Advances in Material Research and Technology

Series Editor

Shadia Jamil Ikhmayies, Physics Department, Isra University, Amman, Jordan

This Series covers the advances and developments in a wide range of materials such as energy materials, optoelectronic materials, minerals, composites, alloys and compounds, polymers, green materials, semiconductors, polymers, glasses, nano-materials, magnetic materials, superconducting materials, high temperature materials, environmental materials, Piezoelectric Materials, ceramics, and fibers.

More information about this series at http://www.springer.com/series/16426

Shadia Jamil Ikhmayies · Hatice Hilal Kurt
Editors

Advances in Optoelectronic Materials

Springer

Editors
Shadia Jamil Ikhmayies
Physics
Jabal El-Hussain
Amman, Jordan

Hatice Hilal Kurt
Department of Physics
Gazi University
Ankara, Turkey

ISSN 2662-4761 ISSN 2662-477X (electronic)
Advances in Material Research and Technology
ISBN 978-3-030-57736-0 ISBN 978-3-030-57737-7 (eBook)
https://doi.org/10.1007/978-3-030-57737-7

This Springer imprint is published by the registered company Springer Nature Switzerland AG
The registered company address is: Gewerbestrasse 11, 6330 Cham, Switzerland

Preface

Optoelectronic materials can be defined as the materials used in the manufacturing of optoelectronic devices. They are divided into three major categories: organic, inorganic, and hybrids. The selection of a material for use in the fabrication of optoelectronic components depends on its properties, especially the optical and electrical properties, stability, and process compatibilities. The optical properties include refractive index, absorption, and emission properties, and electrical properties include mobility and conductivity. Semiconductors are the most used materials in the industry of optoelectronic devices. The majority of the important semiconductors used in optoelectronics industry have diamond or zincblende lattice structures. Group IV semiconductors such as Si and Ge are the most popular diamond lattices, while most of the compounds of group III–V and II–VI have zincblende structures. Some of the most used optoelectronic III–V semiconductor compounds are GaAs, InP, GaN, InN, GaSb and their alloys. And some of the most used optoelectronic II–VI compound semiconductors are CdSe, CdS, CdTe, ZnO, ZnSe, ZnS, ZnTe and their alloys.

The field of optoelectronic materials had attracted much attention due to the significant improvements in material design and purification. In order to suit the materials used in the manufacture of optoelectronic devices, experimental, theoretical, and computational researches can tune the optical, electronic, and electrical properties of these materials. For example, doping the material with selected elements and reducing the size of the crystallites to nanoscale result in tuning the band gap energy. On the other hand, adding dilute quantities of selected ferromagnetic elements results in changing magnetic properties of the material.

This volume of the ebook series "Advances in Material Research and Technology" covers several topics such as deposition and characterization techniques of optoelectronic materials. It presents measurements, computations and methodologies needed for determining optical and electrical properties of advanced optoelectronic materials that are used for emerging advanced electronic and optoelectronic devices. Besides physics, applications of real 2D materials (graphene, MoS_2, WS_2 etc.) for optoelectronics are covered. It also discusses gas sensing properties of nanostructures such as $Zn_{1-x}Sn_xO$ which was found to be sensitive for NO under UV radiation, and the effect of tin doping on ethanol sensing of In_2S_3:Sn. Plasma-semiconductor structures, charge transport mechanisms in porous zeolite, and silver-modified porous zeolite structures are also thoroughly discussed. Moreover, magnetoelectricity in ion-implanted ferroelectric crystals such as $BaTiO_3$ implanted by Co, Fe, and $TlInS_2$ and $TlGaSe_2$ crystals implanted by Co are experimentally studied and thoroughly discussed.

The audience of this book mainly consists of graduate and undergraduate students, physicists, chemists, material scientists and engineers, photonic and optical engineers, and all people working in optoelectronics and semiconductor industries. The beginners, nonspecialists, and new comers to the field of optoelectronic materials will find basics about deposition and characterization methods. The interested reader will find in this book a description of experimental and computational methods with new developments which are thoroughly analyzed and explained. This book can be used as a textbook for several materials related to semiconductors, photonics, advanced materials, thin films, and nanotechnology for undergraduate and graduate students. It is also a rich reference for graduate students, postdoctoral fellows, researchers in the fields of physics, chemistry, material science, and those of several engineering disciplines, where they can find representative examples of experimental and computational work.

This book contains nine chapters that cover wide range of optoelectronic materials, where chapter "A Structural and Optical Look at Functional Materials" by Mustafa Kemal Ozturk is an introductory chapter which presents overviews with examples of the main methods in analyzing, producing, and characterizing functional materials. Chapter "Hybrid Optoelectronic Materials for Photodetection Applications" by Xianguang Yang et al. reported two types of ultraviolet-visible photodetectors based on a single hybrid nanowire and a crossed junction formed with two hybrid nanowires, respectively. These two types of photodetectors would have potential applications and offer novel functionalities in optoelectronic devices and integrated circuits. In chapter "Two-Dimensional (2D) Materials for Next Generation Nanoelectronics and Optoelectronics: Advances and Trends" by Zhi Peng Ling the author provided an overall review of the fascinating world of atomically thin 2D novel materials, the physics underlying these materials, the recent applications surrounding these novel 2D materials and their devices, and the progressive efforts at large-scale commercialization. As examples, graphene, boron nitride (BN), molybdenum disulfide (MoS_2), and phosphorene are overviewed. This is coupled with an insight into the future development trends for such materials.

Chapter "Enhancement of the NO Gas Sensitivity by UV Light in Sn-Doped ZnO Nanostructures" by Irmak Karaduman Er et al. looks at the importance of chemical solution-based technologies and the SILAR method on the gas sensing properties in details. The authors overviewed production methods, dopants, and gas sensing applications of ZnO nanostructures. They produced $Zn_{1-x}Sn_xO$ thin films by SILAR method, processed and characterized the films, and examined them as gas sensors. The films exhibited a good response to very low NO gas and showed an excellent response to low concentrations (20 ppb) of NO gas at room temperature with illumination by UV light of wavelength 365 nm. The authors believe that a combined understanding of theoretical and experimental aspects will open new fields for metal oxide gas sensor and beyond.

Chapter "An Ab-initio Study of Structural, Elastic, Electronic, Vibrational and Optical Properties of Semiconductor NaAlSi Compound for Optoelectronic Applications" by Yasemin O. Ciftci presents an investigation of the structural, electronic, elastic, lattice dynamic and optical properties of NaAlSi alloy, which belongs to half-Heusler family. The author used ab initio density-functional theory (DFT) within the generalized gradient approximation (GGA) for the exchange-correlation potential and found that NaAlSi has indirect band gap, and it should be denoted as a semiconductor. From calculated elastic constants, the author found that NaAlSi is mechanically stable. She also found that further investigation of this alloy is required, and it could be useful for optoelectronic applications.

In chapter "Optoelectronic Properties of GaAs, GaP and ZnSe Cathodes in a Plasma-Semiconductor Cell" by H. Hilal Kurt, the infrared (IR) detectors which are sensitive to infrared wavelengths in a microplasma system were reviewed. This chapter is focused on the optical and electrical characterization of semiconductor materials in a plasma-semiconductor cell, because of their applications in electronic devices and infrared image converter cells. The motivation of this chapter was focused on the electrical gas discharges with various semiconductor electrodes. Ar and air discharge was extensively examined. Experiments were conducted by gas discharge cells with ZnSe, GaAs, and GaP semiconductor electrodes. The major difference of this study from the conventional plasma investigations is that the plasma is produced between the semiconductor electrode and thin-film covered glass electrode for the micro-dimensional discharge gap.

B. G. Salamov in chapter "Charge Transport Mechanisms in the Silver-Modified Zeolite Porous Microstructure" had experimentally investigated non-thermal plasma characteristics in the gas discharge electronic device (GDED) with a zeolite cathode (ZC) up to atmospheric pressure (AP) in air medium. The author used porous zeolite cathodes for plasma light source applications with very low supply voltage and power consumption. The enhanced effect of electric field on the fundamental characteristics of GDED with constant diameter D of ZC, transport of charge carriers, conductivity measurements, and stable discharge glow regions depending the electric field and gas pressure are all investigated. This study has been undertaken to better understand the effect of geometrical (discharge gaps d) and operating (different residual pressures and applied voltages U_0) parameters on the current behavior of the system and operation principle of charge transport in nanoporous materials.

In chapter "Physical Properties and Ethanol Response of Sprayed In_2S_3:Sn Films", the authors Mabrouk Kraini and Noureddine Bouguila presented their study of tin-doped indium sulfide thin films (In_2S_3:Sn) prepared using the chemical spray pyrolysis method on glass substrates, where the molar ratio of Sn:In was varied in the range (0–1% by mol in solution). The material was characterized using X-ray diffraction, optical absorption, Raman and photoluminescence spectroscopies, field-emission scanning electron microscopy, energy-dispersive X-ray spectroscopy, atomic force microscopy, and structural, morphological, vibrational, and optical properties are investigated and discussed. Effect of tin doping on ethanol sensing response was also investigated, and the developed sensor showed promising applications in food control systems and can be used to control drunk driving.

Finally, in chapter "Magnetoelectricity in Ion-Implanted Ferroelectric Crystals" prepared by F. A. Mikailzade, the author reviewed the recent works on the synthesis of nanocomposite materials that are prepared by the inclusions of nanosized magnetic 3D metals in the surface layer of various ferroelectric crystals using ion-beam implantation technique. The different studies showed that the nanocomposite layers of ferroelectrics synthesized by ion-beam implantation have ferromagnetic properties and strong magnetoelectric effects that make them useful for magnetoelectronic applications.

Amman, Jordan — Shadia Jamil Ikhmayies
Ankara, Turkey — Hatice Hilal Kurt

Contents

A Structural and Optical Look at Functional Materials

Mustafa Kemal Ozturk

Abstract This chapter is an introductory chapter which presents overviews of the main methods in analyzing, producing, and characterizing functional materials. A brief idea about deducing optical properties from dielectric function using Kramers–Kronig relations is presented, with $Be_xZn_{1-x}Te$ alloys as an example. The reflectance, transmittance, real and imaginary parts of the dielectric function, refactive index, extinction coefficient, absorption coefficient, and loss function of these alloys are plotted against energy and discussed. Brief ideas about some growth techniques of single crystals such as Bridgman, Czochralski, and Kyropoulos are overviewed. Besides, some deposition methods of films are described. These include molecular beam epitaxy (MBE), metal–organic chemical vapor deposition (MOCVD), and chemical vapor deposition (CVD). Moreover, the main characterization techniques X-ray diffraction (XRD), X-Ray reflectivity (XRR), scanning electron microscopy (SEM), X-ray fluorescence (XRF), small and wide-angle X-ray scattering (SAXS and WAXS), secondary ion mass spectroscopy (SIMS), atomic force microscopy (AFM), spectroscopic elipsometry (SE), photoluminescence spectroscopy, Fourier transform infrared spectroscopy (FTIR), Raman Spectroscopy, UV-Vis spectroscopy, and others are briefly described, and some examples are presented.

Keywords Functional materials · Growth techniques · Characterization techniques · Structural properties · Optical properties

1 Introduction

When we look at the interaction of light and matter in nature, color spectrum formed according to lattice length of material after the light is absorbed by it and is caused by the optical behavior of the material. When we hit also to different materials with our hand slightly, we notice that different sounds come out from the material. This is

M. K. Ozturk (✉)
Photonics Application and Research Center, Department of Physics, Faculty of Science, Gazi University, 06500 Ankara, Turkey
e-mail: ozturkm@gazi.edu.tr

S. J. Ikhmayies et al. (eds.), *Advances in Optoelectronic Materials*,
Advances in Material Research and Technology,
https://doi.org/10.1007/978-3-030-57737-7_1

related with the acoustic property of the material. Furthermore, materials have such properties as having friction very near to zero, high electron mobility, superconductivity at low temperatures, responding to sunlight, sensitivity to light and movement, piezoelectric behavior producing current dependent on movement, producing current by cooling or heating (Peltier) or just the opposite, cooling or heating by current, storing data by spin behaviors, remembering original shape (shape memory effect), mechanical resistivity, and elastic properties. For this reason, the functional properties of the material, today scientists and Investigation and Development workers pay great attention on investigation of material properties. For developing nanoscale optoelectronic devices, related incredible studies take place in literature. For these developments, spectroscopic techniques are used intensively and it gave way to new electronic devices in nano-scale. By the help of these, nanoscale solar cells, paints, drugs, self-cleaning clothes and walls, nanorobots, smart TV, smart phone, and smart glass, and many other industrial products are designed. Studies on these subjects are continuing. After the first lamp transistor used in primitive radio and calculator, developing of diodes and binary system (1 right, 0 false) radios and TVs got smaller and faster and took their place in people's lives with great velocity. Later 4, 8, 16, 32, 64, and 128-bit systems are introduced, and dimensions of the devices got smaller and smaller. Computers, radios, TVs, mobile phones, smart phones, and many other smart devices became a piece of our lives very fast. These developments are usually traced by patents, and till now, many patents are produced. For this reason, mentioning people here who have the most patents will be useful. Making a simple search on net by the year 2018, one can reach this knowledge on many Web sites. Kia, Siver, and Brook took 4732 patents on mostly, digital materials, electronic devices, LCD, security inks, chemistry, DNA, mechanics, and press between 1994 and 2017 years. Shunpei YAMAZAKI follows these with 4315 patents on solar panels, flash memories, and liquid screens. As the third, Paul LAPSTUN took 1278 patents on press, digital paper, electronic, and Internet fields.

During the growth of technology, elements which the material science is focused on are II–VI groups on periodic table and compound semiconductor materials formed by them. Of course polymers and organic materials also take attention. Semiconductors are structurally good crystals. Atoms and molecules form a base on a point of crystal, and this base is called the lattice. 3D settlement of these lattices forms the crystal. The most important property of this crystal is symmetry. And symmetric operations are given as reflection, rotation, and inversion in structure. There are seven different types of lattice; the general lattice kind of these is triclinic due to point symmetry. These lattice types are very important in terms of functional properties of solids. Stable cubic and hexagonal semiconductors are preferred due to their structural, electrical, and optical properties. These materials may present different electrical and optical properties according to their lattice structure. Here, mobility of free orbital electron movement determines the electrical properties. For the atoms which are suitable for disturbing, electronic transitions between energy levels are related with forbidden band gap and optical properties. Besides, smoothness of the surface and to see sun light in colored spectra indicates a successful growth period, else cracked and rough surface structure can be seen. Lattice mismatch and fault structure of the crystal formed on these circumstances change the optical behavior of the structure. Lattice settlement of the crystal is caused by the vibrational phonon movements. Lattice vibrations are called phonons. Phonon movements belonging to structure, elastic constants of the crystal structure, phonon frequency, and stable stress show different properties, and these properties may change by faults. Semiconductors may be in crystal, polycrystalline, or amorphous structure. In further investigations, for growth procedure of the semiconductor, it is desired that it should have good optical and electrical behaviors. With different crystal growth techniques, it is possible to grow more stable epitaxial layers under high vacuum and at high temperatures. Semiconductors may be grown as cubic and hexagonal structures (Si, Ge, GaAs, GaN, ZnO, GaP, GaInP AlGaAs, AlN, GaN, InGaN, AlInN, AlGaN, etc.) in MBE or MOCVD systems with high crystallized epitaxial layers. Light-emitting diode, high electron mobility transistor, photo detector, sensors, and diodes may be formed in heterostructure as quantum wells and multi-quantum wells. Cubic structure crystal semiconductor materials as Si, Ge, InP, GaP, GaAs, AlGaAs, and InGaAs are the first-generation investigation materials. These materials are still used for production of optoelectronic devices in nanotechnology field. But because they have very narrow structural and optical properties, this sector went on looking for new materials. In these investigations apart from the materials mentioned above which do not have cubic crystal structure, interest on GaN-based hexagonal systems which have resistivity against high temperature, frequency, and voltage is increasing. Of course, the first problem noticed is c-oriented lattice mismatch between epitaxial layer and Al_2O_3 substrate which disorders the stability of the material. While lattice mismatch causes lattice relaxation, dislocations increase to a very serious level, and finally, this failure causes cracks on the surface. Furthermore, while cooling from growth temperature to room temperature, bending movements with wide dislocations are seen together with mosaic defects (such as layer tilt angle, twist angle, and crystallized blocks). Because of this, full width at half maximum (FWHM) value for (002) symmetric plane for GaN layer used as buffer layer increases above 400 arcsec ($1^\circ =$ 3600 arcsec). Of course, these problems may be overcome by growing cut-thickness GaN layers in order to decrease the lattice mismatch of AlN and GaN nucleation layers at low temperatures to prevent dislocations moving on c-direction. By the help of these improvements, GaN-based AlGaN/GaN mixed structure field-effect transistors (HFET) , AlGaN/GaN mixed junction bipolar transistors, GaN metal oxide semiconductor field-effect transistors, Schottky and p-i-n rectifiers are used as new generation optoelectronic devices [1–7] . Recently, S.V. Novikov and coworkers succeeded in the growth of GaN and AlGaN epitaxial layers on GaAs with MBE without surface cracking [8] . Although III–V group compounds with nitrogen are a wide research area, ZnS, CdS, HgTe, CdTe, and CdHgTe alloys from II to VI group are a newer research area [9]. Besides, double and triple semiconductors formed by using La, Hf, Ru, Tm, Si, Ge, Li, Pt atoms are a research area for more daily optoelectronic devices [10].

2 Theoretical Optical Approach to Functional Materials

When the material is exposed to electromagnetic waves, as a result of interaction of electrons with wide spectrum, some optical events occur. If the energy which the spectrum has is equal to forbidden band gap E_g, the electron found in orbital is disturbed to a higher energy level. If the energy is lower than the forbidden energy band gap, instead of absorption, spectrum is transmitted, and this material is called transparent [11]. Transmission or absorption of a photon is related with electron configurations of semiconductor, metal, or insulator [12]. Dielectric property of the material is a measure of insulation, and forbidden band gap determines whether the material is an insulator or a conductor. For this reason, by calculating the dielectric constant of the material, one can gain knowledge about optical behaviors of this material. Absorption of incident spectrum by the material has some periods. The reason for the absorption is electron transmissions of optical energy in the material's fundamental forbidden band borders. Absorption of the spectrum is excitation of surface electrons, transmitted between bands with optical energy, in a bulk material. Transition between bands can be seen in every material. That is, optical properties of a material are related with electron density in transition situations and band structures of that material [13]. There are many models for numerical optical behaviors of the material. All these models have better properties and calculation conveniences. One or more of these models can be calculated in detail with many programs—for example, Drude model, Lorentz model, Induced field-effect method, Debye relaxation, Kramer–Kronig equations, Hagen–Rubens equation. Calculations can be done by using the mentioned approximations above [14]. These models given for optical properties are the most frequent used ones. Here, the Kramer–Kronig equations are given, and then, an example of using it is presented.

2.1 *Kramers–Kronig Relations*

Kramers–Kronig relations are used to calculate the real part from the imaginary of a complex function or vice versa. So, the dielectric function ($\varepsilon = \varepsilon_1 + i\varepsilon_2$) which is a function of frequency ω, has a real part $\varepsilon_1(\omega)$ and imaginary part $\varepsilon_2(\omega)$. Any of them can be used to produce the other [14]. For example

$$\varepsilon_1(\omega) = 1 + \frac{2}{\pi}\int_0^\infty d\acute{\omega}\frac{\acute{\omega}^2\varepsilon_2(\acute{\omega})}{\acute{\omega}^2 - \omega^2} \quad (1)$$

Kramers–Kronig relations are also to relate the real and imaginary portions for the complex refractive index $n^*(\omega) = n(\omega) + \mathrm{i}\,\kappa(\omega)$. Also, they allow to calculate the reflectivity, energy loss function $L(\omega)$, and absorption coefficient $\alpha(\omega)$ all as functions of frequency or energy.

Refraction index $n(\omega)$ and extinction coefficient $\kappa(\omega)$ can be determined using the components of dielectric function;

$$n(\omega) = \left[\sqrt{\varepsilon_1^2(\omega) + \varepsilon_2^2(\omega)+\varepsilon_1(\omega)}\right]^{1/2} /\sqrt{2} \tag{2}$$

$$\kappa(\omega) = c\left[\sqrt{\varepsilon_1^2(\omega) + \varepsilon_2^2(\omega)+\varepsilon_1(\omega)}\right]^{1/2} /\sqrt{2} \tag{3}$$

Reflectiivity $R(\omega)$ and absorption coefficient $\alpha(\omega)$ can be calculated using the following equations:

$$R(\omega) = \left[\frac{\sqrt{\varepsilon_1^2(\omega) + j\varepsilon_2^2(\omega) - 1}}{\sqrt{\varepsilon_1^2(\omega) + j\varepsilon_2^2(\omega) + 1}}\right]^2, \tag{4}$$

$$\alpha(\omega) = \sqrt{2\omega}\left[\sqrt{\varepsilon_1^2(\omega) + \varepsilon_2^2(\omega)} - \varepsilon_1(\omega)\right]^{1/2} \tag{5}$$

and energy loss function $L(\omega)$ is identified by the following formula:

$$L(\omega) = \mathrm{Im}\left[\frac{-1}{\varepsilon(\omega)}\right] = \varepsilon_2(\omega)/\left[\varepsilon_1^2(\omega) + \varepsilon_2^2(\omega)\right] \tag{6}$$

2.1.1 Optical Properties of $Be_xZn_{1-x}Te$ Alloys

$Be_xZn_{1-x}Te$ is formed by II–VI group atoms and used for blue-green laser diode production. This structure can be grown in MBE growth system with Be, Zn, and Te sources. In this part, optical properties, static dielectric constants, refraction indexes, absorption coefficients, and plasmon frequency values are calculated according to x alloy ratio as examples [15].

Dielectric Functions

The value for starting absorption of imaginary part of dielectric function for $Be_xZn_{1-x}Te$ gives the splitting $\Gamma_v - \Gamma_{c.}$ Also, it represents the direct optical transition between valance band and conduction band. On the other hand, these values are the same with band gap [16]. Real and imaginary parts of the dielectric function for $Be_xZn_{1-x}Te$ are given in Fig. 1. As shown in this figure for $Be_{1-x}Zn_xTe$, the main peaks for real part of the dielectric function are 3.54, 2.79, and 2.37 eV for $x = 0.25$, 0.50, and 0.75 doping values, respectively. For $\varepsilon_1(\hbar\omega) = 0$, energy values 7.29, 8.43,

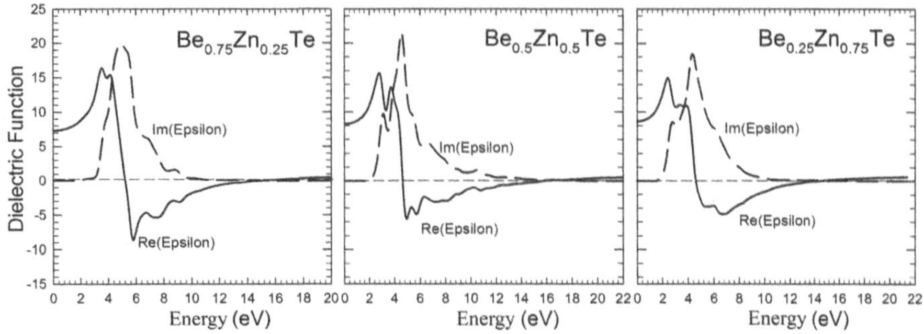

Fig. 1 Real and imaginary parts of dielectric constant for $Be_{0.75}Zn_{0.25}Te$, $Be_{0.5}Zn_{0.5}Te$ and $Be_{0.25}Zn_{0.75}Te$ alloys against energy

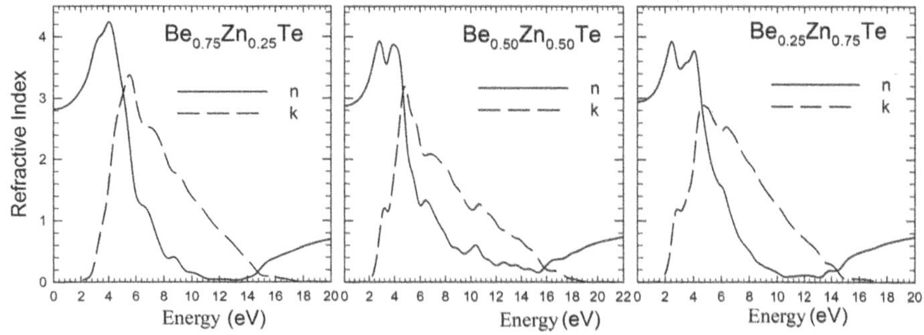

Fig. 2 Refractive index (n) and extinction coefficient (k) for $Be_{0.75}Zn_{0.25}Te$, $Be_{0.5}Zn_{0.5}Te$, and $Be_{0.25}Zn_{0.75}Te$ alloys against energy

and 8.75 eV give static dielectric constant for increasing Be doping values $x = 0.25$, 0.50, and 0.75, respectively. The imaginary part of the dielectric constant begins to increase at about 2.04, 2.11, and 1.58 eV according to doping values. These values are near to band gap energy. This situation represents the optical transition between valance band and conduction band. The alloy acts as a transparent material until the value where dispersion curve started to increase. Here is the low dispersion region. The energy values for the maximum of the imaginary part of the dielectric constant are at 4.96, 4.5, and 4.37 eV for the doping values 0.25, 0.50, and 0.75, respectively. These values correspond to interband transitions [17, 18].

Refractive Index and Extinction Coefficients

Variation of refractive index $n(\hbar\omega)$ and extinction coefficient $k(\hbar\omega)$ for $Be_xZn_{1-x}Te$ are given in Fig. 2 The region where refractive index $n(\hbar\omega)$ is maximum shows the transparent region. The value of refractive index at zero frequency is $n(0)$, and the square of $n(0)$ equals to the isostatic real dielectric coefficient ε_1 (0). It is seen that ε_1

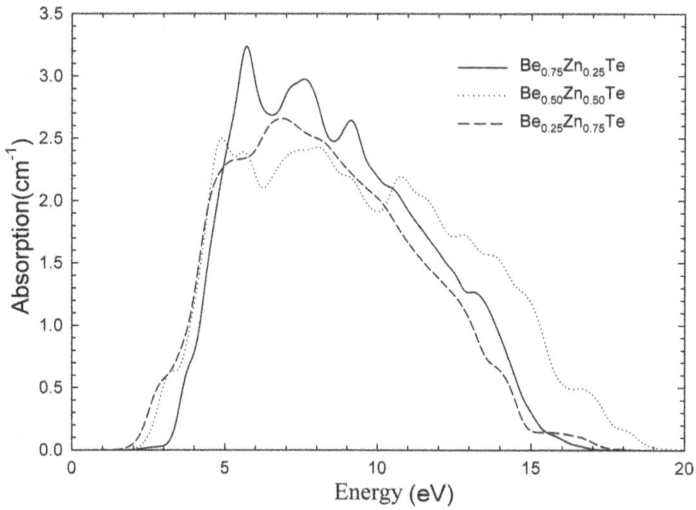

Fig. 3 Absorption coefficient for $Be_{0.75}Zn_{0.25}Te$, $Be_{0.5}Zn_{0.5}Te$, and $Be_{0.25}Zn_{0.75}Te$ alloys against energy

(0) and $\{n\ (0)\}^2$ are in accordance with each other for $Be_xZn_{1-x}Te$. Also, it is seen that maximum value of extinction coefficient of alloys $k\ (\omega)$ and static dielectric constant which ε_1 (0) corresponds to zero energy are in accordance with each other. Dispersion curves of refraction index for $Be_{1-x}Zn_xTe$ are shown in Fig. 2, and values of n (0) are found as 2.85, 2.89, and 2.95 for $x = 0.25$, 0.50, and 0.75, respectively. It is understood that when the value of Zn increases, refractive index also increases. It is seen that real part of dielectric function and refractive index and imaginary part of the dielectric function and extinction coefficient are in agreement with each other

Absorption Coefficient

Absorption coefficient defines absorption of light by the materials per unit length. The limit of the absorption is related to optical transitions between fundamental band gaps. Interband absorption is the excitation of a solid material electron, transmitting optically between bands. Absorption coefficient is obtained from the real and imaginary parts of the dielectric function as shown in Eq. (4), and the results are shown in Fig. 3 for $Be_{1-x}Zn_xTe$ alloys. Absorption coefficient values are starts to increase at 2.06, 2.11, and 1.56 eV for $Be_{1-x}Zn_xTe$, respectively, according to x doping values. It is seen that the point where the absorption coefficient started to increase agrees with those when the imaginary part of the dielectric function and the extinction coefficient began to increase.

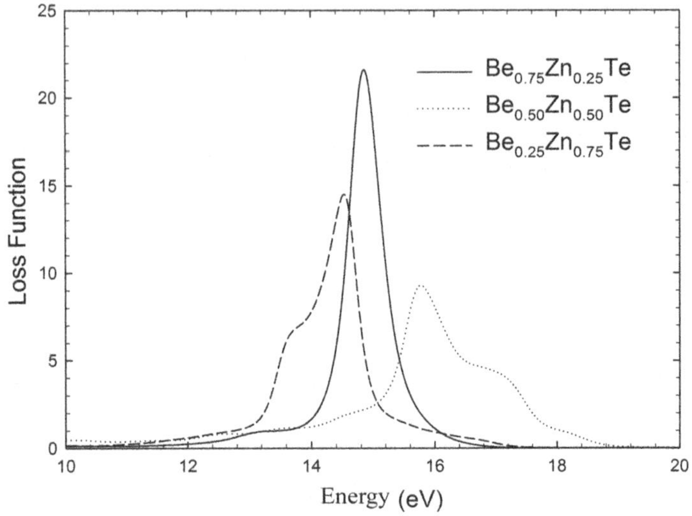

Fig. 4 Loss function L(*w*) for $Be_{0.75}Zn_{0.25}Te$, $Be_{0.5}Zn_{0.5}Te$, and $Be_{0.25}Zn_{0.75}Te$ alloys against energy

Loss Function Calculation

Loss function variation with energy is given in Fig. 4 for $Be_xZn_{1-x}Te$. It is the variation of loss energy while the electron is transmitting from the top of the valence band to the bottom of the conduction band. Here, the main peak value is known as *plasmon frequency*. Alloys act as insulators above plasmon frequency value and as metals below plasmon frequency value.

Loss function of the electron is deduced from the real and imaginary parts of the dielectric function for $Be_{1-x}Zn_xTe$ alloys, and it is shown in Fig. 4. Loss function has some peaks between 12 and 19 eV. The frequency corresponding to the main peak of the loss function is called as plasmon frequency. Imaginary part of the dielectric function is minimum where the peak of loss function is maximum. For the doping values ($x = 0.25$, 0.5, and 0.75), the energies corresponding to plasmon frequency values are 14.80, 15.70, and 14.48 eV. When we look at the plots, it is seen that absorption is too little between 0 and 12 eV. The reason is that there is absorption in this frequency range as can be deduced from the imaginary part of the dielectric function [19].

3 Growth Techniques of Functional Materials

Semiconductor technology has become quite common today. For this reason, the optoelectronic devices desired to be obtained are very much concerned with the formation of single crystalline structures such as group III–V, II–VI for the epitaxial layers. Formation of single crystalline materials has become possible by various crystalline growth methods. Thin films can be grown with sophisticated and simple techniques. Crystalline growth methods are divided into bulk and thin film growth. Bulk growth techniques are Bridgman and Czochralski. Thin film growth techniques are generally molecular beam epitaxy, metal–organic chemical vapor deposition, and chemical vapor deposition techniques.

3.1 Bulk Crystalline Growth Techniques

Techniques for single crystalline growth are solution growth, melt growth, and vapor growth techniques [20]. In cubic or hexagonal systems, crystals oriented in a desired orientation and additive density is grown on a core crystal (AlN, ZnSe, CdSe, GaAs, GaP, Ge, S, GaN, Al_2O_3, etc.). Brief ideas about some of these techniques will be given below.

3.1.1 Bridgman Technique

In Bridgman technique, the crystal is grown in a vacuum quartz tube surrounded by a crystal oven or thermos elements. By the Bridgman method, the crystalline surface, which is solidified by crystallization, is moved in the vertical direction, and the pointed end in the furnace or quartz tube is lowered. The pointed tip is the starting point for the first cooling and acts as a core during the growth. By pulling the pointed end, the crystal in molten state is enlarged [20]. In the Bridgman technique, the oven or quartz tube can be moved by holding the crystal carrier stationary. The crystal carrier can be moved by holding the furnace or quartz tube constant, or it is possible to grow the crystal by different methods by simply changing the temperature without movement.

3.1.2 Czochralski (CZ) Technique

Czochralski is a fast method to grow single crystals. It is widely used for semiconductors, oxides, and fluorides to be used in optical applications. This method is the most widely used technique in the production of single crystalline silicon (Si)), and germanium (Ge) is the Czochralski (CZ) technique [21]. Ultra-high-purity Si is usually dissolved in a carrier which is quartz. If n-type or p-type single crystalline Si is grown, then additive atoms such as boron (B) or phosphorus (P) are added in this step. A Si single crystal (which can be expressed as a seed crystal) in a suitable crystal orientation is immersed in the melt and is slowly rotated and pulled up at the same time. With the CZ method, it is possible to obtain a single crystal in a very large cylindrical shape. This method is often used in the growth of silicon single crystals.

This method produces the most homogeneous and perfect crystals. But it is only applicable to the materials that melt congruently or nearly congruently, because the solid and melt compositions are the same at equilibrium.

3.1.3 Kyropoulos Technique

The Kyropoulos technique was designed to create a smooth crystal at low temperatures in the mid-twentieth century. This technique is one of the most advanced techniques for growing large single crystals before CZ technique. The most important difference between this technique and the CZ technique is that it can form different crystal forms. In Kyropoulos method, a solution in the pot is heated. By slowly cooling the melt, the solid–liquid interface between the melt and the core crystal begins to form. At the beginning of the crystal formation process, the nucleus forms a cone-shaped shape in the crystal aligned to the melt. In this method, instead of pulling up the crystal from the melt, the growth in the crucible is completed by cooling slowly. The size of the resulting crystal is almost equal to the diameter of the pot [22].

3.2 Thin Film Growth Techniques

3.2.1 Molecular Beam Epitaxy (MBE)

MBE is an ultra-high vacuum (UHV) -based thin film growth technique used to produce single-step, controlled, and high-quality epitaxial structures. It has become a widely used technique for the production of semiconductors, superconductors, and metals with a high degree of purity. The underlying phenomenon of the MBE growth system is quite simple: The solid source materials are placed in the evaporation channels to provide angular distributions of atoms or molecules. The substrate is heated to the desired temperature, and the substrate is rotated for growth to be homogeneous. The ultra-high vacuum hits the substrate surface heated in the environment and is bonded to the structure on the enlarged surface. Despite its conceptual simplicity, material purity and slow growth are a bad advantage, but it is a significant advantage that growth can be controlled through the system. In the MBE system, such as GaAs, AlGaAs, InGaAs, InGaP, InP, AlGaN, and AlGaAsP structures can be grown [23].

3.2.2 Metal–Organic Chemical Vapor Deposition (MOCVD)

MOCVD is chemical vapor deposition method for epitaxial growth of materials, exclusively compounds. The growth of crystal structures in the MOCVD technique occurs chemically. As an example, the GaAs film growth is as follows: gallium (Ga) or arsenic (As) compounds, called precursor gases. Usually, a metal–organic compound such as arsenic hydride (AsH_3) for As and trimethyl gallium (TMGa) for Ga is used. The precursor gases are injected into the reactor via the carrier gas (hydrogen or nitrogen). The reactor and substrate are heated to a certain temperature. Temperatures may range from about 500–1500 °C depending on the material to be produced. Reaction in the heated reactor and sub-surface occurs in the gas phase. The particles, together with AsH_3 and TMGa, are transported to the substrate surface where they are deposited on the substrate and the remaining gases are discarded. Within these steps, the GaAs layer grows on the buffer layer. Unreacted molecules are separated from the surface and discharged from the reactor with the aid of carrier gas [24, 25].

As a sample, the high-resistivity GaN layers were grown on 6H-SiC substrates, with different buffer layers structures in MOCVD system. For these structures, trimethylaluminum (TMAl) , TMGa, and ammonia (NH_3) as Al, Ga, and N precursors are used, respectively. Oxide layer in surface of SiC substrate was removed by annealing of the under hydrogen ambient at 1200 °C for ~10 min. For four different samples (called A, B, C and D), growth procedure is applied with different buffer layers. All of the samples nominally contain a 190 Å thick GaN layer after the buffer structures (Table 1).

Sample A contains 100 Å-thick, low temperature growth AlN, 1400 Å-high temperature growth AlN and 500x100 Å GaN/AlN pair super lattice structures under high-resistant GaN layer. In sample B, 100 Å-low temperature AlN and 1400 Å AlN layers were grown as a buffer layer at high temperature. The buffer layers of sample C includes a 100 Å-thick AlN nucleation layer grown at low temperature (705 °C) and 2300 Å-thick AlGaN layer.Fhere, XRD scan was used to obtain the Al content of the $Al_xGa_{1-x}N$ layer, and its value is determined as 67.0%. And 300 Å-thick AlN (high-temperature AlN) is grown at 1100 °C and is used as a buffer layer in sample [26].

Table 1 High-resistant GaN barrier structures designed in the different structural properties

A	B	C	D
High-resistivity GaN layer 190 Å	High-resistivity GaN layer 190 Å	High-resistivity GaN layer 190 Å	High-resistivity GaN layer 190 Å
GaN/AlN superlattice 500 Å	–	Undoped-AlGaN layer 2300 Å	–
High-temperature AlN layer 1400 Å	High-temperature AlN layer 1400 Å	–	High-temperature AlN layer 3000 Å
Low-temperature AlN nuclation 100 Å	Low-temperature AlN nuclation 100 Å	Low-temperature AlN nuclation 100 Å	–
Semi-insulation 6H-SiC substrate	Semi-insulation 6H-SiC substrate	Semi-insulation 6H-SiC substrate	Semi-insulation 6H-SiC substrate

3.3 *Chemical Vapor Deposition (CVD)*

CVD is a process in which gases chemically react and form reaction products on the substrate surface. The basic process steps in CVD are [27]: (I) Transport of reaction materials to the reaction chamber. (II) Chemical reaction at the sample surface. (III) Accumulation of reaction products on the heated surface of the sample. (IV) Separation of residual gases from the sample surface.

3.4 *Vapor Phase Epitaxy (VPE)*

VPE is a chemical vapor deposition method on a substrate for epitaxial solid layers. There are two methods of growing groups III and V with VPE: chloride and hydride. In the chloride method, chlorine is passed through the gallium arsenic metal to form $GaCl_3$. In the hydride method, hydrogen is mixed with arsenic (AsH_3) and $GaCl_3$ gases and sent on the substrate, which is held on a rotating plate at about 1100 K. The gases are trapped on the substrate surface. The enlargement of structures such as GaAs by the VPE method can create a perfect crystal with a better purity [28].

3.5 *Liquid Phase Epitaxy (LPE)*

LPE is based on the decomposition of sub-solid material from a cooling solution layer. The substrate is stored in the growth chamber, and the solution is saturated with the growth medium until the desired growth temperature is reached. The solution is then contacted with the single crystalline substrate surface, allowed to time for the desired growth to occur, and finally, the material is allowed to cool after growth has occurred [29].

4 Characterization Methods of Functional Materials

4.1 *Scanning Electron Microscopy (SEM)*

In this system, electrons are thermionically emitted from an electron gun fitted with a tungsten filament. The emitted electrons which have an energy of 5–30 keV are accelerated with the help of electrical force, which is reversely directed to the electric field, and these electrons are bombed by means of focusing lenses toward the surface of the material, which is coated with one of Au, Pt, Ag, and Cu metals (Fig. 5). Because of this situation, two types of collisions are seen that are elastic and inelastic.

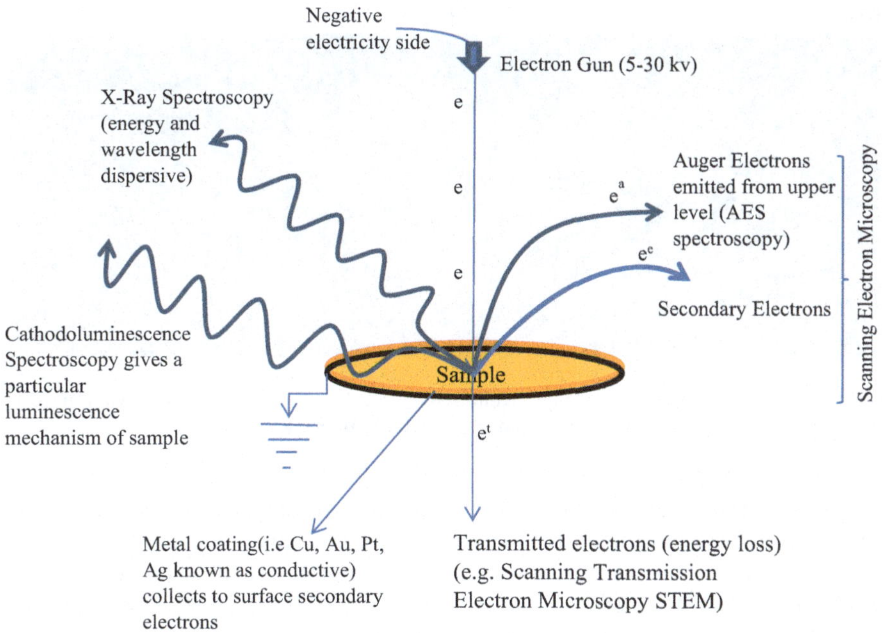

Fig. 5 SEM diagram and spectroscopy techniques by electron–material interactions

SEM and spectroscopic techniques are related to inelastic collisions. In inelastic collisions, electron interacts with the atoms of the sample material. Incident electron rips an electron which may be of 1 s core energy level and passes to the back side of the sample by losing most of its energy. These electrons give the opportunity of scanning transmitted electron microscopy (STEM) image. The electron ripped from 1 s is known as the secondary electron. Secondary electrons may also result from other inelastic interactions between the primary electron beam and the sample. Secondary electrons which are ripped from the atoms approximately 10 nm down the surface of the material are taken to the surface by the help of coated metals, and they radiate on the surface of the material with the kinetic energy they have. Conductor metal collects the electrons on the surface, and so, there are no charges inside. Electrons on the surface make a wide radiation spectrum according to their energy density. Later, these radiations are transformed to image analyzers, amplifiers, and detectors [30].

As an example of scanning electron microscope images, Fig. 6 shows the SEM micrograph representative of the boron-doped polyvinyl alcohol-HfO_2 fibers in different voltages (14.0 and 28.0 kV). It can be observed that the formation of single fiber and cluster structure was the predominant morphology; only a large number of them show irregular fiber orientation, and the average fiber diameter was 670 nm [31].

In order to be stable, 1 s level takes an electron from 2 s or 2p levels. Energy difference between 2 s and 2p and 1 s levels rips an electron from upper levels of

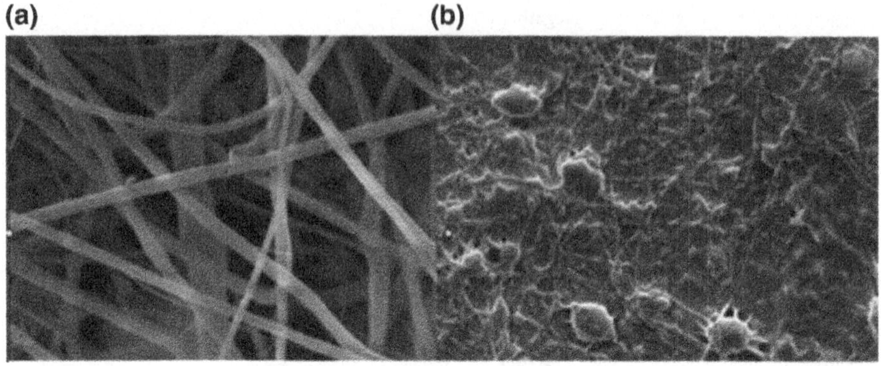

Fig. 6 SEM micrograph of polyvinyl alcohol-HfO_2 doped with boron which is synthesized by electrospinning technique at different voltages. **a** 14.0 kV, **b** 28.0 kV [31]

the atom, and this electron is called as Auger electron (Fig. 5). Auger comes from 1 to 3 nm thickness of the sample surface. Also, this situation causes emission of X-ray spectrum. Auger electrons are not seen in atoms such as lithium, helium, and hydrogen which do not have any electrons upper than p level. Auger electrons give the opportunity for both mineral analysis and surface analysis as (topography, morphology, composition, shape, etc.). This technique is called as Auger electron spectroscopy. X-ray spectroscopy caused from electronic transitions presents very useful analysis of elements. X-ray spectroscopy with energy and wavelength is known as EDX and WDS [32], respectively. Electron transition energies between stable levels are impaired with a database determined before, and one can decide which element these transitions belong to. Mapping can be done. Furthermore, ratio as a percentage of the compound in the composition can be determined. When the transition energies of the unknown samples are needed to be determined, one can use high-resolution fast semiconductor detector.

Figure 7 presents an example on EDX spectra, where the EDX spectra of CaO and SiO_2 are shown. If the CaO sample was examined by WDS, peaks would be sharper, and so, FWHM values would be lower.

Also, with EDX, one can do quantitative analysis, where EDX report gives the composition of the sample. For the example above, the composition of the sample is given in Table 2, where the concentrations of Si, Ca, and O elements are give by weight, atomic percent, and compound percent, respectively.

Cathodoluminescence in Fig. 5 is the term that describes the analysis of radiation emitted due to bombing the sample with electrons. By the help of cathodoluminescence spectroscopy, surface mapping can be done using deep energy values, and surface image can be formed. As an example, Fig. 8 shows the cathodoluminescence spectra of the p-GaAs reference substrate and the porous p-GaAs measured at room temperature. Porous GaAs layers are formed by electrochemical etching on highly doped p-type wafers with (001) crystal orientation by Bioud, et al. [33].

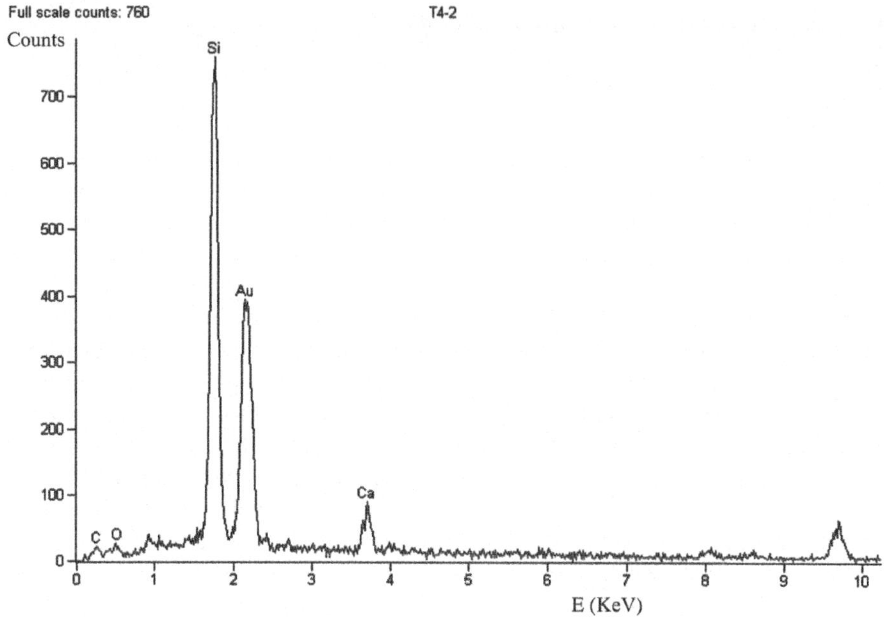

Fig. 7 EDX spectra of CaO and SiO_2

Table 2 EDX analysis for CaO and SiO_2

Element	Weight Conc %	Atom Conc %	Compound Conc %	Formula
O	51.06	65.58	0.00	O_2
Si	42.59	31.16	91.12	SiO_2
Ca	6.35	3.25	8.88	CaO

Fig. 8 Cathodoluminescence spectra at room temperature for the p-GaAs reference substrate and the porous p-GaAs [33]

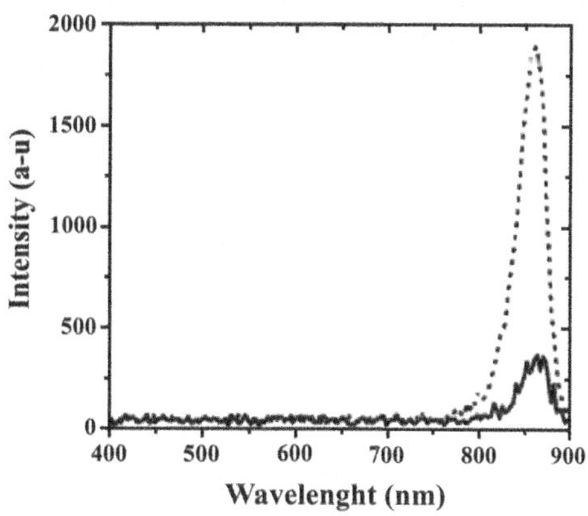

4.2 X-Diffraction and Spectroscopy

4.2.1 X-Ray Types

X-rays can be generated by an evacuated X-ray tube that uses a high voltage (~40 kV) to accelerate the electrons released by a hot cathode to a high velocity. The high-velocity electrons collide with the anode which is a metal target, such as Mo, Cu, Co, Fe, or Cr. X-rays are generated via interactions of the accelerated electrons with electrons of the anode. There are two types of generated X-ray: characteristic X-rays and bremsstrahlung one. The characteristic X-ray is produced when a high-energy electron collides with an inner shell electron (1S, 2S, …), and both are ejected from the atom leaving a "hole" in the inner shell. This is filled by an outer shell electron with a loss of energy emitted as a monochromatic X-ray photon. If the inner shell is 1S (called K shell), and the outer shell is 2S, 2p (called L shells), then the emitted X-ray is called $K_{\alpha 1}$, but if the outer shell is 3s, 3p, 3d (called M shells), then the emitted X-ray is called $K_{\alpha 2}$. But if the inner shell is L shell, and the outer is M shell, then the X-ray is called $K_{\beta 1}$. If the outer shell in this case is 3s, 3p, 3d (called N shells), then the emitted X-ray line is called $K_{\beta 2}$, and so on. In other words, K_α is between L-K, K_β is between M-K, and K_γ is between N-K. In X-ray diffraction techniques, the sharpest peak, K_α, is used. K_β peak is absorbed by some filters such as Zr, Ni, Fe, Mn, and V. The second type of generated X-ray is produced when electrons pass near the nuclei in the anode material, where they slow down and loss their kinetic energy continuously. During their deceleration they give the continuous X-ray radiation known as a bremsstrahlung X-ray. As a result of characteristic and bremsstrahlung radiation, the generated X-ray will be similar to the one shown in Fig. 9. Characteristic X-ray peaks are the sharp peaks labeled as K_α and K_β. This

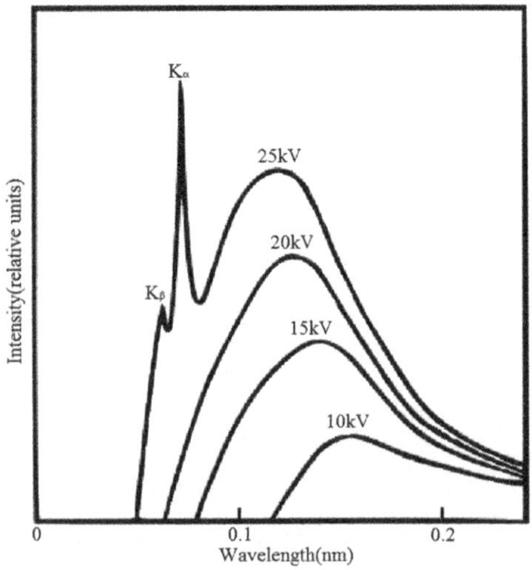

Fig. 9 Molybdenum X-ray spectrum with different potentials between the anode and cathode [34]

spectrum can be manipulated by changing the X-ray tube, current, or voltage settings, or by adding a filter to get monochromatic X-ray.

4.2.2 X-Ray Diffraction (XRD)

The X-ray diffractometer consists of three basic parts: an X-ray tube, a sample holder, and an X-ray detector, and it usually operates at known voltages and current. When a monochromatic beam of X-rays such as (K_α) is incident on a sample as shown in Fig. 10, it will be scattered from the atoms within the sample. To get monochromatic X-ray, a filter is required. $K\alpha$ consists of $K_{\alpha 1}$ and $K_{\alpha 2}$, where $K_{\alpha 1}$ has a slightly shorter wavelength and twice the intensity as $K_{\alpha 2}$, so a weighted average of the two is used. If the material is crystalline, then the scattered X-rays will undergo constructive and destructive interference. This process is called diffraction, and the maxima of the intensity of diffracted X-rays, which correspond to constructive intereference, are called diffraction peaks, and they satisfy Bragg's law:

$$2d \sin \theta = n\lambda \tag{7}$$

where d is the lattice spacing, θ is the diffraction angle, which is the angle between the incident X-ray and the scattering plane as shown in Fig. 10a. n is the order, and λ is the wavelength of incident X-rays. These diffracted X-rays are then detected by the detector, processed, and counted. The produced XRD pattern (diffractogram) is a fingerprint of the material under investigation. When a powder or polycrystalline

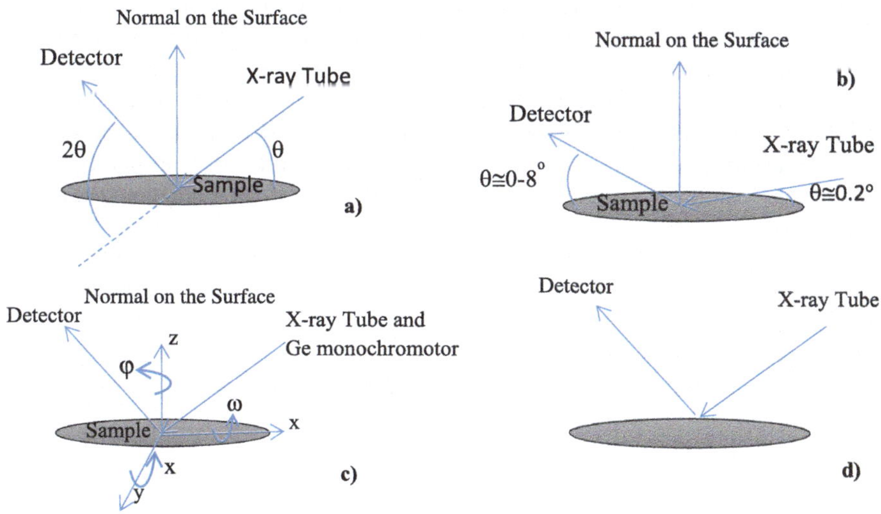

Fig. 10 **a** XRD measurement device scheme. **b** XRR measurement device scheme. **c** HRXRD, texture, and stress measurement device scheme. **d** XRF measurement device scheme

material with randomly oriented crystallites is placed in the X-ray beam, the beam will see all possible interatomic planes. By scanning the sample through a range of 2θ angles, all possible diffraction directions from the sample will be detected. The positions of the diffraction peaks can be converted to d-spacings, which allow the identification of the material by comparison with standard reference patterns [35].

X-ray diffraction is now a common technique for the study of crystal structures and atomic spacing. This technique provides information about structures, phases, preferred crystal orientations (texture), average grain size, crystallinity, strain, and crystal defects, and it is used for analyzing minerals, alloys, spreading phases, corrosion products, polymers, zeolites, explosives, and cement like materials. The parafocusing (or Bragg–Brentano) diffractometer is the most common geometry for diffraction instruments.

4.2.3 X-Ray Reflectivity (XRR)

The basic principle of X-ray reflectivity (XRR) is to reflect a beam of X-rays at grazing angles of incidence or low angles (typically, from 0 to 4° from grazing incidence) as shown in from a flat surface (Fig. 10b) and then to measure the intensity of reflected X-rays in the specular direction (reflected angle equal to incident angle) as a function of angle. The total external reflection occurs below the critical angle θ_c, which is very small, and depends upon the electronic density of the material. The higher the angle of incidence of the X-ray relative to θ_c, the deeper the X-rays penetrate into the material. Below the critical angle of total external reflection, X-rays penetrate only a few nanometers into the sample. At every interface where the electron density changes, a part of the X-ray beam is reflected. The interference of these partially reflected X-ray beams results in the oscillatory pattern observed in reflectivity experiments. From these reflectivity curves, layer parameters such as thickness (until 1000 nm), density (with %1 error), interface, and surface roughness (until 5 mm) for single and multilayer stacks on semiconductor wafers can be determined, regardless of the crystallinity of each layer (single crystal, polycrystalline, or amorphous). Also, morphology can be determined by spreading beam, correlation lengths, and fractal parameters.

Figure 11 shows a fit curve modified with dynamical equations of XRR scan for AlN crystal grown on SiC substrate by MOCVD method. The figure shows reflected intensity of X-rays against $2\theta/\theta$, where θ is the incidence angle and 2θ is the angle between the detector and incident X-rays. Here, by using fringes width thickness is found as 61.340 nm and by using amplitude of fringes roughness is found as 1.121 nm and density is found as 3.17 g/cm^3.

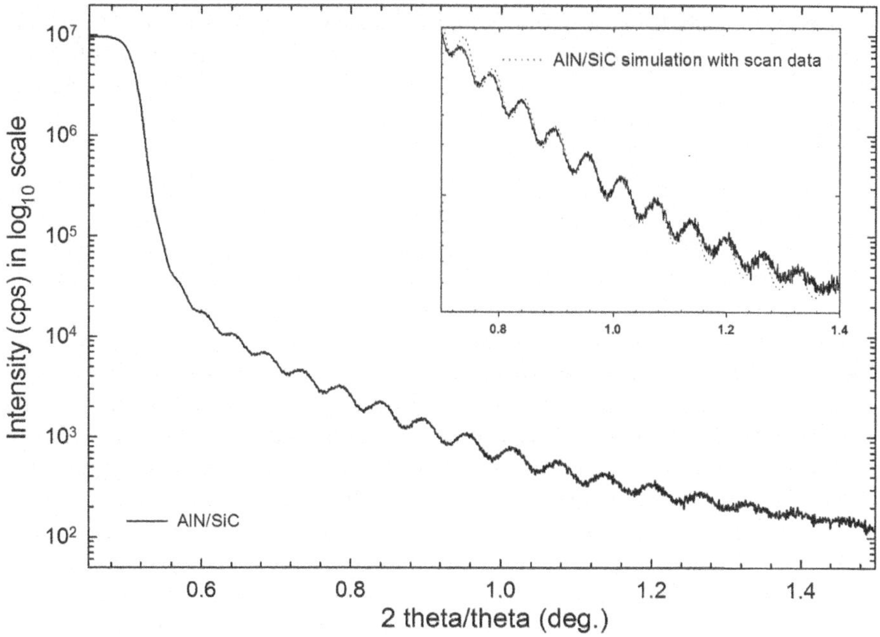

Fig. 11 Reflectivity scanning of AlN buffer layer grown on SiC by MOCVD. It is given with Inset fit value

4.2.4 High-Resolution X-Ray Diffraction (HRXRD)

HRXRD is an indispensable technique for semiconductor and polymer-semiconductor materials. HRXRD can be obtained using a laboratory-based triple-axis (four-circle) diffractometer. It requires a sealed X-ray tube or rotating anode, a mirror with a four-bounce duMond–Hart–Bartels (dHB) design monochromator to isolate $K_{\alpha 1}$ in the incident beam path, because in this method only $K_{\alpha 1}$ spectral line is used as X-ray source, and a two- or three-bounce monochromator in front of the detector. Usually, a Ge(220) crystal monochromator is used. A Ge(440) crystal monochromator can be used to further improve the resolution at the expense of intensity. Before describing the work principle of this method, it is essential to emphasize that the sample should be prepared very carefully. For an efficient $K_{\alpha 1}$, Ge monochromator, slit and absorber apparatus should be used. While inserting the sample and adjusting the detector, detector and $K_{\alpha 1}$ source tube should stand at the same axis level. Sample is attached to the Cradle attachment centered by the help of such methods as vacuum, clamp, glass paste, gum, or magnet. $K_{\alpha 1}$ beam in the detector is seen in full intensity value. During the z or height adjustment of the sample at the first time, when z approaches to beam, the total intensity value of the beam goes to zero. Here, the zero value of the total intensity, tube, and the X-ray beam direction behind the sample meet in attachment surface. Here, we shall adjust the direction of the X-ray beam toward the middle of the side section thickness of the sample. For this reason, the middle of the getting zero full intensity value is calibrated as z value. Secondly, θ is scanned between -2 and $+2$, and the vision should be a triangle. Here, the aim is searching whether the surface of the sample has tilt or not. Sample adjustment is over if maximum peak point of the scan is at zero and it corresponds to the half of the full intensity value; otherwise, Z and θ scanning should be made again. Z axis-directed preferred planes are primarily examined by $\omega - 2\theta$ scanning for substrate, epilayer, heterostructure, MQW, and such nanotechnological devices. Data is taken from all layers and differentiated from each other. Among these planes, mostly c-directed symmetric (002), (004) and (006) planes are scanned. For example, while examining GaN layer, symmetric scans for these planes can be made.

In order to see crystal planes, sample will not be changed to powder, but it should be put into the system completely. For this reason, besides θ and 2θ, there should also be x, y, z, ω, χ, and φ axes (Fig. 10c), where ω is the angle between incident X-ray beam and the sample. HRXRD usually measures intensity of scattered X-rays as a function of ω and/or 2θ. There are different types of scans: first, a rocking curve which is a plot of X-ray intensity versus ω with the sample tilted, but the detector is set in an angle, second, a detector scan in which X-ray intensity is plotted against 2θ without changing ω (fixed source), and third, a coupled scan in which scattered X-ray intensity is plotted against 2θ, but if sample rotates, the detector also rotates, or in other words, ω changes in a way that is linked to 2θ. In this scan, the source is fixed, and $2\theta = 2\omega +$ offset. It shall be noticed that ω scan and 2θ scan are needed in the alignment process, and coupled scan is usually used to show the final result.

The diffraction planes can be classified to be symmetric planes and asymmetric planes. The angle between diffraction plane and incident angle is ω, while the diffracted angle is 2θ. Symmetrical planes are the planes that are parallel to the surface of the sample, while asymmetrical planes are the planes that have an angle to the surface of the sample (obtained by tilting the sample). Symmetrical and asymmetrical scans are both needed to be able to investigate different properties of materials. Symmetrical scans only provide information about in plane (parallel) lattice parameter. Among these planes, mostly c-directed symmetric (002), (004), and (006) planes are scanned. But (102), (111), and (121) are asymmetrical planes. Symmetrical plane measurement was used to find the tilt of the epitaxial layer, while asymmetrical plane measurement was used to find lattice mismatch, strain, and relaxation. Also, c-lattice length is determined from symmetric plane peaks, and a-lattice length is determined from asymmetric plane peaks. For example, while examining GaN layer, symmetric scans for these planes can be made. χ and φ axes are also used with ω and 2θ axes for (102), (111), and (121) asymmetrical planes. In this kind of analysis, ω, 2θ and χ values of crystallographic asymmetric planes or tilt angle should be determined with a software or by using vector operations. After calibrating these values on the device, φ scanning is made. After calibration of φ, determined from scanning peaks, sample is scanned in 2θ angle from right and left. For example, $\pm(1, 2$and 3) degrees.

For example, Fig. 12a shows a coupled scan plot, in which intensity is plotted against 2θ, and ω is changing. The figure shows the $\omega - 2\theta$ scans of AlxGa$_{1-}x$N/GaN ($x = 0.3$) structures with and without a high-temperature AlN interlayer and AlN buffer layer on sapphire substrates for the three high-electron mobility transistor

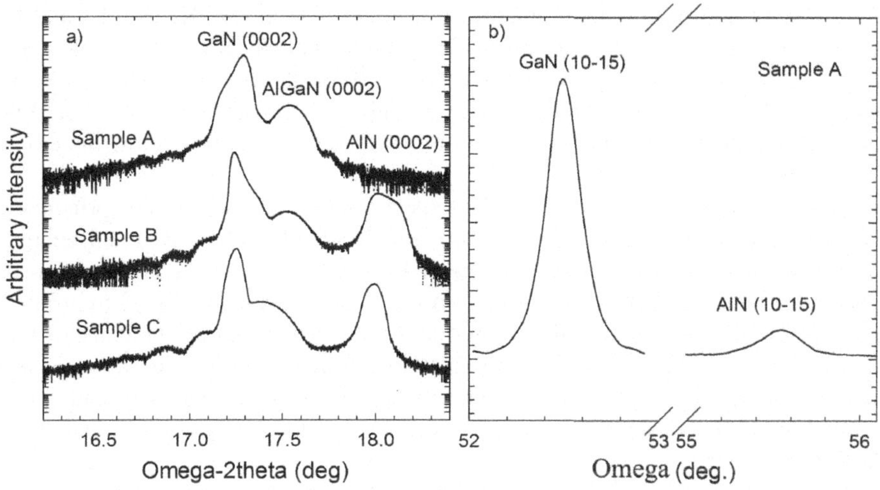

Fig. 12 **a** High-resolution Bragg reflections near the (0002) reflection of samples A, B and C. **b** The $(10\bar{1}5)$ rocking curve of the GaN layers and the AlN interlayer in sample A [36]

(HEMT) samples [36]. As shown in Fig. 12a, besides the AlGaN (0002) and AlN (0002) reflections, an intense GaN (0002) reflection along with its strong interference with AlGaN was observed, in turn indicating that the GaN, AlN, and AlGaN layers possess high-quality wurtzite structures. Here, miller planes of hexagonal structure are indicated by (hk(-h-k)l). In the skew geometry of XRD, the ω peak of the asymmetric $(10\bar{1}5)$ plane in the interlayer was the only one found and was shown by the ω scan of the GaN $(10\bar{1}5)$ reflection, as shown in Fig. 12b.

Qualitatively, the determination of the quality of crystal can be taken based on the full width at half maximum (FWHM) of the peaks. A perfect crystal will have a very sharp peak. A broadening of the curve can be related to the presence of mosaicity, dislocation, or curvature in the sample. Plane quality is determined by fitting the ω turning peak half height width (FWHM) to the Gaussian function. The quality of the FWHM increases when approaching to the FWHM value of 0.00435° (15arcdeg) gained from silicon calibration of Ge monochramator.

4.2.5 X-Ray Florescence (XRF)

The mechanism of this technique is easy to understand if compared with XRD. In XRD, only K_{α} source spectra are used, but in XRF, all the spectrum coming from the tube is used which is called primary. In this technique, irradiation of a sample by bombarding with primary X-ray beam from an X-ray tube causes emission of fluorescent X-rays with discrete energies characteristic of the elements present in the sample. It can be seen in Fig. 10d that when the whole beam comes over the sample, in all types of atoms forming the sample, the electrons in inner shells like

K, L, and M are excited, and luminescence effect is seen. So, K_α, K_β, L_α and L_β spectras are seen for every element in the sample. The spectra of these elements are compared with database, which was prepared for all elements, and constituent elements of the sample under study are determined. This determination should be made before the compound determination in XRD. It is useful to know the atoms forming this compound in XRD.

Analysis is made using one of two spectrometers of two design principles: wavelength-dispersive X-ray fluorescence spectrometers (WD-XRF) and energy-dispersive X-ray fluorescence spectrometers (ED-XRF). WD-XRF spectrometer uses a crystal monochromator as a grating to diffract the fluorescent X-rays to produce a spectrum. So, just a single wavelength at a time is received by the detector that measures their intensity. But the ED-XRF spectrometer permits the fluorescence radiation of all wavelengths emitted to reach the detector simultaneously, and it uses a pulse height discriminator to classify the energy of the X-ray photons that strike the detector.

All the elements till uranium can be analyzed by this technique. Geological and underground good samples are commonly examined with XRF technique. Especially for underground investigations, there are portative XRFs put in a bag or shaped like a gun. A historical self-portrait painted by Sir Arthur Streeton (1867–1943) has been studied with fast-scanning X-ray fluorescence microscopy using synchrotron radiation given in Fig. 13 figured by Hovard et al. [37]. Also, XRF technique is commonly used in the powder analyses with XRD [38].

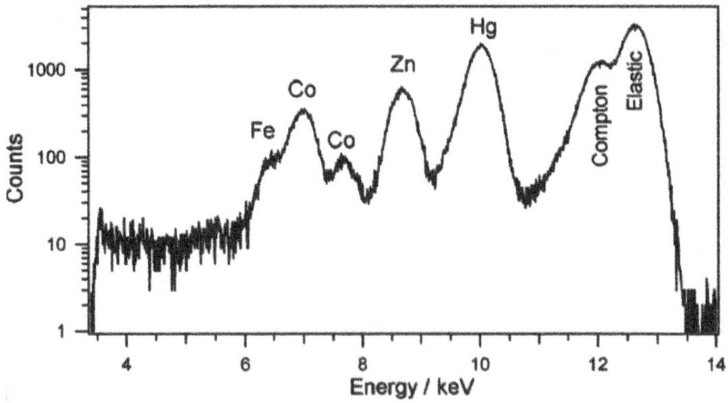

Fig. 13 Representative X-ray fluorescence spectrum of a 50 mm^2 area of the painting near the chin collected by 1 of the 384 detector elements of the Maia detector array. Several major elemental fluorescence lines and the scattering components are labeled [37]

4.3 Secondary Ion Mass Spectroscopy (SIMS)

In this technique, an ion gun is used to produce a precisely focused energetic primary ion beam (typically between 25 and 30 keV kinetic energy) that sputters the surface of the sample. Because of this primary ion sputtering, a collision cascade is induced in the upper surface layers of the sample. Due to energy dissipation resulting from these collisions, bonds will be broken and secondary particles will be emitted. The emitted secondary ions (positive or negative) are collected and analyzed by mass spectroscopy, where they are separated from each other according to their charge and mass ratios. Three types of signals are taken from the scattered ions. These are ion images, depth profiles, and mass spectra, respectively. So, 3d mapping of the elements and molecules can be made. As an example, SIMS depth profile of gallium-doped zinc oxide (GZO) thin films which were successfully deposited onto n-Si and glass substrates by RF magnetron sputtering at room temperature using a power of 200 W [39] is presented in Fig. 14. Ga and Zn homogeneity in the layers give very good results.

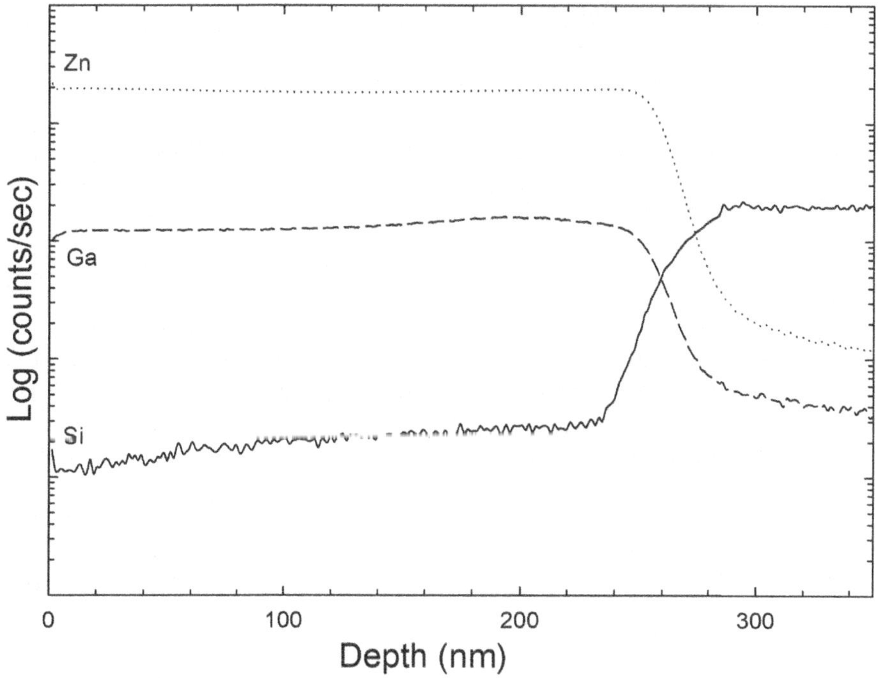

Fig. 14 SIMS depth profile of GZO thin films onto n-Si and glass substrates [39]

4.4 Stress Measurements

In stress measurements, there are two modes: the "iso inclination" and the "side inclination." In the iso-inclination mode, the 2θ is measured by aligning the χ plane (the plane that connects the normal of the sample surface to the normal on lattice plane) to the detector scanning plane (2θ scan). In the side inclination mode, the 2θ is measured by setting the detector scanning plane perpendicular to the χ plane. This mode permits the use of large values of the angle χ to measure the bottom or curved sections of a gear, which are difficult to measure by the iso-inclination method.

Using XRD and Bragg's law, stress measurements can be performed for semiconductor and metal material by measuring detector ane 2θ while varying the angle χ, where χ is formed by the normal of the sample surface and the normal to the lattice plane. In these measurements, if there is stable stress in any direction, the diffraction peak of this direction is used. Setting the angle of the X-ray beam at fixed value, the value of d at the position of the diffraction peak can be found by substituting $n = 1$ in Bragg's law, $2d \sin(\theta) = y$. For a symmetric and asymmetric peak in the ω scans, the peak will be determined through scanning in a narrow range. The value of d and data corresponding to changing χ value is used the stress equation;

$$d = d_0 \frac{1+\nu}{E} \sigma \sin^2(\chi) - d_0 \frac{\nu}{E} (\sigma_{11} + \sigma_{22}) \tag{8}$$

According to this equation, the plot of d versus $\sin^2(\chi)$ will be linear. So, by performing a linear fit, and using the slope m obtained from the fit, the stable stress can be calculated from the formula:

$$\sigma = \frac{m}{d_0} \frac{E}{1+\nu} \tag{9}$$

where Young's modulus E, Poisson ratio ν, and lattice spacing in stress-free conditions d_0 are known for the material. As an example, d versus $\sin^2(\chi)$ curve is given in Fig. 15 for Ge semiconductor. χ angle is taken between 0 and 4° with steps of 0.1° for stress measurements by using side inclination mode. By using symmetric (004) ω-scans we find d of the peak with the help of Eqs. (8) and (9) as mentioned above.

Stress curve is plotted. The equation of the regression straight line is found from the plot as: $y = -0.0004x + 1.3663$, where $x = \sin^2(\chi)$. Using the slope of the straight line, which is $m = -0.0004$, Young module, for Ge $E = 150$ GPa, Poisson ratio $\nu = 0.3$, and universal $d_0 = 1.3577$ values, the stable stress was calculated and found to be −34 MPa.

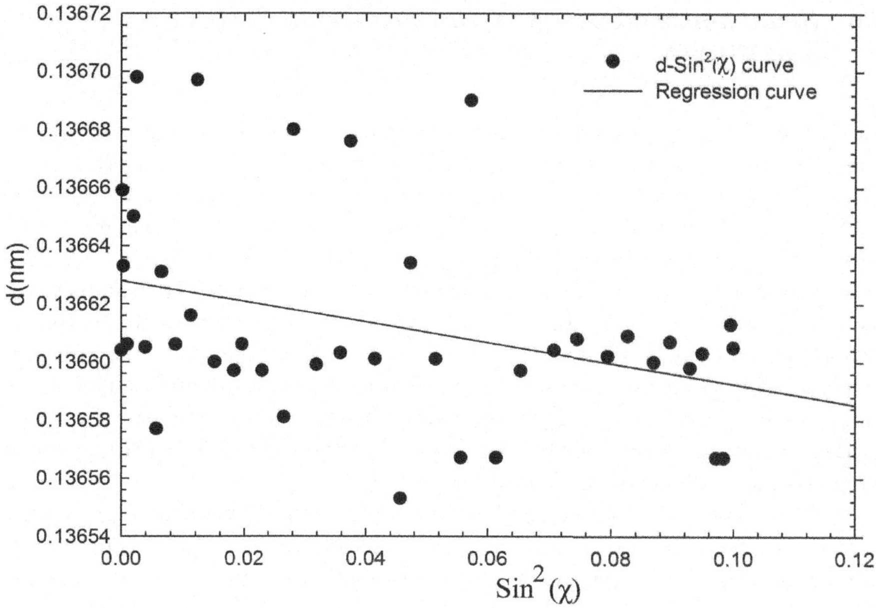

Fig. 15 *d* versus $sin^2(\chi)$. plot for Ge semiconductor

4.5 Texture Calculation

In this method as in XRD, $K_{\alpha 1}$ and subdued $K_{\alpha 2}$ spectral peaks are used. Generally, pinhole apparatus is used, and pole images are made. Primarily in a textured material, diffraction peaks should be seen, and when the Bragg condition is maintained with θ, the range giving the *d* values is kept constant. After this, diffraction peaks are tried to be seen again and again with Cradle attachment and χ tilt and φ turn angles in a certain range. After this, for analyzing the data peak intensities which have d values gained from every χ (tilt) and ϕ (rotation) angle scans, three-dimension pole (χ, ϕ, and intensity) is plotted. Texture mapping can be given as an example to this situation. In texture analysis, as standard measurement ranges, 2θ range is 2.5–70°, χ tilt angle range is 0–85° or 0–350°, and ϕ turning angles can be used in every 6-degree steps. Pole graphs need wide data analysis because of fingerprints of the X-ray beam by sintilization detectors. Texture analysis became more convenient with fast pixel detectors. In texture analysis, spreading of the lattice plane direction is shown in a half sphere. Sample may be weak, heavy, and normal texture. In investigations, direction of a plane can be determined with 2θ scan without using pole graph. In the degree of direction, texture percentage of crystal lattice plane is determined by the ratio of single peak area to total area of peaks.

4.6 Small and Wide-Angle X-Ray Scattering (SAXS and WAXS)

Small-angle X-ray scattering (SAXS) is an analytical technique that measures the intensities of X-rays scattered by a sample as a function of the scattering angle. Measurements are made at very small angles, typically in the range of 0.1–5°. A SAXS signal is observed whenever a material contains structural features on the length scale of nanometers (1–100 nm). SAXS method is one of the most versatile techniques for the structural characterization of nanomaterials. The samples may be solid objects, powders, gels, or liquid dispersions, and they may be amorphous, crystalline, or semi-crystalline. SAXS also gives knowledge about the structural dynamics. Measurements can often be done in situ, and only, minimal sample preparation is required. It is essential to use a detector which has a high linearity, a high dynamic range, and negligible intrinsic noise. For high-end SAXS instrumentation, a high spatial resolution of the detector is beneficial.

Wide-angle X-ray scattering also known as wide-angle X-ray diffraction (WAXD) is the same technique as small-angle X-ray scattering (SAXS), except that the distance between the sample and the detector is smaller, and thus, diffraction maxima at larger angles are observed for WAXS. Wide-angle X-ray scattering (WAXS) probes the material for the much smaller length scale (interatomic distances). It is used to determine the crystalline structure of polymers and to determine the chemical composition, phase, or texture of films. Wide-angle X-ray scattering can be measured simultaneously with SAXS.

Small-angle X-ray scattering and wide-angle X-ray scattering (SAXS and WAXS) are complementary techniques. WAXS usually covers angular 5–60°, while SAXS covers much smaller range, which is up to 5°. WAXS and SAXS methods do not harm the sample. As an example, spherical polyelectrolyte brushes analysis with SAXS is given in Ref. [40].

4.7 Atomic Force Microscopy (AFM)

AFM is a very high-resolution scanning probe microscope, with a resolution of fractions of Angstrom. It is used widely for examining organic and inorganic sample surfaces. It is in the first plan for examining structural properties of surfaces of optoelectronic devices. AFM contains a cantilever with a sharp tip (probe) at its end that is used to scan the specimen surface at the nanoscale level. Fundamentally, it depends on the interaction of the tip with atoms (Fig. 16). AFM has three different modes. These are contact mode, where the tip touches the surface of the sample; noncontact mode, where the tip does not touch the surface; and the tapping mode, where the tip hits the sample surface. Tips are made of different materials for different aims. Tips are made of Si, Si_3N_4, Au, and Pt, or other molecular materials coated with them. For this reason, in each mode, different tips are used. Here, forces coming out from

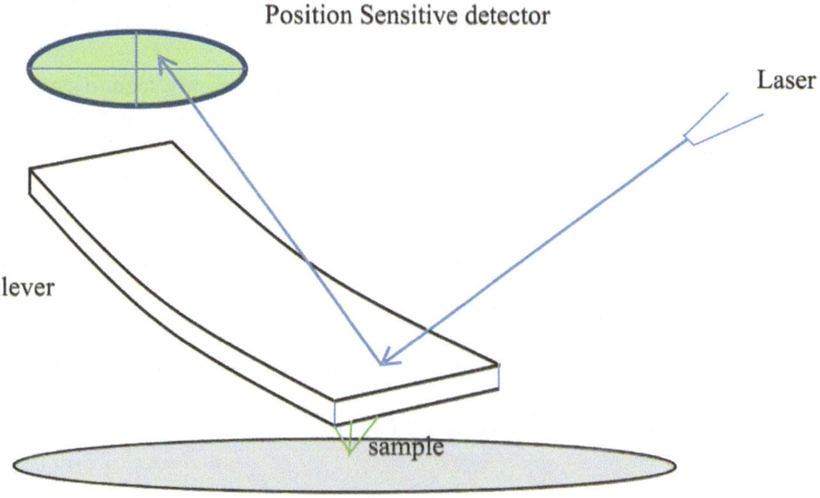

Fig. 16 Schematic of atomic force microscopy (AFM)

Wan der Waals interaction (F = kz) transforms to motion on the tip made up of Si_3N_4 and SiO_2 materials. This motion transforms to data with the help of a photodiode which makes laser trace. While the tip is approaching to the surface pulling and pushing, Wan der Waals interacts and a resonance period starts. Roughness, friction, electrical properties, magnetic properties, nanomechanical properties, and corrosion can be determined about the sample.

Surface images of the AlInN/AlN/GaN/Al_2O_3 HEMT (high-electron-mobility transistor) structures grown by MOCVD with various thicknesses of GaN top layer and AlInN layer having different indium compositions are given in Fig. 17 [41], where the scan area of the AFM images of GaN top layer of the samples is 4.0 × 4.0 μm^2. The thickness values of the buffer AlN layer for samples A, B, and C are 14.0, 8.0, and 14.0 nm, respectively, and for GaN, cap layer are 1.0, 2.0, and 4.0 nm

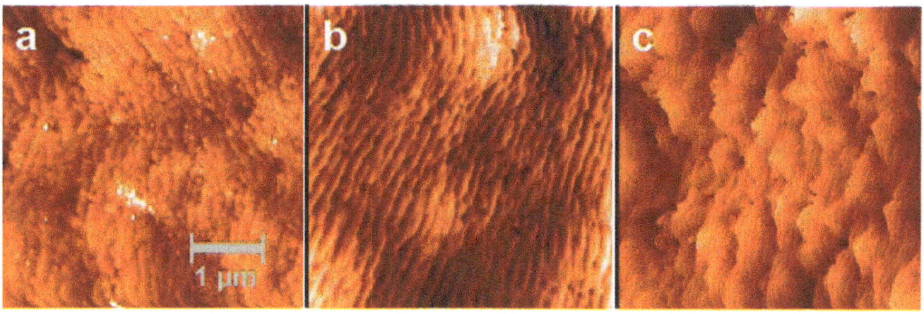

Fig. 17 AFM images of AlInN/AlN/GaN heterostructures with different AlN Buffer Thickness [41]

for samples A, B, and C, respectively. As it is clearly seen in these images, surfaces are similar and are composed of steps and terraces.

Not withstanding the pure atomic steps that are commonly seen in sample C, the black points, which are typical for GaN surfaces [42–46], still appear on the surface image of the top layer. The average root mean square (RMS) roughness of samples A, B, and C was attained as 0.450, 0.270, and 0.570 nm in a scan area of $4 \times 4\ \mu m^2$.

Lattice mismatch between sapphire and AlN increases the dislocation density in the AlN layer, and dislocation is carried to the surfaces causing cracks. In this case, the nucleation of AlN and GaN interlayers can be grown to prevent surface cracking. As a result, the surface morphology looks good [43].

4.8 Spectroscopic Elipsometry (SE)

Elipsometry method is widely used for the examination of the films over substrates for measuring the optical activity in surface investigations [47], where optical activity of the films is known as polarimetry. It is also used for determining the thickness of unknown films, optical properties of materials, and surface density of overlayers on reflective materials. But it must be noticed that materials properties and other physical parameters determined by ellipsometry are inherently dependent on the physical model assumed by the user. SE method is a harmless method for determining optical properties of semiconductors, metals, and insulators. It is also used for determining the boxing of the reflected light from the surface of the film [48].

Ellipsometry in general makes use of the fact that the polarization state of light may change when a light beam is reflected from the surface of a sample. Ellipsometry analyzes this change of the state of polarization and yields information about the layers of the thin film that are often even thinner than the wavelength of the probing light. A basic ellipsometer consists of a source of light, a polarization state generator (PSG) placed in the path of the beam before the sample, a polarization state analyzer (PSA) placed in the path of the beam behind the sample, and a photo detector. The PSG controls and analyzes the polarization state of the probing light, while the PSA controls and analyzes the polarization state of the reflected. There many different types of PSGs and PSAs exist. In ellipsometry, the probing beam hits the sample under an oblique angle of incidence. Consequently, the phase change caused by the probing beam to the reflectivity of the sample is different for light with polarization parallel and perpendicular to the plane of incidence, respectively (p- and s-polarizations). Ellipsometry measures both the ratio of the reflectivity and the relative phase change of the p- and s-components and yields the ellipsometric angles ψ and Δ:

$$\tan \Psi = \frac{|R_p|}{|R_s|} \tag{10}$$

$$\Delta = \Delta_p - \Delta_s \tag{11}$$

Table 3 Solar cell structure

n-GaAs Cap layer 50 Å
n-GaInP 2000 Å
p-GaInP 2000 Å
p-GaAs 150 Å
n-GaAs 150 Å
n-GaAs 2000Å
p-GaAs buffer layer 3000 Å
3″ p-GaAs (100) substrate

where R_p and R_s are reflectivity of p- and s-components, respectively, while Δ_p and Δ_s are the induced phase change on these p- and s-components, respectively. By using elipsometric equations, elipsometric ratio is calculated by the following formula;

$$\rho = \left| \frac{R_P}{R_S} \right| = e^{i(\Delta_p - \Delta_s)} = \tan \psi e^{i\Delta} = \frac{R_p}{R_s} \tag{12}$$

These equations are fitted with suitable models based on Fresnel equations with elipsometric data, so they give the optical parameters of the solid material [49]. With this SE method, primarily, refractive index, film thickness, extinction coefficient, and crystallite and alloy ratio can be determined.

SE analysis for the $Ga_{0.516}In_{0.484}P/GaAs$ solar cell structure (Table 3) was measured at an angle of incidence 70° by Kinaci, et al. [50]. Figure 18a shows the real and imaginary parts of the dielectric function against photon energies in the range 0.6–4.7 eV. It should be noticed that (E_0, $E_0 + \Delta_0$, E_1 and $E_1 + \Delta_1$) in Fig. 18a are the critical point energies. We have calculated numerically the second derivative spectra of the real part of the experimental dielectric function to perform the line–shape analysis, which was given in Fig. 18b. The experimental curves were fitted to determine the critical point energies. So, E_0, $E_0 + \Delta_0$, E_1 and $E_1 + \Delta_1$ were found as 1.92, 2.05, 3.25, and 3.45 eV, respectively [50].

4.9 Photoluminesance Spectroscopy (PL)

Photoluminesance spectroscopy (PL) is an optical method used in characterization of materials. In this technique, sample is exposed to electromagnetic radiation and the energy is absorbed and the material is transferred to a higher energy state. The release of energy through the emission of radiation during relaxation of the material to return to lower energy states is called photoluminescence (PL). PL is a very simple method, where the intensity of the incident wave and spectrum can be determined very fast, and it does not deform, or heat the sample. PL may be used in determining doping ratio of III–V group semiconductors, structural defects, and band

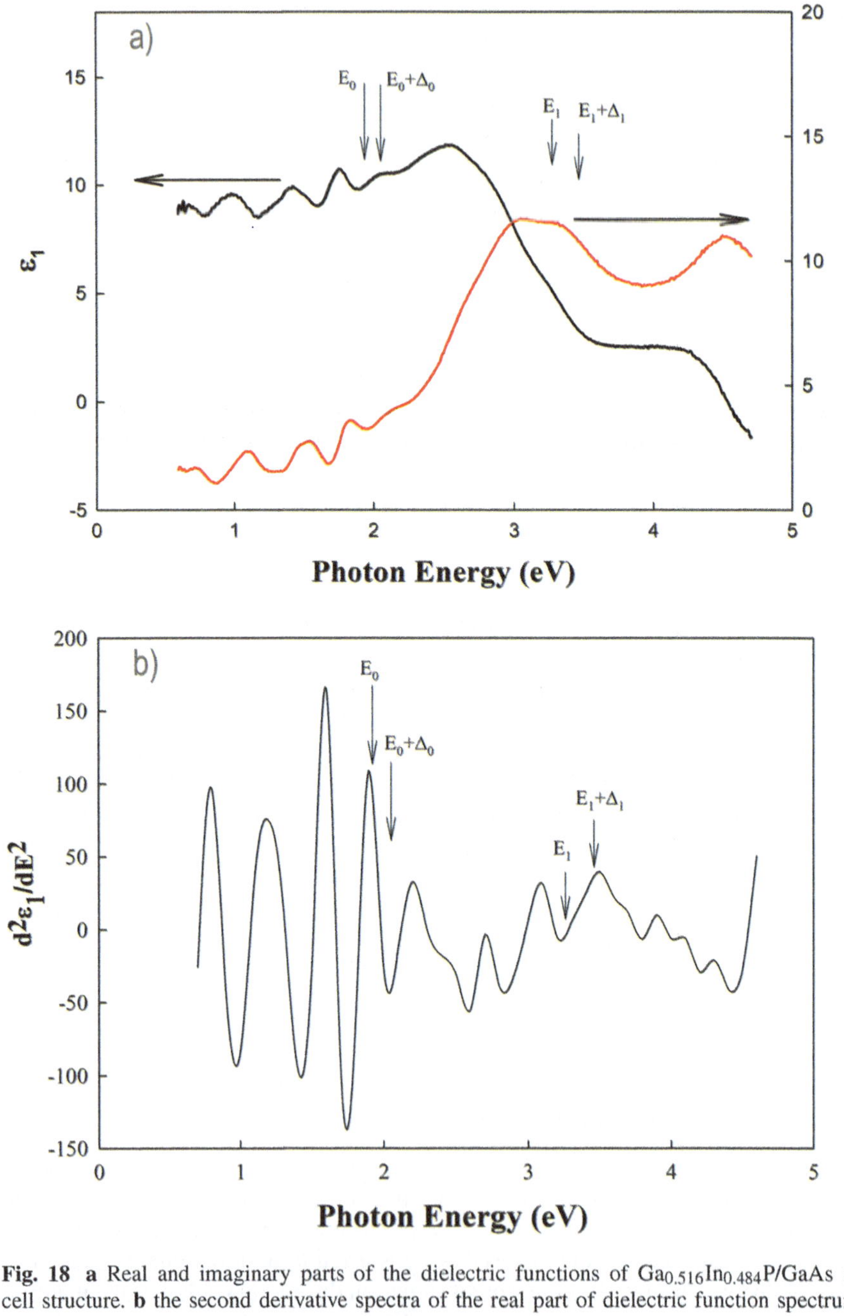

Fig. 18 **a** Real and imaginary parts of the dielectric functions of $Ga_{0.516}In_{0.484}P$/GaAs solar cell structure. **b** the second derivative spectra of the real part of dielectric function spectrum of $Ga_{0.516}In_{0.484}P$/GaAs solar cell structure [50]

gap. Intensity and spectral property of PL give knowledge about different properties of the sample [51]. It is an efficient method in determining forbidden band gap and different behaviors. PL is sufficient in determining forbidden band gap at room temperature in helium atmosphere; even it is more stable, for project studies. When producing LED, detector, and photodetector, it is a routine spectrum technique [52]. By using PL measurements, it is possible to control conditions which come out with surface variations. It gives optical knowledge about deformed region near to surface under high pressure. It can examine optical behaviors of interfaces. Surface area of the sample which is exposed to PL may decrease the PL intensity. According to other spectrum devices, sample surface roughness or structural irregularity is also not limiting [53]. In semiconductor materials which have indirect band gap, the PL has less efficiency. Irradiative transitions cause relaxation of excited populations in these samples. This situation may increase population because of structural defects such as surface oxidation. For this reason, fast and irradiative excitations may occur. PL peak may be seen even in this condition. This peak may be characterized with both radiative and irradiative excitations. With PL, it is not possible to determine irradiative traps directly. But the traces of these traps may be seen clearly in some measurement types of PL. It is a difficult investigation to examine interface and doping conditions by using PL spectrum. If the mentioned situations are radiative mechanism situations, they can be examined in PL spectrum and the size of the corresponding PL peaks permits to examine the doping condition of the sample. But it is difficult to measure the density of these levels, and it is exhausting that the exciting intensity of PL peaks is in ratio [52].

As an example, the PL measurements of hetero-junction GaAs and solar cell are taken at 300 K (room temperature), and it is shown in Table 4 as an example. Forbidden band gaps and alloy ratios(x) of semiconductor materials forming GaAs solar cell are gained from PL spectra peak positions. PL intensity versus wavelength plot is given in Fig. 19 for hetero-junction GaAs solar cell structure. Forbidden band gap of GaAs semiconductor material is calculated as 1.42 eV by using these peak positions. $Al_xGa_{1-x}As$ window layer's forbidden band gap is found as 1.865 eV. Aluminum composition ratio in $Al_xGa_{1-x}As$ alloy is found as %36.00 by the help of forbidden band gap [54].

Table 4 Schematic view of hetero-junction GaAs solar cell grown at 300 K

p+GaAs (Cap) 20 nm
p-AlGaAs 30 nm
p-GaAs 500 nm
n-GaAs 500 nm
n + GaAs (buffer)
3" n-GaAs (100) substrate

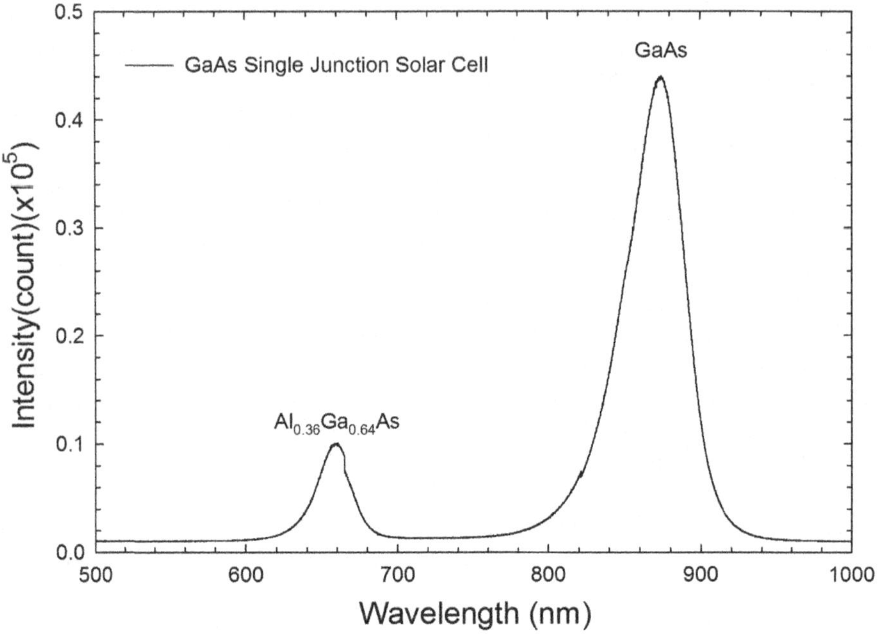

Fig. 19 PL intensity versus wavelength for hetero-junction GaAs solar cell

4.10 Raman Spectroscopy

At the beginning of twentieth century, first, Raman spectroscopy was performed for determining solutions and organic materials by Raman [55]. The measurements gained by the help of this method can be taken by not deforming the material. Raman spectroscopy method is used as material characterization by the vibrations belonging to the material. In the middle of the twentieth century, it is started to be used commonly in semiconductor technology. When a monochromatic light beam (usually laser beam) excites a material, most of the scatters over the material come out elastically. In these scatterings, no energy variation is seen among atoms and molecules. For this reason, the incident electromagnetic radiation scatters in the same wavelength. Scattering in this wavelength may be seen together with less intensity of non-elastical scattering. When the energy difference between incident and scattered light is investigated, it is observed that the energy of inelastic scattered light is different from material to material [56], which means that vibrational energies differ from material to material or each material has its fingerprint. If the scattering is elastic [57], there is no difference in the energy between incident and scattered photons, and it is called Rayleigh scattering. If the scattering between the molecule and the photon is inelastic, it is called Raman scattering.

Vibrational energy after the collision and the wavelength of the inelastically scattered photon are inversely proportional with each other. Momentum and total energy are conserved during scattering. After inelastic scattering, the molecule moves to a new energy state, and the scattered photon shifts to a different energy, or to a different frequency. The energy difference of the scattered photon equals to the energy difference between the initial and final energy states of the molecule. If the final energy state of the molecule is higher than the initial one, or in other words the vibrational energy of the molecule increased after the collision, then the scattered photon will be shifted to a lower frequency (lower energy) and that the total energy of the system remains constant. This shift in frequency is called a Stokes shift, or downshift. If the energy of the final state of the molecule is lower than the initial state, or in other words the vibrational energy of the molecule decreased after the collision, the scattered photon will be shifted to a higher frequency, which is called an anti-Stokes shift, or upshift, and in this case, the molecule should be excited [56, 58].

Some positive and negative aspects of Raman spectra:

(a) Completive data may be gained by different adjustments.
(b) It is easy to apply.
(c) It does not require material preparation.
(d) Experimental procedure may continue repeatedly.
(e) Adding some noise to the spectrum may cause complications.

Raman spectra of the high-resistivity GaN epilayers are given in Fig. 20a grown on 6H-SiC substrate with different buffer structures which were recorded at room temperature [26]. As shown in Fig. 20a, in the back-scattering geometry measurements, the allowed E_2 (high) and the A_1 (LO) phonon modes of the c-direction-oriented GaN and the E_2 (TA), A_1 (LA), E_2 (TO), and A_1 (LO) phonon modes of the 6H-SiC substrate can be fitted as Ref. [59]. In Fig. 20b, GaN E_2 (high) and A_1 (LO) mode's Raman spectras are seen separately [26].

For a detailed Raman analysis, a second sample can be given. Raman spectra for samples A and B of details in Table 5 are shown in Fig. 21, where spectra are gained using micropublisher 5.0 camera software. During these measurements, the 785 nm wavelength laser is used for 60 s for both samples.

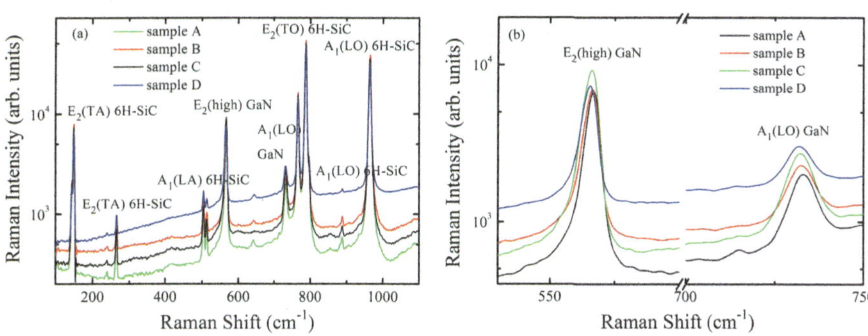

Fig. 20 **a** Raman spectra obtained at $T = 300$ K for the HR-GaN epilayers grown on 6H-SiC substrate with different buffer structures, for back-scattering (surface) geometry ($\bar{z}(x, x)z\bar{z}(x, x)z$). **b** E_2 (high) and A_1 (LO) phonon modes in the HR-GaN epilayers [26]

Table 5 Schematic diagram of (a) Sample A and (b) Sample B [60]

Sample A	Sample B
p-InGaN layer 20 nm	p-InGaN layer 20 nm
p-InGaN layer 50 nm	p-InGaN layer 50 nm
Graded InGaN layer 25 nm	–
p-InGaN layer 200 nm	InGaN layer 220 nm
n-GaN layer 1.9 μm	n-GaN layer 1.9 μm
Ud–GaN layer 1.9 μm	Ud–GaN layer 1.9 μm
LT–GaN 2.45 min	LT–GaN 2.45 min
4H–Al_2O_3 substrate	4H–Al_2O_3 substrate

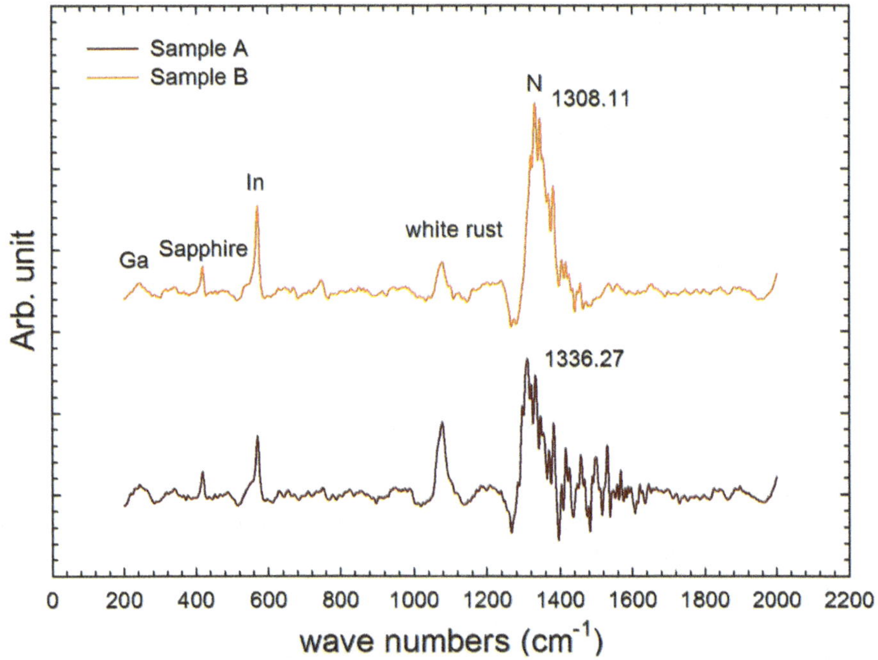

Fig. 21 Raman spectras for Sample A and B [60]

Raman spectra can be used for determining, crystallite size, identification of material, and stress. In Table 6, one can see the results for peak center, peak height, peak area, and full width at half maximum values (FWHM) values for InGaN/GaN structure. There is also one more layer detected by Raman measurement that is white rust formed by touching the air for a long time.

Amount of materials in the samples are estimated by peak intensity ratio and peak area ratios that are shown in Table 6. Crystallite size can be estimated using FWHM data in Raman spectra. Atomic radius of materials can be calculated by

Table 6 Raman analysis data for two InGaN LED structures

	A					B				
	Ga	Al_2O_3	In	N	White rust	Ga	Al_2O_3	In	N	White rust
Peak center	244	419	571	1336.27	1078	244	419	571	1308.11	1078
Peak area ratio(%)	3.28	1.95	8.20	17.60	3.24	3.21	2.62	9.47	17.39	6.92
Peak height ratio(%)	2.22	5.83	10.74	76.03	5.16	5.82	7.98	18.89	45.53	21.76
FWHM	34.09	8.91	12.48	26.69	–	33.83	9.54	10.65	25.37	–

equation $r = r_0.A^{1/3}$ [61], where r is the atomic radius, A is atomic number, and r_o equals 1.2 fm. If FWHM values of Ga and N are compared with their atomic radii, it is seen that they are in good agreement. FWHM values of sample A are generally bigger than those of sample B. This means that crystal quality is better for sample A. As mentioned before stress can be calculated from Raman spectra. During the calculation of stress values, unstressed peak center values must be found from RRUFF database. These values are subtracted from experimental values and divided by calibration coefficients to gain stress in MPa. This relation is shown with equation $\sigma(\text{MPa}) = \Delta w(\text{cm}^{-1})$ [62].

In Raman calculation, N peak center value is used because it is the most intense peak in both samples. Stress values from Raman are calculated as 4.27 and −23.89 Mpa. There is one more layer detected by Raman that is white rust ($3Zn(OH)_2ZnCO_3H_2O$) . Peak intensities of white rust are different in both samples, which is related to different cleaning durations.

4.11 Fourier Transform Infrared Spectroscopy (FTIR)

FTIR spectroscopy is related to the measurement of interatomic bond vibrations of the (IR) absorbing material. IR spectrum is modulated with strain, shrinkage, and twist bonds in the structure. Modulation of positions in these interatomic bond vibrations is taken into resonance with unstable positions in IR region. Every functional group has its own vibrational frequency and spectrum [63].

IR spectroscopy is a fast and simple technique for determining different functional groups. As in all spectroscopic methods, IR spectroscopy is also dependent on the interaction of atoms or molecules with electromagnetic radiation. Infrared (IR) spectroscopy is the technique to explain inorganic, organic, and biological structures [64–67]. Shortly, IR spectroscopy is the technique, which is used for measuring vibrational spectra, that is absorbed in a wide range dependent on wavelength. IR measurements can be done using transmission or reflection [68], but transmission situation is more common. The wavelength range of IR spectrum is between 0.78 and 1000 μm (12500–10 cm^{-1} wave number). This may be grouped as far IR (100–10 cm^{-1}), middle IR (4000–100 cm^{-1}) [69, 70], and near IR (12500–4000 cm^{-1}) [71]. Near and middle IR are the most commonly used regions [59].

IR rays cause atoms to vibrate with an increasing amplitude around bonds connecting atomic groups. IR light does not have sufficient energy to excite electrons as ultraviolet or high-energy radiations do. Because functional groups of organic molecules are formed by ordering of connected atoms belonging to functional group, absorption of IR energy by organic molecules comes out typically by atoms found in determined functional groups. Vibrations coming out like this are the vibrations, which are quantized by the absorption of IR energy by compounds. The position of an IR absorption band is measured as the reverse of the centimeter (cm^{-1}), and it is indicated by wave number (ν) or wavelength (λ) . Wave number is the vibration number of the wave in a unit length, while the wavelength is the distance between two tops of these vibrations. It is given by $\bar{\upsilon} = 1/\lambda(\text{cm})$ or $\bar{\upsilon} = 10.000/\lambda(\mu\text{m})$.

Vibration of covalent bonds resembles tiny springs, which hold atoms together. Atoms may vibrate as they are adjusted in certain frequencies. Excitation of the molecule from one vibrational energy to the other happens by absorption of IR rays at a certain energy (because $\Delta E = h\nu$).

Molecules may vibrate in two different types. These are;

1. Strain vibrations
   a. Symmetric strain; atoms at two sides of the bond approach and get further at the same time.
   b. Asymmetric strain; one of the bonds gets longer and the other gets shorter at two sides or just the inverse of this.
2. Bending vibrations
   a. Inplane bending (shearing, shake)
   b. Out-of-plane bending (twist, shake front, and back)

Strain frequencies in IR spectra may have two reasons. These are masses of the bonded atoms (heavy atoms have less vibrational frequency if compared with light ones) , and the hardness of the bond. Trio bonds vibrate at higher frequencies than double bonds, and they are harder. The same relation is present between mono and double bonds. IR spectra of a compound are like its fingerprint. For this reason, if there are two different IR spectras, it is certain that they are different compounds. IR spectra of the same compounds are also the same [72]. Results in IR spectra are taken as absorption versus frequency plot.

In FTIR systems, no monochromater crystal is used. Spectrum is gained in time scale instead of frequency scale if all the frequencies interact with the sample. Data gained in time scale are named by interferogram. Interferogram is the Fourier transform of the absorption spectrum. Inverse Fourier transform operations are changed to data in interferogram frequency scale with the help of computer. Therefore, the well-known absorption spectra are obtained. Another function of the computer is gaining the spectrum in time scale many times, storing this data, and later measuring the total signal without noise. Here, a system called Michelson Interferometer is used [73]. FTIR spectroscopy is faster than IR spectroscopy, where the spectrum is registered in a few seconds. Sensitivity of the spectrum does not change because there are no solids and a high distorted spectrum is gained. Using FTIR spectroscopy, chemical bond and molecular structure of the top of a surface can be determined in a few microns. It is important to use IR spectroscopy in investigating organic materials with covalent bonds. These type of materials are analyzed in 2500-25000 nm wavelength range.

As an example, FTIR spectras for flash conductive-bridging rem application, SiO_2 grown on Si/W and Cu doped SiO_2 films are given in Fig. 22 [74]. The Si–O–Si bonding and the Si–O bonding were observed at 1060 cm^{-1}, 835 cm^{-1}, and 461 cm^{-1}, respectively, in the SiO_2 film. For the Cu-doped SiO_2 film, there was additional Cu–O bonding appearing at 606 cm^{-1} [74].

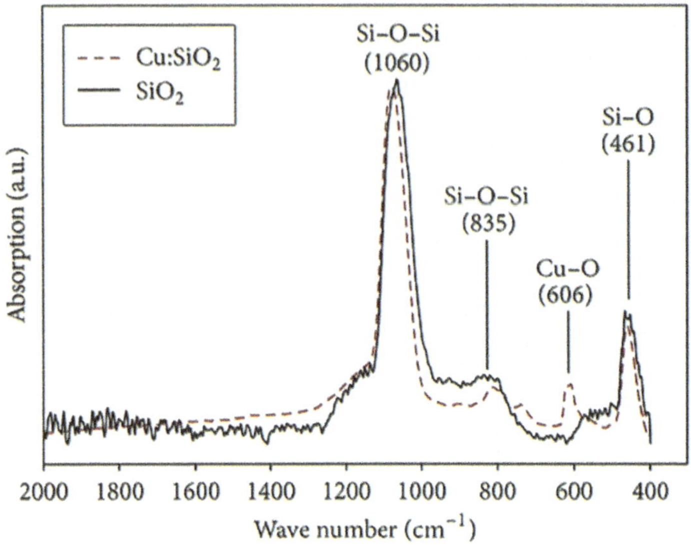

Fig. 22 FTIR analysis of the deposited SiO_2 and $Cu{:}SiO_2$ films [74]

4.12 Ultraviolet–Visible Spectroscopy (UV-Vis)

The principle of UV-visible spectroscopy is based on the absorption or reflection of ultraviolet or visible light light of wavelength in the range 160–780 nm by a material, which results in the production of absorption spectra, and it depends on the interaction between light and matter. UV-Vis spectroscopy is mostly used for quantitative determination of molecules in a solution, or inorganic ions, or complexes. Most molecules absorb UV or VIS wavelengths, and a spectrum formed by absorption bands shows the structure of the molecule.

Charge transfer transitions among σ π, and n molecular orbitals (both in organic molecules and in complexes) are seen. Also, there are transitions between d and f orbitals (in complexes) [75].

In organic compounds, it is difficult to examine excitations due to absorbing light at low wavelengths ($\lambda < 185$ nm). For this reason, vacuum is required for work on UV region. Electrons causing the absorption in organic molecules are electrons making bonds (electrons in bond orbitals π and σ) and electrons in non-bonding orbitals like sulfur, oxygen, nitrogen, and halogens which are found around atoms sharing electron pairs.

In thin film semiconductor structures, refractive index, energy band gap, and such optical coefficients can be determined by analyzing absorption and transmission spectra. Depending on the fact that a semiconductor material absorbs photons with larger energy than the forbidden band gap, and transmits the ones of smaller energy, it may be suggested that edge of the absorption spectra may correspond forbidden energy value.

Using the percent transmittance (T) measured with the spectrometer, the optical absorption coefficient of a thin film material with thickness d is determined by;

$$\alpha = -\frac{\ln(T)}{d} \tag{13}$$

In direct band gap semiconductors such as CdS and ZnO, the relationship between optical absorption coefficient and band gap energy is given as;

$$\alpha h\upsilon = B(h\upsilon - Eg)^{1/2} \tag{14}$$

Here, B is a constant, h is Blanck's constant, ν is the radiation frequency, $h\nu$ is the energy of the incident photon, and E_g is the band gap energy [76]. Optical band gap of films may be determined by fitting the linear region of $(ah\upsilon)^2$ versus $h\upsilon$ plot. By this method, when $(ah\upsilon)^2 = 0$, the intercept point of the energy axis corresponds to the value of forbidden band gap energy. When the semiconductor material has indirect band gap, exponential term ½ in equations is replaced by 2 [76].

As can be understood from the curve energy versus transmittance shown in Fig. 4.19 for n-type silicon, transmittance is about %90. Also, it is possible to measure forbidden band gap by using transmittance. For this operation, Tauc's plot is used given above. Changing of optical transmittance and $(ah\upsilon)^2$ value versus photon wavelength/energy is given in Fig. 4.19. In this graph, x-axis intercept of the plot is known as forbidden band gap (Fig. 23).

Using Eq. (13), the absorption coefficient can be calculated from the transmittance. The plot of the absorption coefficient against radiation energy is given for n-type silicon in Fig. 24. In the plot, there can be seen a broad large peak with two smaller peaks. The position of the maximum peak for n-type silicon is seen around 2.33 eV.

UV-VIS can give wide optical data about the material. Using transmittance and reflectivity, the refractive index can be found. The following equations can be used;

$$n = \frac{\sum_{\lambda=\lambda_0}^{\lambda_m}\left(\frac{1+\sqrt{(1-T^2(\lambda))}}{T(\lambda)}\right)}{m} \tag{15}$$

$$n = \frac{\sum_{\lambda=\lambda_0}^{\lambda_m}\left(\frac{1+R(\lambda)-\sqrt{R(\lambda)})}{1-R(\lambda)}\right)}{m} \tag{16}$$

In these equations, m is the number of transmitters and reflectors for every wavelength value. Transmittance and reflectance spectra for of Corning glass are given in Fig. 25. By taking a certain λ value after the absorption edge, and using T(λ) /100 and R(λ)/100 at this selected wavelength, then substituting in Eqs. (15) and (16) given above, the refraction index is found as 1.530 from transmittance and 1.553 from reflectance. The difference between two situations is instability between 200 and 360 nm. It will be more suitable not to take this region.

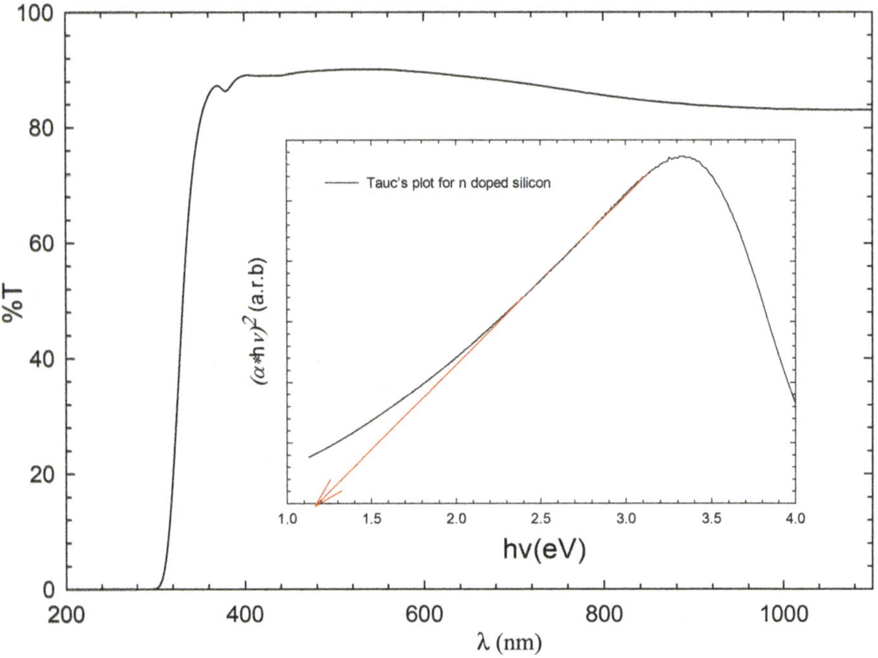

Fig. 23 Transmittance versus wavelength plot for n-type silicon, and $(ah\upsilon)^2$ versus photon energy for silicon film (inset)

5 Conclusions

As a conclusion, there are ways to examine wide number of materials and their properties including every part of the electromagnetic spectrum with spectroscopy techniques. These techniques are developing continuously. Using spectral mapping, visualizing in 3D is an opportunity to improve characterisation. It is doubtless that development of nanotechnology is dependent on the development of these techniques. Besides, some of the questions that will be asked in the future are as "there is nanotechnology but why there is no Pico technology?" and qubit instead of 0 and 1 maintained by spin motions nowadays. That is 0 and/or 1 at the same moment, began to be mentioned for quantum computers for electronic bazaars. Although traditional devices are rapidly developing because their physics is not changing using frequency, bazaar percentage is too large. The life time of a device is changing between 15 and 30 years, because structural, electrical, and optical properties of the device are important, and these types of devices have become always popular.

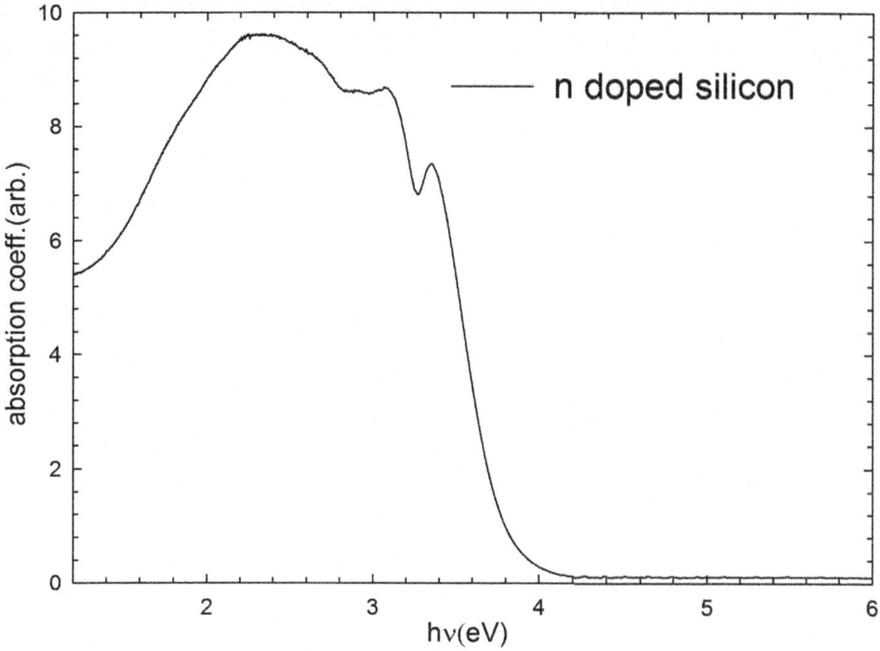

Fig. 24 Plot of the absorption coefficient against radiation energy for n-type silicon

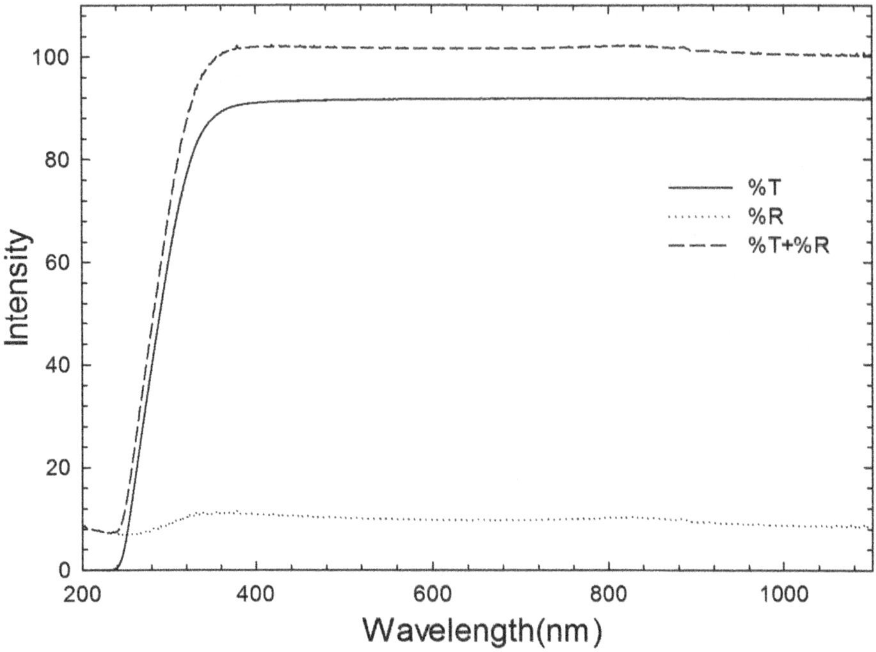

Fig. 25 Transmittance and reflectance spectra for Corning glass

Acknowledgements I would like to thank our many colleagues and also Director Prof. Dr. Süleyman Özçelik, Dr. Ömer Akpınar, Dr. Gürkan Kurtuluş, Dr. İlknur Kars, Dr. Nihan Akın Sönmez, Dr. Tuğçe Ateşer, and Dr. Neslihan Akçay, at Photonics Application and Research Center for their assistance in the work outlined in this short review. I would also like to thank Dr. Uğur Özdemir, Dr. Engin Arslan, Dr. Barış Kınacı, and Dr. Ahmet Kürşat Bilgili for his help in this the study.

References

1. A.K. Bilgili, O. Akpinar, M.K. Ozturk, C. Baskose, S. Ozcelik, E. Ozbay, Appl. Phys. A-Matter. **125**(1), 36 (2019). https://doi.org/10.1007/s00339-018-2338-2
2. E. Arslan, M.K. Ozturk, S. Ozcelik, E. Ozbay, Philos. Mag. **99**(14), 1715 (2019)
3. T. Ataser, M.K. Ozturk, O. Zeybek, S. Ozcelik, Acta Phys. Pol., A **136**(1), 21 (2019)
4. O. Akpinar, A.K. Bilgili, M.K. Ozturk, S. Ozcelik, E. Ozbay, Appl. Phys. A-Mater. Sci. Process. **125**(2), 112 (2019). https://doi.org/10.1007/s00339-019-2402-6
5. A.K. Bilgili, O. Akpinar, M.K. Ozturk, S. Ozcelik, Z. Suludere, E. Ozbay, J. Mater. Sci.-Mater. El. **30**(11), 10391–10398 (2019). https://doi.org/10.1007/s10854-019-01379-w
6. M. Demirtas, B. Tamyurek, E. Kurt, I. Cetinbas, M.K. Ozturk, J. Electron. Mater. **48**(11) (2019)
7. U. Ozdemir, S. Korcak, A. Gultekin, M.K. Ozturk, Mater. Res. Exp. **6**(9) (2019)
8. S.V. Novikov, A.J. Kent, C.T. Foxon, Prog. Cryst. Growth Ch. **63**(2), 25 (2017)
9. A.L. Rogach, M.T. Harrison, S.V. Kershaw, A. Kornowski, M.G. Burt, A. Eychmuller, H. Weller, Phys. Status Solidi B **224**(1), 153 (2001)

10. H.M. Tutuncu, E. Karaca, G.P. Srivastava, Phys. Rev. B. **95**(21) (2017)
11. D.A Young, Phase Diagr. Elem. (1991)
12. S.Q. Wang, H.Q. Ye, Phys. Rev. B. **66**(23) (2002)
13. M. Fox, *Optical Properties of Solids*, 2nd edn. (Oxford Master Series in Physics, 2010)
14. H. Akkus, Dissertation, University of Cukurova, 2007
15. A. Gultekin, Dissertation, University of Gazi, 2016
16. D.X. Li, Y. Haga, H. Shida, T. Suzuki, Y.S. Kwon, G. Kido, J. Phys.-Condens. Mat. **9**(48), 10777 (1997)
17. S.M. Hosseini, Phys. B-Condens. Matter. **403**(10–11), 1907 (2008)
18. R. Khenata, A. Bouhemadou, M. Sahnoun, A.H. Reshak, H. Baltache, M. Rabah, Comput. Mater. Sci. **38**(1), 29 (2006)
19. Z. Hongsheng, Y. Tao, W. Deqi, J. Ningxia Univ. (Natural Science Edition) **33** (2012)
20. A. Molchanov, J. Friedrich, G. Wehrhan, G. Muller, J. Cryst. Growth **273**(3–4), 629 (2005)
21. D.A. Neamen, *Semiconductor Physics and Devices Basic Principles* (Mc Graw Hill, 2003), pp. 16
22. G. Singh, B. Ghosh, R.Y. Deshpande, Cryst. Res. Technol. **16**(11), 1239 (1981)
23. M.A. Herman, H. Sitter, Microelectr. J **27**(4–5), 257 (1996)
24. E.C. Kendrick, Dissertation, University of Canterbury, New Zealand, 2008
25. J.M. Olson, M.M. Aljassim, A. Kibbler, K.M. Jones, J. Cryst. Growth **77**(1–3), 515 (1986)
26. E. Arslan, M.K. Ozturk, E. Tiras, T. Tiras, S. Ozcelik, E. Ozbay, J. Mater. Sci.-Mater. El. **28**(4), 3200 (2017)
27. S.L. Wong, H.F. Liu, D.Z. Chi, Prog. Cryst. Growth Ch. **62**(3), 9 (2016)
28. S.H. Baxamusa, S.G. Im, K.K. Gleason, Phys. Chem. Chem. Phys. **11**(26), 5227 (2009)
29. F. Kawamura, M. Morishita, K. Omae, M. Yoshimura, Y. Mori, T. Sasaki, Jpn. J. Appl. Phys. **42**(8A), L879 (2003)
30. R.A. Strading, P.C. Klipstein, *Growth and Characterisation of Semiconductor* (IOP Publishing Ltd., 1991)
31. I. Uslu, T. Tunc, M.K. Ozturk, A. Aytimur, Polym-Plast Technol. **51**(3), 257 (2012)
32. S. Rajeshkumar, J. Genetic Eng. Biotechnol. **14**(1), 195 (2016)
33. Y.A. Bioud, A. Boucherif, A. Belarouci, E. Paradis, D. Drouin, R. Ares, Nanoscale Res. Lett. **11**(1), 446 (2016). https://doi.org/10.1186/s11671-016-1642-z
34. C. Suryanarayana, M. Grant Norto, *X-Ray Diffraction A Practical Approach* (1998)
35. B.D. Cullity, *Elements of X-Ray Diffraction* (Addison-Wesley Publishing Company, 1956)
36. M.K. Ozturk, H. Altuntas, S. Corekci, Y. Hongbo, S. Ozcelik, E. Ozbay, Strain **47**, 19 (2011)
37. D.L. Howard, M.D. de Jonge, D. Lau, D. Hay, M. Varcoe-Cocks, C.G. Ryan, R. Kirkham, G. Moorhead, D. Paterson, D. Thurrowgood, Anal. Chem. **84**(7), 3278 (2012), https://doi.org/10.1021/ac203462h
38. M.S. Shackley, *X-Ray Fluorescence Spectrometry (XRF) in Geoarchaeology* (2010)
39. N. Akin, B. Kinaci, Y. Ozen, S. Ozcelik, J. Mater. Sci.-Mater. El. **28**(10), 7376 (2017)
40. X.J. Yu, W.H. Wang, L. Li, X.H. Guo, Z.M. Zhou, F.C. Wang, Chinese J. Polym. Sci. **32**(6), 778 (2014)
41. S. Corekci, S. Dugan, M.K. Ozturk, S.S. Cetin, M. Cakmak, S. Ozcelik, E. Ozbay, J. Electron. Matter. **45**(7), 3278 (2016)
42. S. Corekci, M.K. Ozturk, M. Cakmak, S. Ozcelik, E. Ozbay, Mat. Sci. Semicon Proc. **15**(1), 32 (2012)
43. J. Bai, T. Wang, P.J. Parbrook, K.B. Lee, A.G. Cullis, J. Cryst. Growth **282**(3–4), 290 (2005)
44. N. Faleev, C. Honsberg, O. Jani, I. Ferguson, J. Cryst. Growth **300**(1), 246 (2007)
45. P. Gay, P. Hirsch, A. Kelly, Acta Metall. **1**(3), 315 (1953)
46. T. Ide, M. Shimizu, X.Q. Shen, K. Jeganathan, H. Okumura, T. Nemoto, J. Cryst. Growth **245**(1–2), 15 (2002)
47. D.J. Wolford, G.D. Gilliland, T.F. Kuech, L.M. Smith, J. Martinsen, J.A. Bradley, C.F. Tsang, R. Venkatasubramanian, S.K. Ghandi, H.P. Hjalmarson, J. Vac. Sci. Technol. B **9**(4), 2369 (1991)
48. H.G. Tompkins, E.A. Irene, *Handbook of Ellipsometry*, 1st edn. (Springer, Berlin, 2005)

49. S.S. Cetin, B. Kinaci, T. Asar, I. Kars, M.K. Ozturk, T.S. Mammadov, S. Ozcelik, Surf. Interface Anal. **42**(6–7), 1252 (2010)
50. B. Kinaci, Y. Ozen, T. Asar, S.S. Cetin, T. Memmedli, M. Kasap, S. Ozcelik, J. Mater. Sci.-Mater. El. **24**(9), 3269 (2013)
51. C. Sancar, Dissertation, University of Gazi, 2004
52. S. Perkowitz, *Optical Characterization of Semiconductors: Infrared, Raman, And Photoluminescence Spectroscopy* (Academic Press, London, San Diego, 1993)
53. S. Ozkaya, Dissertation, University of Gazi, 2005
54. T. Ataser, Dissertation, Balıkesir University, 2017
55. J.R. Ferraro, K. Nakamoto, *Introductory Raman Spectroscopy*, 1st edn. (Academic Press, San Diego, 1994)
56. D. Long, *Raman Spectroscop* (Mc Graw Hill, New York, 1971)
57. J. Bardeen, W.H. Brattain, Phys. Rev. J. Archive **74**(2), 230 (1948)
58. J.F. Suyver, Dissertation, University of Utrecth, 2003
59. M. Gallignani, M.D. Brunetto, Talanta **64**(5), 1127 (2004)
60. A.K. Bilgili, O. Akpinar, G. Kurtulus, M.K. Ozturk, S. Ozcelik, E. Ozbay, XRD vs raman for InGaN/GaN structures. J. Polytechnic. (2019). https://doi.org/10.2339/politeknik.537733
61. K. Heilig, Hyperfine Interact. **24**(1–4), 349–375 (1985)
62. C.P. Constable, D.B. Lewis, J. Yarwood, W.D. Münz, Surf. Coat. Technol. **184**(2–3), 291–297 (2004)
63. G.B. Kilic, A.G. Karahan, J. Food **35**(6), 445 (2010)
64. K. Nakamoto, *Infrared and Raman Spectra of Inorganic and Coordination Compounds*, 5th edn. (A Wiley-Interscience Publication, New York, 1997)
65. N. Puviarasan, V. Arjunan, S. Mohan, Turk. J. Chem. **26**(3), 323 (2002)
66. H.U. Gremlich, *Infrared and Raman Spectroscopy of Biological Materials* (New York, 2001)
67. M.G. Rockley, D.M. Davis, H.H. Richardson, Science **210**(4472), 918 (1980)
68. B. Stuart, *Infrared Spectroscopy: Fundamentals and Applications* (Wiley, Hoboken, 2004)
69. L.I. Maklakov, G.G. Suchkova, Spectrochim Acta A. **71**(1), 238 (2008)
70. A.R.W. McKellar, D.R.T. Appadoo, J. Mol. Spectrosc. **250**(2), 106 (2008)
71. B.J. Reddy, R.L. Frost, S.J. Palmer, Spectrochim Acta A. **71**(2), 430 (2008)
72. T.W. Graham Solomons, *Craig Barton Fryhle Organic Chemistry,* 7th edn. (Wiley, London, 2001)
73. A. Yildiz, Ö. Genç, *Instrumental Analysis* (Hacettepe University Publications, 1993)
74. J.Y. Lin, B.X. Wang, Adv. Mater. Sci. Eng. **6**, 594516 (2014). http://dx.doi.org/10.1155/2014/594516
75. H.H. Perkambus, *UV-VIS Spectroscopy and Its Applications* (Springer Laboratory, Berlin, 1992)
76. S. Kumar, V. Gupte, K. Sreenivas, J. Phys.-Condens. Mat. **18**(13), 3343 (2006)

Hybrid Optoelectronic Materials for Photodetection Applications

Xianguang Yang, Qin Chen, and Baojun Li

Abstract One-dimensional hybrid optoelectronic materials based photodetectors are highly desirable because low-dimensional nanostructures are fascinating platforms for manipulating electrons and photons at the sub-wavelength scale. In this chapter, we reported two types of ultraviolet–visible photodetectors based on a single hybrid nanowire and a crossed junction formed with two-hybrid nanowires, respectively. For a single nanowire device, the high-performance is attributed to the high density of trapping states at the organic-inorganic interfaces. For crossed junction device, the excellent performance is mainly attributed to the intermediate re-absorption process. These two types of photodetectors would have potential applications and offer novel functionalities in optoelectronic devices and integrated circuits.

Keywords Hybrid nanowire · Interface · Detectors

1 Introduction

Photon detection in the ultraviolet and/or visible waveband shows various applications, such as biological imaging, chemical sensing, environmental monitoring, optical information processing, and missile launching [1, 2]. Commercial detectors are usually fabricated with silicon and/or silicon carbide for photon detection in visible and/or ultraviolet wavebands, respectively. It's desirable to fabricate a low-cost, multi-waveband detector in a single device. Several types of detectors have been fabricated [3–7], among which hybrid detectors have recently attracted great attentions [6, 7]. Compared with other types of detectors, organic-inorganic hybrid detectors have the advantages of organic-based detectors including easy-formation

X. Yang (✉) · Q. Chen · B. Li (✉)
Institute of Nanophotonics, Jinan University, Guangzhou 511443, China
e-mail: xianguang@jnu.edu.cn

B. Li
e-mail: baojunli@jnu.edu.cn

S. J. Ikhmayies et al. (eds.), *Advances in Optoelectronic Materials*,
Advances in Material Research and Technology,
https://doi.org/10.1007/978-3-030-57737-7_2

and function tune-ability [8, 9], but also the merits of inorganic based detectors, such as broadband absorption and high carrier mobility [10, 11]. Meanwhile, the hybridization of organic and inorganic material can avoid their own disadvantages [12]. On the other side, the unique interaction at the organic-inorganic interface is highly beneficial to the detection applications [13, 14].

In view of the above advantages, several prototypes of organic-inorganic hybrid detectors have been fabricated and characterized for optoelectronic applications [6, 7, 12–16]. Nevertheless, some of them were in three-dimensional bulk materials and/or two-dimensional film systems [12, 14–16]. Owing to the large aspect ratio and the high surface-to-volume ratio, one-dimensional nanomaterials have enormous surface trap states which could prolong the lifetimes of charge carriers. Moreover, the low dimensionality of one-dimensional nanomaterials could short the transit times and confine the effective area, which leads to high photoconductive gain in one-dimensional nanomaterial-based detectors [17, 18]. Hybrid nanowire could be a rational choice, because it combines the advantages of one-dimensional nanomaterials and the enormous trap states of organic-inorganic interfaces [19]. In contrast to two-dimensional and three-dimensional nanomaterials, one-dimensional organic–inorganic hybrid nanowire is the optimal dimension to investigate the transport mechanism for carriers.

In this chapter, we reported an organic-inorganic hybrid broadband detector based on a single hybrid nanowire with organic-polyaniline and inorganic-quantum-dots. Polyaniline polymer was selected as the organic component because of the transparent conductivity [20]. Quantum dots were chosen as the inorganic component because of theoptoelectronic properties [21]. The hybrid nanowire was fabricated by drawing method at low-cost. Compared with the high-cost detector with sandwich structure, our device structure does not need the use of transparent electrodes and allow the external incident light to efficiently illuminate the photo-active region. The constructed detector exhibits high responsivity of 10^5 A/W and external quantum efficiency of $10^7\%$ across the whole waveband of 350–700 nm. Furthermore, we also reported an ultraviolet-visible detector based on a single crossed heterojunction assembled with two types of hybrid nanowires. The device performances of 9 ms response/recovery time and $10^5\%$ external quantum efficiency were benefited to the abundant inorganic-organic interfaces and the 10^6 nm^2 heterojunction area formed with the crossed nanowires, which leads to the efficient light absorption. The operation waveband could be easily tuned to the desired waveband by altering the emission properties of the quantum dots. Since the synthetic methodology has developed, several conductive polymers and colloidal quantum dots would have been synthesized. This platform would build a new class of organic-inorganic hybrid nanowire detectors for optically and electrically active detection applications.

2 Materials and Methods

Inorganic nanocrystal of CdSe/ZnS core/shell quantum dots and transparent conductive organic-polyaniline were commercially available from Zkwy Bio-Tech (Beijing, China) and Alfa Aesar (Tianjing, China), respectively. Hybrid nanowires were fabricated by using a physical drawing method. First, 900 mg of organic-polyaniline was dissolved into 1 mL of dimethylbenzene to form a homogenous solution containing dimethylbenzene and polyaniline. Second, organic-polyaniline and inorganic-quantum-dots blend solution was obtained by adding quantum-dot-dimethylbenzene solution with concentration of 8 μM/L into the homogenous polyaniline-dimethylbenzene solution. The quantum dots concentration of the polyaniline-quantum-dot blend solution was tuned from 0 to 60 wt% with respect to the content of organic-polyaniline by changing the added volume for quantum dots solution. Then, the polyaniline-quantum-dot blend solution was stirred at room temperature for 180 min, and followed by 50 min of ultra-sonications to have a homogeneous solution at an appropriate viscosity for physical drawing. Third, the silica fiber tip was immersed into the polyaniline-quantum-dot blend solution for 1–5 s and then removed with a velocity of 0.1–5 m/s, leaving an organic–inorganic hybrid wire extending between the fiber tip and the solution with the rapid evaporation for the dimethylbenzene solvent. The diameters of the hybrid nanowires are ranging from 200 to 500 nm.

The surface morphology and diameter of fabricated organic–inorganic hybrid nanowires could be inspected roughly and selected preliminarily under optical microscopy. After the examination, the preliminarily selected organic–inorganic hybrid nanowires were transferred into a preliminarily cleaned SiO_2 substrate. Then, the SiO_2 substrate was spin-coated by poly(methyl methacrylate) and methyl methacrylate, and the patterns of electrodes were defined by using the electron beam lithography. The electrodes of 25/85 nm Cr/Au were prepared by using the metal evaporation. The electrical characterization for the single nanowire detector were conducted by using the semiconductor characterization system and the equipped probe station under a clean box at room temperature. As a comparison, a similar detector based on a single pristine nanowire was also prepared and characterized by the same process. The monochromatic light illumination was achieved from a xenon lamp with power of 200 W by using a monochromator. The light power for optical illumination was measured by the use of a power meter (OPHIR NOVA). The mechanically chopped light illumination was obtained with frequency modulated chopper for the measurements onthe temporal photo-responsivity.

For thejunction device, the emission wavelengths for the p-type and n-type oil-soluble CdSe/ZnS core/shell quantum dots are 580 and 650 nm, respectively. The crossed heterojunction formed with two types of hybrid nanowires could be assembled by using a commercial micromanipulator (50 nm-resolution) equipped with a tungsten probe (tip diameter of 300 nm) under optical microscopy. After the junction assembly, the crossed heterojunctions were transferred into a preliminarily cleaned SiO_2 substrate. Then, the SiO_2 substrate was spin-coated by using the poly (vinyl alcohol) and poly (methyl methacrylate), while the electrode patterns were defined by using the electron beam lithography. The 15/65 nm Cr/Au electrodes were successfully prepared by using the metal evaporation. The electrical characterization for the crossed heterojunction and homojunction detectors were conducted by using a semiconductor characterization (Keithley 4200) system.

Optical characterization for the single organic-inorganic hybrid nanowires with and without incorporating quantum dots were carried under a CRAIC Micro-Spectrophotometer (20/20 PV). The optical absorption and transmission spectra for single organic-inorganic hybrid nanowires were conducted by using the respective modules equipped in the CRAIC Micro-Spectrophotometer. A continuous work laser at 473 nm wavelength was used to excite the single organic-inorganic hybrid nanowires. The 473-nm excitation laser was focused on a 50 μm spot via an objective at 80× with numerical aperture of 0.6. The photoluminescence signals were collected with the same objective and then directed via a dichroic filter. The filtered light was splitting with a beam splitter and then directing into a charge coupled device camera and a spectrometer for imaging and spectrum measurements, respectively. In addition, the blue light with a wavelength of 473 nm was also coupled evanescently to the organic–inorganic hybrid nanowires by the use of a fiber taper for wave-guiding excitation with high-efficiency.

3 Results and Discussion

3.1 Nanowire Detector

Figure 1a shows a scanning electron microscope (SEM) image of a 300-nm-diameter hybrid nanowire (i.e. quantum dots in polyaniline nanowire) . To inspect the distribution of quantum dots in the polyaniline nanowire, energy-dispersive X-ray spectroscopy (EDS) and transmission electron microscopy (TEM) were performed, the obtained results are shown in Fig. 1b. For comparison, TEM image of a 300 nm-diameter pristine nanowire (i.e. without quantum dots in polyaniline nanowire) is shown at the bottom inset of Fig. 1b. As can be seen from the two TEM images that the CdSe/ZnS core/shell quantum dots were successfully doped into the 300 nm-diamter polyaniline nanowire. The EDS analysis verifies the elementary composition of S, Zn, Se, and Cd elements. The estimated mass ratio of quantum dots in 300 nm-diamter polyaniline nanowire is about 40 wt% relative to the polyaniline content.

Because of the quantum size effect and the broadband absorption, quantum dot is one of the building blocks for fabricating nanoscale photodetectors [2, 4, 10, 21]. In this work, CdSe/ZnS core/shell colloid quantum dot is as a photo-active component, meanwhile the size and optical characterization of quantum dot are shown in Fig. 2. Figure 2a shows the TEM image of experimentally used quantum dots. The corresponding distribution histogram of diameter is shown in the inset. It gives

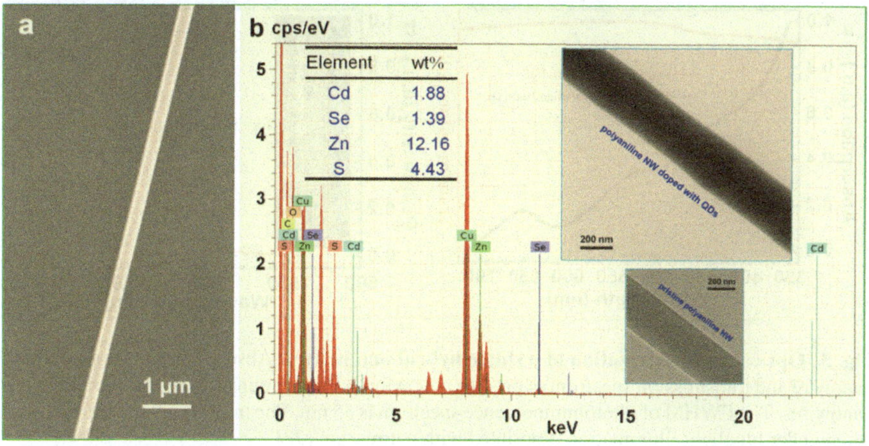

Fig. 1 SEM, TEM, and EDS analysis of a single hybrid nanowire. a SEM image. **b** EDS spectrum and TEM image. Bottom inset shows the TEM image of a pristine nanowire. Reprinted with permission [22]

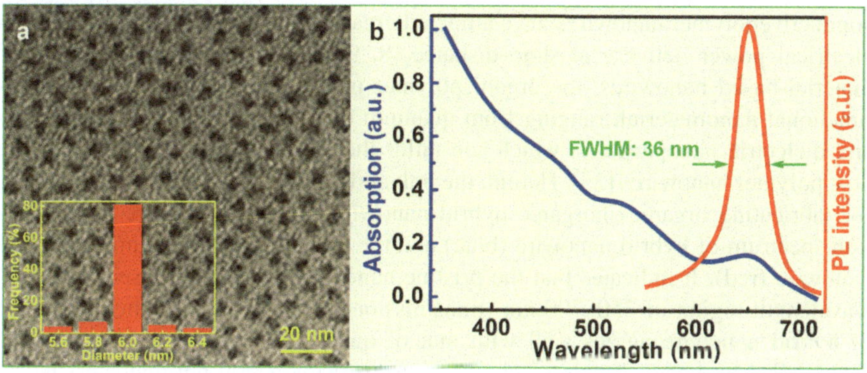

Fig. 2 TEMand optical characterization of CdSe/ZnS quantum dots. a TEM image of quantum dots with average diameter of 6 nm. The inset shows the distribution histogram of diameter. **b** Optical absorption and photoluminescence spectra of quantum dots. Reprinted with permission [22]

that the frequency of quantum dots in 6-nm-diameter is larger than 80%. Semiconductor quantum dot exhibits intriguing optical properties, which can absorb and emit photons at the nanometer scale (1–10 nm). Figure 2b shows the optical absorption (blue) and photoluminescence (red) spectra of quantum dots. The absorption intensity is monotonically decreasing in the wavelength region of 350–600 nm, whereas the 650 nm absorption peak is attributed to the re-absorption effect of quantum dots. The re-absorption peak is corresponding to the center wavelength of photoluminescence emission band. The full width at half maximum (FWHM) of the photoluminescence emission is about 36 nm. As the optical properties of quantum dots are strongly

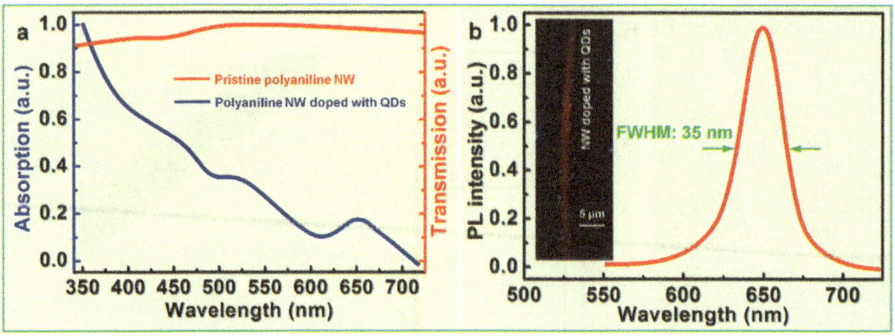

Fig. 3 Optical characterization of a single hybrid nanowire. a Absorption spectrum of hybrid nanowire and transmission spectrum of pristine nanowire. **b** Photoluminescence spectrum of hybrid nanowires. The FWHM of Photoluminescence spectrum is 35 nm. The inset shows the optical image taken under bluelight illumination. Reprinted with permission [22]

dependent on the size, thus indicating an opportunity to tailor the optoelectronic properties of quantum-dot-based photodetectors by tuning the quantum dot sizes.

Owing to the flexible conductivity and the optical transparency, conjugated conductivepolymer nanowires have attracted great potentials for optical energy and electrical power delivery at short-distance [8, 9, 20]. Compared with inorganic material-based nanowires, the organicpolymer matrix is compatible with various functional nanomaterial, ranging from quantum dots and molecular dyes to metal and dielectric nanoparticles, which can tailor the optoelectronic properties of the host polymer nanowires [23]. Herein, the polyaniline acts as organicpolymer matrix for fabricating organic–inorganic hybrid nanowires. Figure 3a shows the absorption spectrum of hybrid nanowire (blue) and the transmission spectrum of pristine nanowire (red). It indicates that the pristine nanowire is optical transparent in the wavelength region of 350–700 nm (transmission intensity> 90%). The absorption of hybrid nanowire agrees well with that of quantum dots, as given in Fig. 2b. Figure 3b gives the photoluminescence spectrum of a 500-nm-diameter hybrid nanowire, excited by blue light illumination at 473-nm wavelength.

To construct a nanowire detector, the fabricated single hybrid nanowire was first placed on a pre-cleaned substrate. Then, two electrodes were deposited thermally on the two end of the nanowire. Figure 4a shows *I*–*V* characteristics of 300-nm-diameter hybrid nanowire (blue) and pristine nanowire (red). Insets give the measurement configurations. They were measured under 350-nm-wavelength light illumination with intensity of 20 mW/cm^2. As can be seen from Fig. 4a that the photo-induced current is enhanced by quantum dots in hybrid nanowire. For wavelengths of 350, 395, 473, 532, 600, 650, and 700 nm light illumination, the *I*–*V* characteristics of nanowire detector are given in Fig. 4b. To evaluate the device performance, the optical spectral responsivity (*R*) and external quantum efficiency (EQE) are usually calculated. Responsiviy (*R*): the photo-induced current generated by per unit power of light. EQE: the ratio of the number of photo-generated carriers to the number of incident photons. They can be calculated according to the equations [24]:

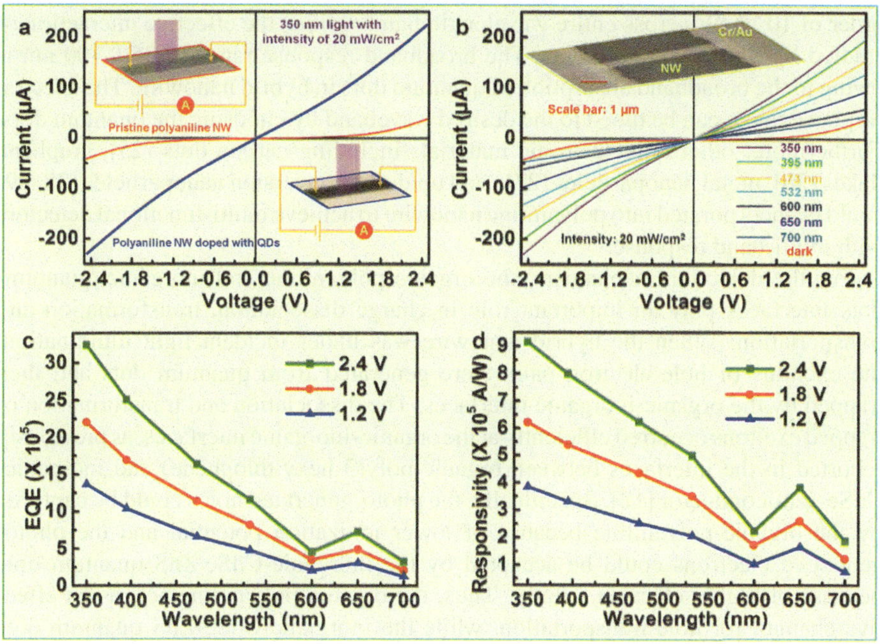

Fig. 4 **Electrical characterization of nanowire detectors.** **a** *I–V* characteristic curves of detectors based on a pristine nanowire (red) and a hybrid nanowire (blue). **b** *I–V* characteristic curves of a detector based on a hybrid nanowire under illumination at seven discrete wavelengths. Inset: SEM image of the nanowire detector. **c** EQE and **d** Responsivity versus wavelength under three voltages. Reprinted with permission [22]

$$R = \frac{\Delta I}{\mathrm{PS}} \tag{1}$$

$$\mathrm{EQE} = R \times \frac{\mathrm{hc}}{e\lambda} = \frac{\Delta I}{\mathrm{PS}} \times \frac{\mathrm{hc}}{e\lambda} \tag{2}$$

where I is the difference between the photo-induced current and dark current, P is the light power density, and S is the illumination area. The constants: h is Planck's constant with a value of 6.6×10^{-34} J·s, e is the electronic charge with a value of 1.6×10^{-19} Coulombs and c is the velocity of light in vacuum with value of 3.0×10^{8} ms. Figure 4c gives the relationship between EQE and wavelength under bias voltages of 2.4, 1.8, and 1.2 V. The maximum EQE of 350 nm-wavelength measured under 2.4 V reaches up to $3.25 \times 10^{8}\%$. Even the minimum EQE of 700 nm-wavelength measured under 1.2 V still reaches up to $1.33 \times 10^{7}\%$. The organic-inorganic interfaces benefits the enhanced light-nanowire interaction and the high density of trap states, which can facilitate the generation and trapping of excitons. Then, the dissociation of trapped excitons is efficiently occurred, thus reducing the recombination of photo-generated carriers. Figure 4d gives the relationship between responsivity and wavelength under bias voltages of 2.4, 1.8, and 1.2 V. The large responsivity with magnitude at the order of 10^5 A/W across entire wavelength band verifies the effective interaction of studied light with quantum dots. The broadband response range of 350–700 nm is owing to the broadband absorption of quantum dots in hybrid nanowire. The spectral response range can be tuned to the desired waveband by choosing the quantum dots. Furthermore, other optoelectronic materials including carbon dots [25], graphene flakes [26], metal nanoparticles [27] , and up/down conversion nanoparticles [28, 29] could be incorporated into polyaniline nanowire to achieve multi-functional detectors with multi-band response.

For the detection mechanism, the organic-polyaniline and inorganic-quantum-dots interfaces play an important role in charge dissociation, transformation and transportation. When the hybrid nanowire was under incident light illumination, the excitons of hole-electron pairs were generated from quantum dots and then trapped by the organic-inorganic interfaces. The dissociation and transformation of trapped excitons occurred efficiently at the organic-inorganic interfaces, as previously reported in the interfaces between organic-poly(3-hexylthiophene) and inorganic-CdSe-semiconductor [12]. Specifically, the photo-generated holes could be captured by the organic-polyaniline because of lower ionization potential and the photo-generated electrons could be accepted by the inorganic-CdSe/ZnS quantum dots because of higher electron affinity. Thus, the organic-polyaniline acts as an effective channel for hole transportation, while the inorganic-CdSe-ZnS quantum dots act as the channel for electron transportation. Consequently, the recombination of photo-generated holes and electrons were reduced largely. Moreover, in the hybrid nanowire, the CdSe/ZnS quantum dots are dispersed highly in the organic-polyaniline matrix, leading to the formation of one-dimensional interconnected networks. Such a network structure results in abundant organic-inorganic interfaces for exciton trapping and charge separation.

3.2 *Junction Device*

Figure 5a gives a field emission SEM image of a straight hybrid nanowire at a diameter of 630 nm. The inset shows a corresponding high-resolution SEM image. Figure 5b gives the SEM image of a crossed heterojunction assembled by two-hybrid nanowires at diameters both of 500 nm. To carefully see the distribution of quantum dots in the polyaniline nanowire, TEM operating at 200 kV, and EDS analysis were performed, the obtained results are given in Fig. 5c, d, respectively. Figure 5c gives the TEM image of a hybrid nanowire at 500-nm-diameter, which indicates that the CdSe/ZnS (Fig. 5d) core/shell quantum dots (red arrows) were successfully incorporated into the polyaniline nanowire.

To obtain a crossed heterojunction, colloidal CdSe/ZnS core/shell quantum dots at emissions of 650 and 580 nm were incorporated into the polyaniline nanowires. Figure 6a gives a TEM image of quantum dots at the emission of 650 nm, the white circles show the individual quantum dots with 6 nm-average-diameter. As a comparison, a TEM image of quantum dots at the emission of 580 nm was also given

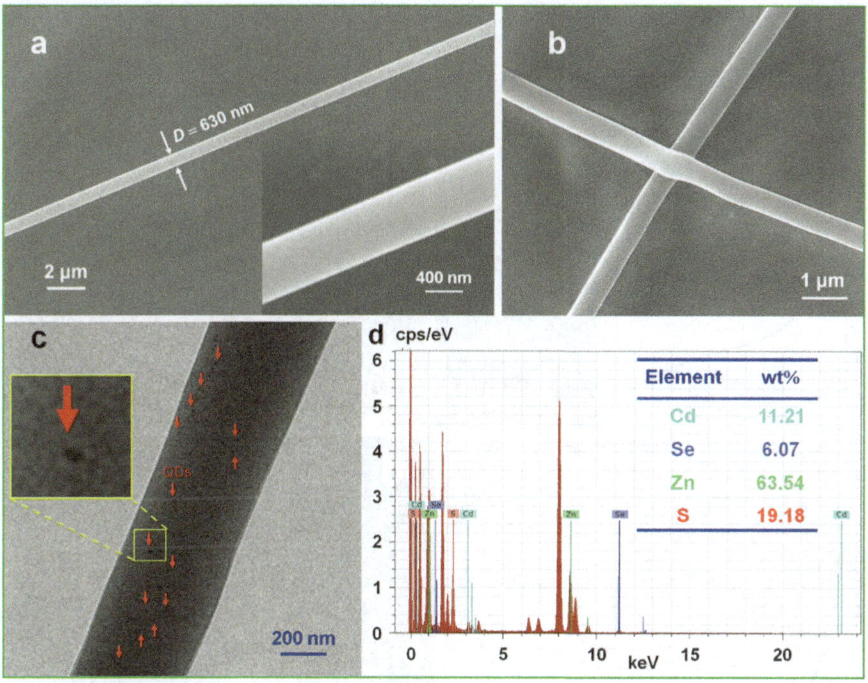

Fig. 5 **a** SEM image of a straight hybrid nanowire and a high-resolution SEM image (inset). **b** SEM image of a crossed heterojunction. **c** TEM image of a hybrid nanowire and zoom-image (inset). The quantum dots were indicated by red arrows. **d** EDS spectrum of the hybrid nanowire shown in (**c**). Reprinted with permission [30]

in Fig. 6b. The insets show the optical absorption and photoluminescence spectra of the quantum dots. To carefully see the quantum dot diameter, Fig. 6c gives a high-resolution TEM image for the quantum dots of (b), and the zoom-in image was also given as inset. As nanocrystal, the lattice fringes could be clearly seen from the white and yellow circles, indicating an average diameter of 4.5 nm. Figure 6d gives the SAED (selected area electron diffraction) pattern for the quantum dots in (c), and the Fourier transform pattern was correspondingly given as inset.

The above characterizations show the size effect of optical properties for the colloidal CdSe/ZnS core/shell quantum dots, providing an efficient method to control the optical properties via the size tuning. Figure 7a gives the bright field optical micrograph for a 630 nm-diameter hybrid nanowire at the emission of 580 nm. The tip diameter of the fiber taper is about 500 nm, and without launching light. When the 473 nm light was launched from the fiber taper into the hybrid nanowire, the incorporated quantum dots would be excited and emitted 580 nm photons. Figure 7b gives a dark field micrograph for emitted 580 nm photons guided along the nanowire. As a comparison, Fig. 7c gives the bright field micrograph for a 500 nm-diameter hybrid nanowire at emission of 650 nm, while Fig. 7d gives the dark field micrograph

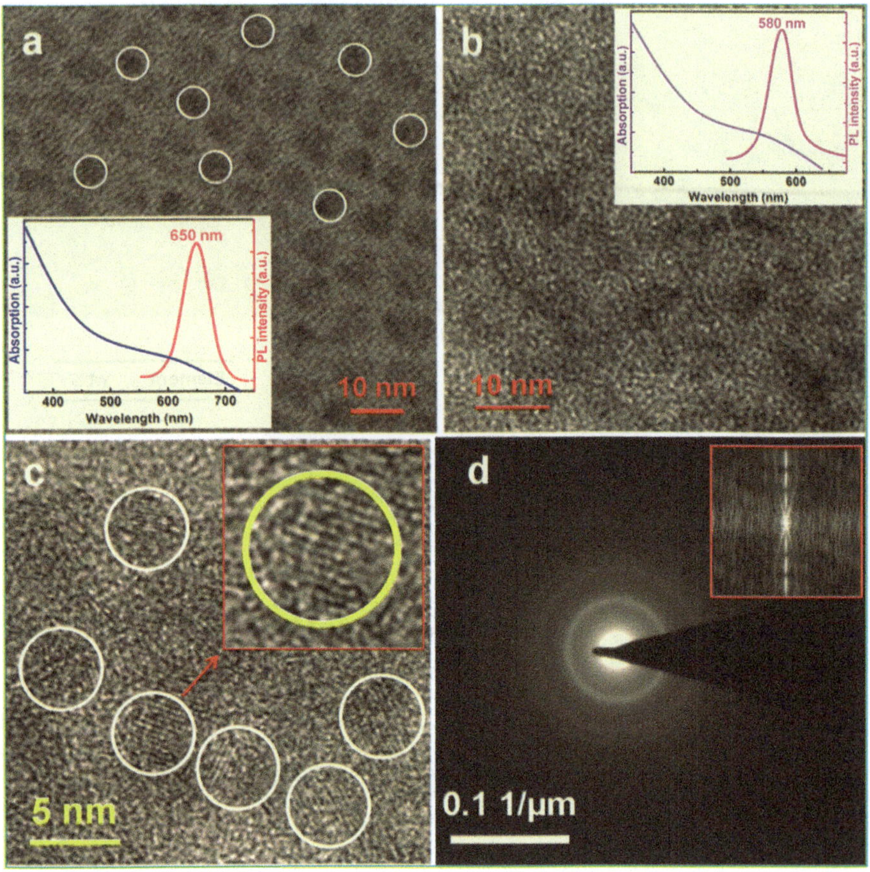

Fig. 6 TEM images of quantum dots at emission wavelengths of 650 (**a**) and 580 (**b**) nm. Insets: the corresponding absorption and photoluminescence spectra. (c) High-resolution TEM image of the quantum dots given in (**b**). The white circles in (**a**) and (**c**) indicate the individual quantum dots. **d** SAED pattern for the quantum dots shown in (**c**). Inset: the Fourier transform pattern. Reprinted with permission [30]

for emitted 650 nm photon guided along the nanowire. To analysis the photoluminescence properties for the hybrid nanowires, Fig. 7e gives the photoluminescence spectra for two types of nanowires given in (b, blue) and (d, red). The results further demonstrated that two types of quantum dots at the emissions of 580 and 650 nm were successfully incorporated into polyaniline nanowires. These two types of hybrid nanowires were used to assemble the crossed heterojunction: one was at emission of 580 nm and the other was at emission of 650 nm. Once the crossed heterojunction was under 473 nm light irradiation, two different colors of photoluminescence emissions could be observed, as given in Fig. 7f. The two photoluminescence colors also verify the formation of the crossed heterojunction. Figure 7g gives the optical absorption spectra for a hybrid nanowire (blue) and a pure nanowire (red). The inset shows the

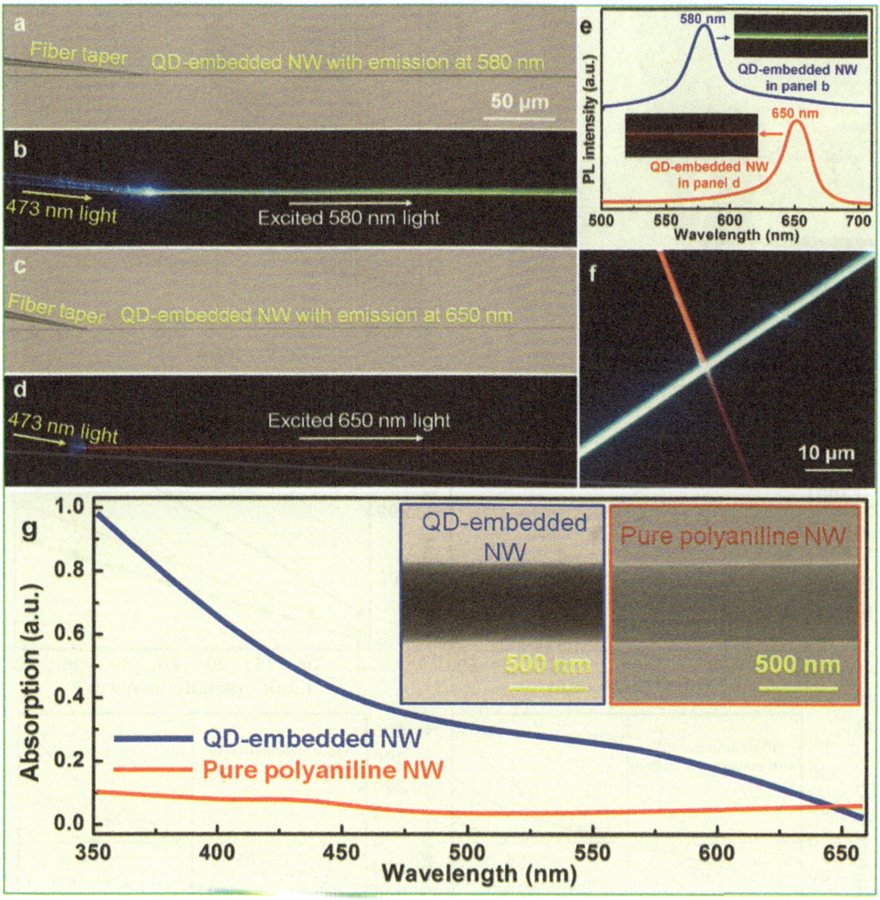

Fig. 7 Bright field optical micrographs of **a** a 630 nm-diameter hybrid nanowire at emission of 580 nm and **c** a 500-nm-diameter hybrid nanowire at emission of 650 nm. The dark field micrographs were taken under 473-nm light excitation, given as (**b**) and (**d**). Scale bar is applicable to (**a**–**d**). **e** Photoluminescence spectra of hybrid nanowires at emission of 580 (blue) and 650 (red) nm. **f** Dark field micrographs of a crossed heterojunction with 473 nm light irradiation. **g** Absorption spectra of a hybrid nanowire (blue) and a pure nanowire (red). Insets: TEM images for each nanowires. Reprinted with permission [30]

TEM images for the nanowires, respectively. As can be seen from Fig. 7g that the absorption intensity for a pure nanowire is smaller than 10%, while the absorption for a hybrid nanowire derives from the incorporated quantum dots. Thus result agrees well with the absorption spectra for colloidal CdSe/ZnS core/shell quantum dots given as insets of Fig. 6a, b.

Figure 8a gives the TEM image of a crossed heterojunction constructed by two types of hybrid nanowires both with 680 nm-diameter: one is at emission of 580 nm

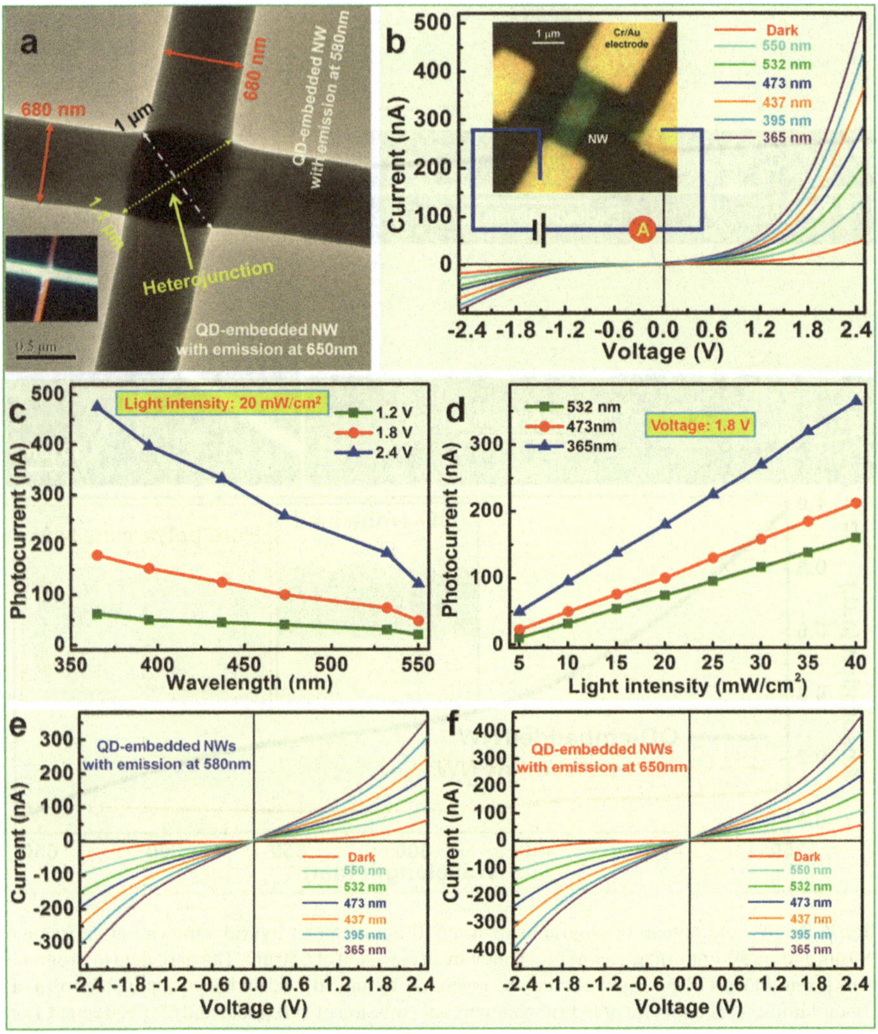

Fig. 8 Electrical characterization of the junction device. a TEM image of a crossed heterojunction assembled by two-hybrid nanowires at 680 nm indiameter. Inset: dark field micrograph. **b** Representative $I-V$ characteristics of the junction detector with different-wavelengths light illumination at intensity of 20 mW/cm^2. Inset: false-color SEM image of the junction detector. **c** Wavelength versus photocurrent for the detector working at 1.2, 1.8, and 2.4 V. **d** Light intensity versus photocurrent for different wavelengths. **e, f** Representative $I-V$ characteristics of the junction device constructed by two-hybrid nanowires at the same emission wavelengths of 580 nm (**e**) and 650 nm (**f**), respectively. Reprinted with permission [30]

and the other is at emission of 650 nm. The diagonal lines of the crossed heterojunction are 1.1 and 1 μm long, as indicated by the yellow and white lines, respectively. The crossed heterojunction is a promising candidate for fast responsivity and high photo-sensitivity due to the enormous interfaces between inorganic-quantum-dots and organic-polyaniline, enhancing the light absorption significantly. The enormous interfaces can avoid the aggregation of quantum dots and make them fully dispersible in organic-polyaniline, leading to the efficient light absorption. On the contrary, for the case of without organic-polyaniline, the quantum dots deposited directly on substrate would aggregate easily so that they are not sensitive to light, which may reduce the light absorption dramatically. As a more efficient light absorption by quantum dots, more electron-hole pairs would be generated and a larger photocurrent would be obtained under the external field. Consequently, the device performance could be significantly improved with respect to external quantum efficiency and spectral responsivity (see the details below). Inset of Fig. 8a gives the dark field micrograph, which was taken under the irradiation excitation. The junction device was readily fabricated by using the as-constructed hybrid nanowire crossed heterojunction. The device morphology of false-color SEM image was given as inset of Fig. 8b. Figure 8b gives the representative $I-V$ characteristics of the hybrid nanowire crossed heterojunction detector. The inset schematically shows a diagram for the device measurements, which are all conducted at room temperature in air. The obtained $I-V$ characteristics were conducted under dark conditions (without light illumination) and under the illuminations of monochromatic light (intensity of 20 mW/cm^2) with wavelengths at 550, 532, 473, 437, 395 and 365 nm. The curves indicate that the junction device exhibits distinct rectification features, which could be attributed to the formation of a p-n junction within the hybrid nanowire crossed heterojunction. The hybrid nanowire at emission of 580 nm is p-type, while the other hybrid nanowire at emission of 650 nm is n-type. The rectification features derive from the p–n junction, locating at the transition area proximal the interface of two nanowires. In addition, the results indicate that the electrical conductance for the junction device under dark conditions (dark current) was relatively weak, while the electrical conductance under light illumination was significantly increased for all of the investigated wavelengths. In fact, the dark current was related to the intrinsic conductivity of organic-polyaniline.

To investigate the spectral response at differentwavelength, the wavelength-related photocurrents were measured for incident light at wavelengths varying from 365 to 550 nm and a fixed light intensity of about 20 mW/cm^2 under operating voltages of 1.2, 1.8, and 2.4 V. The measured results were given as Fig. 8c, which indicates that the junction device exhibits a monotonically decreasing in spectral response, i.e., the photocurrent decreased as increasing wavelength. This spectral response occurs because the short-wavelength light (high energy photons) could excite simultaneously both two types of nanowire and would induce more photo-generated carriers than that of the long-wavelength light (lower energy photons). These phenomena were also in agreement well with the optical absorption spectrum for the hybrid nanowire, given as Fig. 7g. When the wavelength of incident light was larger than 550 nm, the quantum dots could not be excited effectively due to the low efficiency at optical absorption. Thus, the optical cut-off edge for the spectral response was determined by the absorption properties of the quantum dots, which were incorporated into the polyaniline nanowires. Therefore, the operation range for spectral response could be controlled via the emission wavelength changing for preselected quantum dots. The changing of emission wavelength could be easily achieved via altering the physical sizes of the colloidal CdSe/ZnS core/shell quantum dots. In addition, the photo-sensitivity of the junction device at spectral response was increased as the bias voltage increasing from 1.2 to 2.4 V, given as Fig. 8c.

To assessment the capability of optical signals converting to electrical signals under different wavelengths of light, the EQE of junction device could be calculated by the following equation [24]:

$$\mathrm{EQE} = \frac{I}{e} \times \frac{\mathrm{h}\nu}{P} \tag{3}$$

where I stands for the photocurrent, e stands for the elementary charge, hν stands for the photon energy, and P stands for the optical power. The calculated results for an operating voltage of 2.4 V give that the EQE could reach up to 1.4, 2.1, 3.4, 4.7, 6.2, and 8.1 $\times$ 10^5% under wavelengths of 550, 532, 473, 437, 395 and 365 nm, respectively. Figure 8d gives the light intensity-related photocurrents at light wavelengths of 532, 473, and 365 nm for an operating voltage of 1.8 V. The result for photocurrent and light intensity exhibits a good linear relationship. The photo-responsivity (R), under different wavelengths could be calculated by the following equation [31]:

$$R = \frac{I}{P} = \frac{\mathrm{EQE} \times \lambda(\mathrm{nm})}{1240}\ \mathrm{A/W} \tag{4}$$

where λ stands for the light wavelength in nm. By using the measured EQE under 2.4 V operating voltage, the calculated R under wavelengths of 532, 473, and 365 nm for this ultraviolet-visible detector was 901, 1297, and 2384 A/W, respectively. The photo-responsivity of this junction device was larger than that of other nanostructured devices, including graphene [32] and alloy nanoribbon [33].

As a comparison, single cross homojunction constructed by the same type of two nanowires at the same emission wavelengths was also investigated. The cross homojunction was under light illumination at different wavelengths with the same intensity of 20 mW/cm^2 or without light illumination, i.e., under dark conditions. Figure 8e gives the representative $I-V$ characteristics for a single cross homojunction assembled from two-hybrid nanowires at the same emission of 580 nm. When the 365 nm light illumination, the generated photocurrent was about 340 nA for an operating voltage of 2.4 V and the calculated EQE is of 5.8 $\times$ 10^5%. Analogously, Fig. 8f gives the representative $I-V$ characteristics for a single cross homojunction assembled from two-hybrid nanowires at the same emission of 650 nm. When the 365-nm light illumination, the generated photocurrent was about 424 nA for an operating voltage of 2.4 V and the calculated EQE is of 7.2 $\times$ 10^5%. It is different from the result for the cross heterojunction of Fig. 8b, there are no rectification behaviors observed in the cross homojunction with the same type ofnanowires. The same type of nanowires means that they have the same band-gaps, and then could not form a heterojunction but just a homojunction. On the other side, when the 365 nm light illumination, the generated photocurrent for the cross heterojunction of Fig. 8b was about 476 nA for an operating voltage of 2.4 V and the calculated EQE is of $8.1 \times 10^5\%$. The generated photocurrent of 476 nA in the cross heterojunction was higher than that of the weighted average in the two cross homojunctions (382 nA). The resulted difference of 94 nA shows that the emitted 580-nm light could be re-absorbed and could then induce additional photo-generated carriers for photocurrent generation.

In a conventional detector based on the single band-gap semiconductors, the photon-to-electron conversion efficiency was not very high since the photo-generated carriers would recombine and then emit photons via photoluminescence process. That is to say, a few holes located at the top of the valence band would deplete some electrons located at the bottom of the conductance band. This radiative recombination loss would reduce the detection efficiency for the incident photons. Fortunately, this issue could be largely resolved in the heterojunction detectors. The energy of radiative photons from a hybrid nanowire at 580 nm emission could be re-absorbed by another hybrid nanowire at 650 nm emission since the photon energy of the former was larger than the direct band-gap of the latter (i.e., 2.14 eV > 1.91 eV). The re-absorption process for radiative photons could induce the additional photo-generated carriers. Thus, this intermediate re-absorption for radiative photons leads to a larger photocurrent and photo-responsivity in the heterojunction detectors compared with that of single band-gap detectors. Moreover, as an important factor for a detector, the EQE stands for the photo-generated number of electron–hole pairs excited by the incident photons. The calculated EQE of $10^5\%$ for our heterojunction detector was one order of magnitude larger than that of $10^4\%$ for the CdS nanoribbon- and In_2Se_3 nanowire-based detectors [34, 35]. This result further verifies that the intermediate re-absorption for radiative photons in the heterojunction has a significant role for the excellent performance of cross heterojunction detectors.

3.3 Temporal Responsivity

The temporal photo-responsivity is a key parameter for the device performances. Figure 9a gives the time-related photocurrents for the single nanowire detector, which was measured under the 350, 473, and 650 nm light illumination with 20 mW/cm^2 intensity. The incident light was turned on and off repetitively under an operating voltage of 2.4 V. The nanowire detector shows a high and fast responsivity to the investigated 350, 473, and 650 nm light, and the dynamic responsivity indicates that the photo-responsivity is reproducible and time- stable for practical device applications. The switching on/off ratios was 5.5, 3.8, and 2.3 for the 350, 473, and 650 nm light, respectively. This level of ratio was comparable to that (1.15 – 3.5)

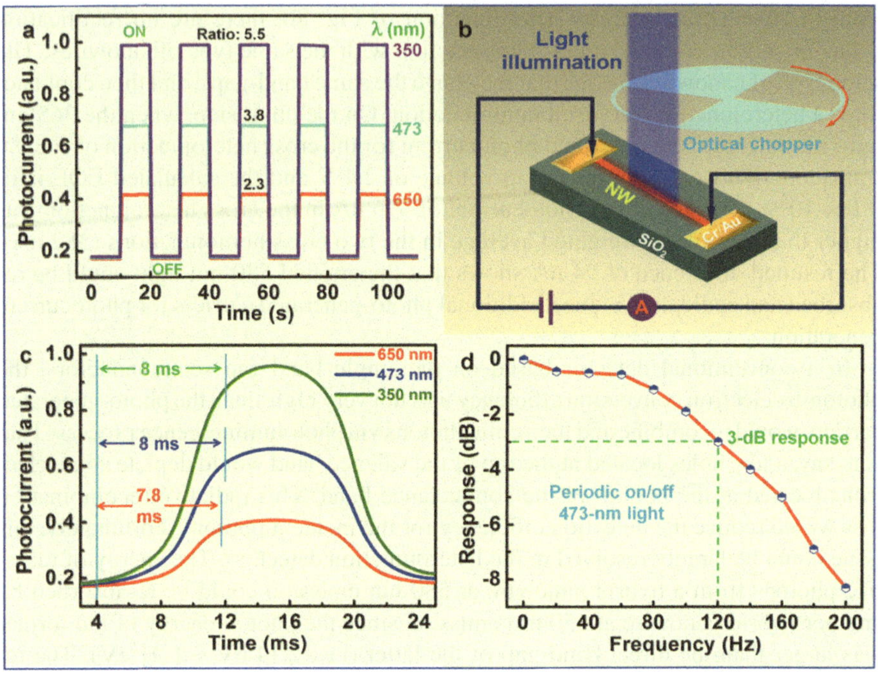

Fig. 9 **Temporal responsivity of the nanowire detector**. **a**Switching on and off the 350-nm, 473, and 650 nm light illumination with five cycles at an operating voltage of 2.4 V. **b** Scheme diagram for the device measurements of responsive speed. **c** Photo-responsivity with periodically switching on and off the 350, 473, and 650 nm light illumination via an optical chopper at 50 Hz frequency. **d** Photo-responsivity as a function of the chopper frequency, which shows a 3 dB responsivity of 120 Hz. Reprinted with permission [22]

of previously reported works [16, 36] and less than those (100 − 180) of previously reported works [6, 7, 14, 37, 38] with respect to the organic-inorganic hybrid detectors. To investigate the responsive speed, Fig. 9b gives the schematic configuration for device measurements, where an optical chopper with tunable frequency was experimentally used to modulate the light illumination. Figure 9c gives the photo-responsive results measured with periodically switching on and off the 350, 473, and 650 nm light illumination via the optical chopper at 50 Hz frequency. The photo-responsive times for nanowire detector were of 8, 8, and 7.8 ms for the investigated 350, 473, and 650 nm light illumination, respectively. The photo-responsive time was calculated by that the device photo-responsivity reaching up to 90% of the maximum photocurrent. The photo-responsive time of 8 ms was faster than that (100 ms) of the organic-poly(3-hexylthiophene) and inorganic-CdSe nanowire detector [38]. While ournanowire detector has a relatively slower photo-responsive speed compared with that of the single-crystalline nanobelts detector [24], it was still faster than that of the all-inorganic nanowire detectors [17, 39]. The relatively high photo-responsive speed was related to the photo-induced charge transfer [40] between the polyaniline and QDs in the organic–inorganic hybrid NW. Moreover, Fig. 9d gives the photoresponsivity versus the chopper frequency, indicating a 120 Hz of 3-dB responsivity. Thisperformance was measured by periodically switching on and off the 473-nm light illumination with tunable frequency.

For the detection application of the organic–inorganic hybrid nanowires, the recombination of photo-generated carriers was not wanted since it would cut down the detection sensitivity for incident photons. Figure 10a gives the photocurrents versus light intensity under an operating voltage of 2.4 V. The obtained results for the 350, 473, and 650 nm light illumination were nearly linear. This results indicate that the photo-excitation efficiency for photo-generated carriers was proportional to the photon flux. Since the high density of trap states [19] around the enormous organic-inorganic interfaces from the hybrid nanowire, this detector could significantly cut down the recombination of photo-generated carriers. Therefore, for the

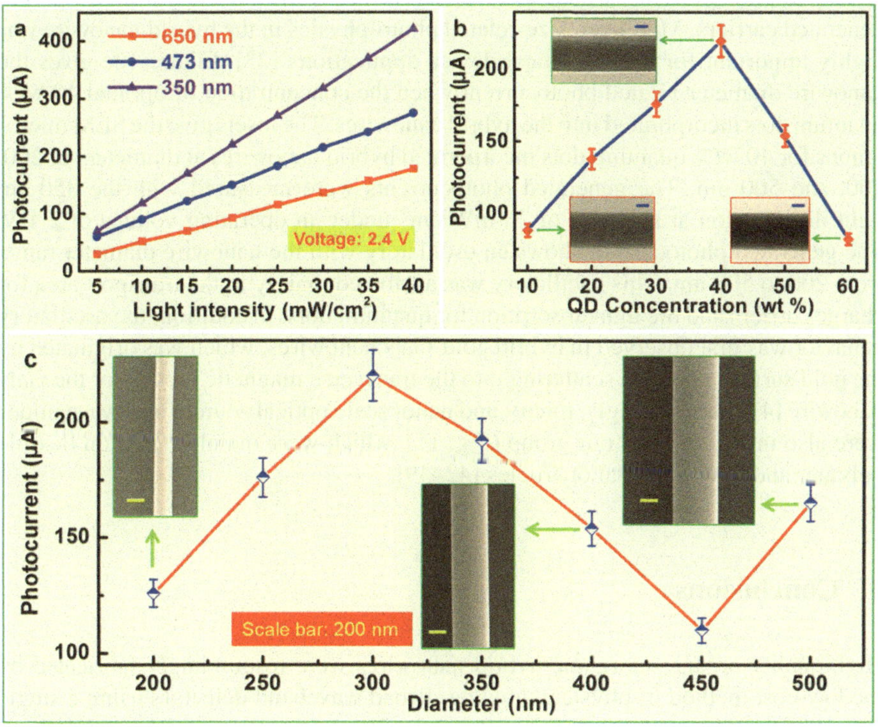

Fig. 10 Light intensity, quantum dots concentration, and nanowire diameter-related photocurrents. **a**Light intensity-related photocurrents under the 350, 473, and 650 nm light illumination at an operating voltage of 2.4 V. **b** Quantum dots concentration-related photocurrents. Insets: TEM images of three hybrid nanowires at different concentrations of quantum dots. Scale bar: 100 nm. **c**Nanowire diameter-related photocurrents. Insets: SEM images of three hybrid nanowires with different diameters. Reprinted with permission [22]

carriers' depletion, the photoconductive channel was preferred while the recombination channel was stopped largely. As the quantum dots are photo-active materials in ournanowire detector, the quantum dots concentration would largely affect the photo-generated currents. Figure 10b gives the photocurrent versus quantum dots concentration. The insets show the TEM micrograph of three 300 nm-diameter hybrid nanowires at 10, 40, and 60 wt quantum dots concentrations. The generated photocurrents were measured by the 350 nm light illumination at intensity of 20 mW/cm^2 with 2.4 V operating voltage. The generated photocurrent has a maximum with the quantum dots concentration of 40 wt%. When the quantum dots concentration was larger than 40 wt%, the incorporated quantum dots would aggregate and could not absorb effectively the incident photons, leading to the reduced photocurrents. In the case for quantum dots concentration was less than 40 wt%, the incorporated quantum dots could be excited effectively by the incident photons. Nevertheless, the low concentration of incorporated quantum dots may induce the low level of generated photocurrents, which was resulted from the low concentration for photo-generated carriers. Moreover, size-related photo-physics in the hybrid nanowire was highly important for their practical device applications [23]. Figure 10c gives the nanowire diameter-related photocurrent when the concentration of optimal 40 wt% quantum dots incorporated into the hybrid nanowires. The insets give the SEM micrographs for 40 wt% quantum dots incorporated hybrid nanowires at diameters of 200, 400, and 500 nm. The generated photocurrents were measured with the 350-nm light illumination at intensity of 20 mW/cm^2 under an operating voltage of 2.4 V. The generated photocurrent shows an oscillatory with the nanowire diameter range from 200 to 500 nm. This oscillatory was attributed mainly to the transport area for charge carriers and the light absorption for quantum dots. In addition, the oscillatory behavior was first observed in hybrid gold-GaN nanowires, which was originated by the gold surface plasmon scattering into the transverse magnetic modes for the GaN nanowire [41] . Interestingly, micro- and nano- scale optical sources and waveguides were also investigated in our group (Fig. 11), which were mainly based on flexible polymer and functional nanoparticles [42–49].

4 Conclusions

High-quality organic-inorganic hybrid nanowires were resoundingly fabricated by the low-cost method of physical drawing. Broad waveband detectors using a single hybrid nanowire and a crossed heterojuntion on the rigid substrate were prepared and characterized. The detector has shown an excellent photo-responsivity in the broad waveband of 350–700 nm. Moreover, the waveband range could be controlled by the size changing for quantum dots. The external quantum efficiency for the devices could reach up to 10^6, the photo-responsivity could reach up to 10^5 A/W, and the responsive time could be down to 8 ms. These indexes for device performances were kept in the high level across the whole waveband, which was resulted from enormous interface traps. The time-resolved photo-responsivity shows the good stability

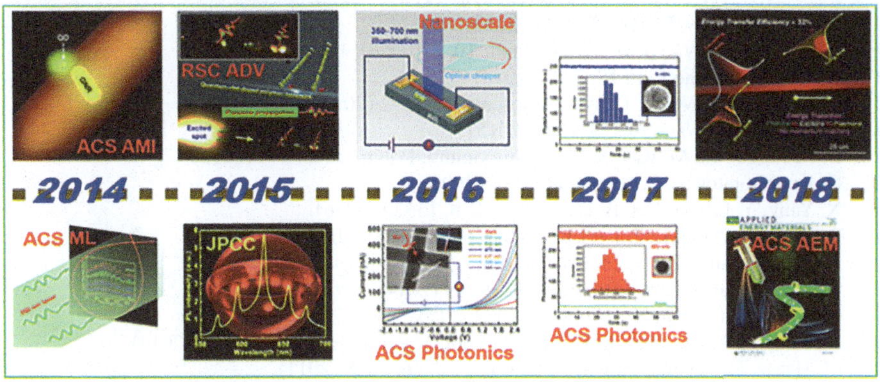

Fig. 11 Roadmap for device development from 2014 to 2018. First authored articles published in ACS Appl. Mater. Interfaces, ACS Macro Lett. ACS Photonics, ACS Appl. Energy Mater. Nanoscale, etc.

and the device reversibility. The bandwidth for device 3-dB responsivity was about 120 Hz. Different from the nanowire detector, the junction detector has the distinct rectifying characteristics associated with p-n junction. The enhanced characteristics for device performances could be attributed to the intermediate re-absorption for radiative photons and the abundant inorganic-organic interfaces. These two types of detectors would have various applications on the integrated optoelectronic devices for advanced on-chip information photonics.

Acknowledgements We thank the financial supports provided by the National Natural Science Foundation of China (Grants 11804120 and 61574158), the Natural Science Foundation of Guangdong Province (Grant 2017A030313026), the Fundamental Research Funds for the Central Universities (Grant 21617334), and the Research Projects from Guangzhou (Grant 201804010468).

References

1. L. Zang, Acc. Chem. Res. **48**, 2705 (2015)
2. G. Konstantatos, E.H. Sargent, Nat. Nanotechnol. **5**, 391 (2010)
3. X. Gong et al., Science **325**, 1665 (2009)
4. R. Saran, M.N. Nordin, R.J. Curry, Adv. Funct. Mater. **23**, 4149 (2013)
5. H. Yuan et al., Nat. Nanotechnol. **10**, 707 (2015)
6. D.J. Xue et al., Adv. Mater. **23**, 3704 (2011)
7. G. Chen, X. Xie, G. Shen, Nano Res. **7**, 1777 (2014)
8. G.A. O'Brien, A.J. Quinn, D.A. Tanner, G. Redmond, Adv. Mater. **18**, 2379 (2006)
9. Y. Che et al., J. Am. Chem. Soc. **132**, 5743 (2010)
10. S.A. Mcdonald et al., Nat. Mater. **4**, 138 (2005)
11. T. Zhai et al., Chem. Soc. Rev. **40**, 2986 (2011)
12. W.U. Huynh, J.J. Dittmer, A.P. Alivisatos, Science **295**, 2425 (2002)
13. A.L. Briseno et al., Nano Lett. **10**, 334 (2009)
14. F.-X. Wang, J. Lin, Y.-Q. Liu, H.-D. Wu, G.-B. Pan, Org. Electron. **15**, 844 (2014)

15. Y. Xie et al., Adv. Mater. **25**, 3433 (2013)
16. Y.R. Tao, X.C. Wu, W.W. Xiong, Small **10**, 4905 (2014)
17. J. Huh et al., Nano Lett. **15**, 3709 (2015)
18. Y.-R. Tao, J.-J. Wu, X.-C. Wu, Nanoscale **7**, 14292 (2015)
19. Y.-Q. Yu et al., Nano Res. **8**, 1098 (2015)
20. B. Aronggaowa, M. Kawasaki, T. Shimomura, Polym. J. **45**, 819 (2013)
21. G. Konstantatos et al., Nature **442**, 180 (2006)
22. X. Yang, Y. Liu, H. Lei, B. Li, Nanoscale **8**, 15529 (2016)
23. P. Wang, Y. Wang, L. Tong, Light Sci. Appl. **2**, e102 (2013)
24. L. Li et al., Adv. Mater. **22**, 3161 (2010)
25. G. Zhang et al., Chem. Commun. **50**, 10244 (2014)
26. C. Meng et al., Light Sci. Appl. **4**, e348 (2015)
27. X. Yang, R. Xu, D. Bao, B. Li, A.C.S. Appl, Mater. Interfaces **6**, 11846 (2014)
28. M. Haase, H. Schäfer, Angew. Chem. Int. Ed. **50**, 5808 (2011)
29. R. A. Ze'ev, A. Niv, X. Zhang, J. Appl. Phys. **109**, 114905 (2011).
30. X. Yang, D. Bao, Y. Zhang, B. Li, ACS Photonics **3**, 1256 (2016)
31. M.A. Khan et al., Appl. Phys. Lett. **60**, 2917 (1992)
32. T. Mueller, F. Xia, P. Avouris, Nat. Photonics **4**, 297 (2010)
33. P. Guo et al., Adv. Mater. **26**, 2844 (2014)
34. L. Yingkai, Z. Xiangping, H. Dedong, W. Hui, J. Mater. Sci. **41**, 6492 (2006)
35. T. Zhai et al., ACS Nano **4**, 1596 (2010)
36. N. Maity et al., J. Mater. Chem. A **3**, 20736 (2015)
37. J.-J. Wang, J.-S. Hu, Y.-G. Guo, L.-J. Wan, NPG Asia Mater. **4**, e2 (2012)
38. X. Wang et al., Adv. Funct. Mater. **23**, 1202 (2013)
39. X. Wang et al., Nanoscale **6**, 12009 (2014)
40. G. Itskos et al., Adv. Energy Mater. **3**, 1490 (2013)
41. J.P. Sundararajan et al., Nano Lett. **12**, 5181 (2012)
42. X. Yang, D. Bao, B. Li, RSC Adv. **5**, 60770 (2015)
43. X. Yang, D. Bao, B. Li, J. Phys. Chem. C **119**, 25476 (2015)
44. X. Yang, Y. Li, Z. Lou, Q. Chen, B. Li, A.C.S. Appl, Energy Mater. **1**, 278 (2018)
45. X. Yang, B. Li, ACS Macro Lett. **3**, 1266 (2014)
46. X. Yang, Y. Li, B. Li, ACS Photonics **4**, 2669 (2017)
47. X. Yang, B. Li, in *Plasmonics*. ed. by T. Gric (IntechOpen, London, 2018), p. 21
48. X. Yang, B. Li, in *Asia Communications and Photonics Conferenc*e (OSA, 2017), pp. Su2C.5
49. X. Yang, B. Li, in *International Conference on Optical Instruments and Technology 2017*,vol **10622** (SPIE, 2018), p. 10

Two-Dimensional (2D) Materials for Next-Generation Nanoelectronics and Optoelectronics: Advances and Trends

Zhi Peng Ling

Abstract Atomically thin two-dimensional (2D) materials has been reported in the International Technology Roadmap for Semiconductors to be one of the potential technological enablers to achieve scaling sizes of 5 nm and lower, digressing away from the conventional silicon material. Given the uniqueness of these 2D materials, with properties ranging from semi-metal to semiconductors to insulating, research scale efforts have demonstrated promising new opportunities for both rigid and flexible nanoelectronics and optoelectronics applications, in numerous fields such as imaging and sensing, medical, industrial, environmental, biological, field-effect transistors, solar cells, batteries, thermoelectrics, light emitting diodes, lasers and ultrafast optical communications. It is the focus of this chapter to provide an overall review of this fascinating world of atomically thin 2D novel materials, the physics underlying these materials, the recent applications surrounding these novel 2D materials and their devices, and the progressive efforts at large scale commercialization.

Keywords 2D materials · 2D devices · Transition Metal Dichalcogenides · Graphene · Boron nitride

1 Introduction

Given the ubiquitous growth of the consumer electronic market, there is a strong demand for devices with more functionality, smaller form-factors, faster performance, improved durability and lower costs among other requirements. To meet these multifaceted demands, the international technology roadmap for semiconductors (ITRS) [1] has been established as a guide, whereby research institutes partner closely with industry players to develop devices with ever-shrinking dimensions which meet or exceed the roadmap plans. Inevitably, this comes with a diverse set of challenges that need to be addressed. These include the search for alternative high

Z. P. Ling (✉)
Department of Electrical and Computer Engineering, National University of Singapore, 4 Engineering Drive 3, Block E4, Singapore 117576, Singapore
e-mail: E0374210@u.nus.edu

S. J. Ikhmayies et al. (eds.), *Advances in Optoelectronic Materials*, Advances in Material Research and Technology,
https://doi.org/10.1007/978-3-030-57737-7_3

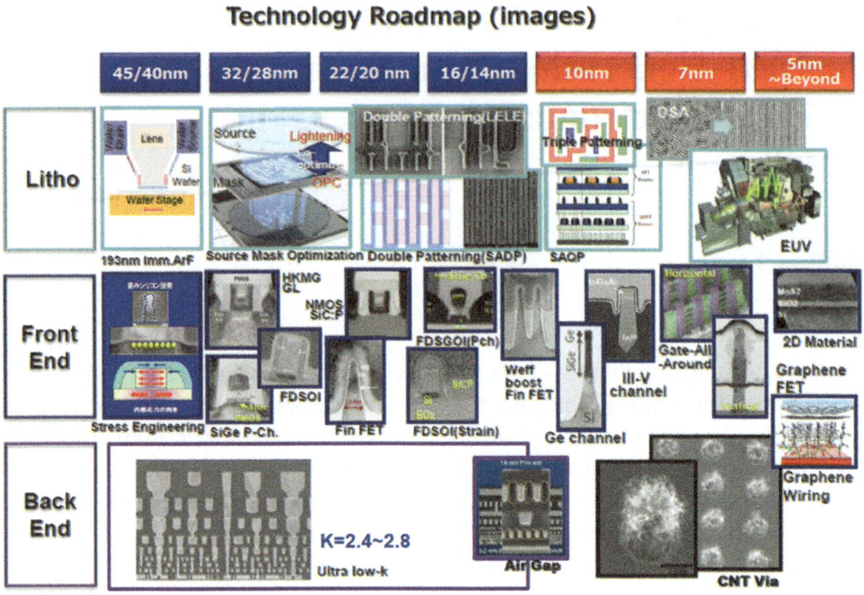

Fig. 1 Technology roadmap for the scaling of semiconductors and the efforts placed on both front-end and back-end processing, according to the 2015 ITRS reports. Figure eproduced from Ref. [1]

mobility channels, achieving low contact resistivity, low parasitic resistance and capacitance, continual effective oxide thickness scaling, and the integration of low bandgap channel materials such as Ge, III-V, and two-dimensional (2D) materials among others. Figure 1 shows an excerpt from the 2015 ITRS workshop on device scaling strategy and existing/potential future solutions [1].

Interestingly, atomically thin 2D materials were highlighted as potential candidates to integrate on next-generation electronic products starting from the year 2019 to achieve dimensions smaller than 5 nm. This is a result of cumulative research and fundamental discovery of unique material, electrical and optical properties from their corresponding bulk counterparts, sparking an immediate interest from academics and industry partners alike. Given the uniqueness of these 2D materials, with properties ranging from semi-metal to semiconductors to insulating, it will be of relevant interest to the reader to gain an overall review of this fascinating world of atomically thin 2D novel materials, the physics underlying these materials, the latest research advances both in terms of the understanding of the material properties and the development of novel 2D devices for innovative applications. This coupled with an insight into the future development trends for such materials will be the focus of this book chapter.

2 History—2D Materials

The interest in atomically thin 2D materials originated from graphene (a monolayer equivalent of bulk graphite), in which theoretical studies conducted more than seventy years ago [2–4] revealed intriguing information about the structure of the electronic energy bands, Brillouin zones, conduction-electrons, magnetic susceptibility among others. Despite that, the experimental realization of such layers only took place much later (i.e. year 2004) by Novoselov and team [5, 6], in which graphene and several other free-standing 2D atomic crystals (such as boron nitride, dichalcogenides, and complex oxides) had been obtained via micromechanical exfoliation. Figure 2 presents an illustration of a typical micromechanical exfoliation process [7] which involves four steps: (i) exfoliation of 2D crystal with low adhesion tape, (1) transfer of 2D crystal flakes from low adhesion tape to polydimethylsiloxane (PDMS) substrate, (2) transfer of 2D crystal flakes from PDMS substrate to target substrate, and (3) peeling off the PDMS substrate slowly to transfer the 2D crystal flakes onto the target substrate.

To build on this simple popular approach, many other approaches to obtain monolayers or few-layers ultra-thin 2D materials had been reported [8], such as liquid-phase exfoliation, molecular assembly, laser ablation [9], electron beam irradiation [10, 11], ultrasonication [12, 13], variants of chemical vapor deposition (CVD) [14–21], atomic layer deposition [22] , and magnetron sputtering [23, 24] among others. Given the vast possibilities to acquire the ultra-thin 2D layers, the unique material, electrical and optical properties can also differ significantly across different approaches as well, thus creating immense opportunities for both academics and industry partners. The progress in this field is further accelerated by the wide range of commercially available 2D materialsin various forms (solution, solid crystals, powders) [25–28], and with properties ranging from semi-metal, semiconducting, and insulating. Although primitive it seems, the ease of access to this plethora of 2D

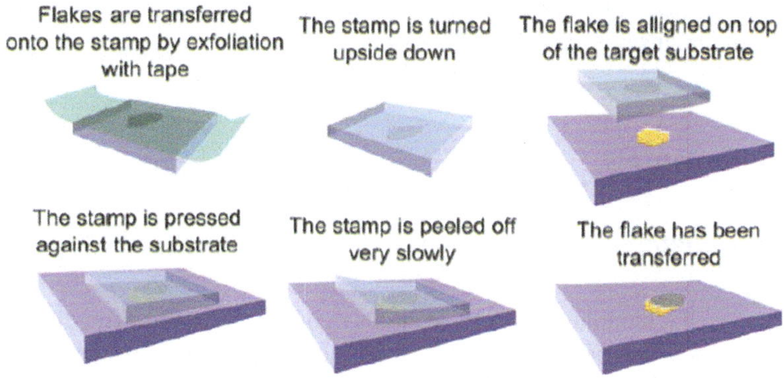

Fig. 2 Schematic of a micromechanical exfoliation and transfer process of 2D crystals' flakes to the target substrate. Figure Reprinted by permission from [7] © 2014 IOP Publishing Ltd. (2014)

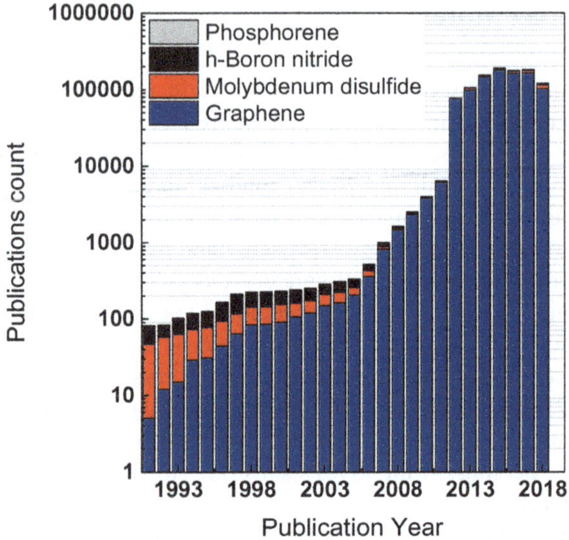

Fig. 3 The ability to obtain atomically thin 2D materials has sparked intense research interest, evident from the exponential growth of publications related to these 2D materials. This chart only shows the publication counts related to graphene (semi-metal), molybdenum disulfide(semiconducting), phosphorene (semiconducting) and hexagonal boron nitride (insulator), extracted from the Web of Science database up to the year 2018. Given the much wider range of commercially available 2D materials, the actual publication count is expected to be much higher

material candidates has led to an exponential growth of 2D materials related publications as seen in Fig. 3. Figure 3 features a summary of the research publications count pertaining to some of the more commonly investigated 2D materials, such as graphene (semi-metal), hexagonal boron nitride (insulator), molybdenum disulfide (semiconductor), and phosphorene (semiconductor). With the intense research progress on varied 2D materials, fulfilling the semiconductor technology roadmap with <5 nm dimensions appears to be an attainable goal.

3 Fundamentals

3.1 Graphene

Graphene is a flat monolayer of carbon atoms tightly packed into a two-dimensional (2D) honeycomb lattice (see Fig. 4), which can be subsequently processed into multi-dimensional structures such as zero-dimensional buckyballs, one-dimensional nanotube, and three-dimensional graphite. Pristine graphene is a zero-gap semi-metal, which exhibits a unique temperature-independent (up to 300 K) ambipolar

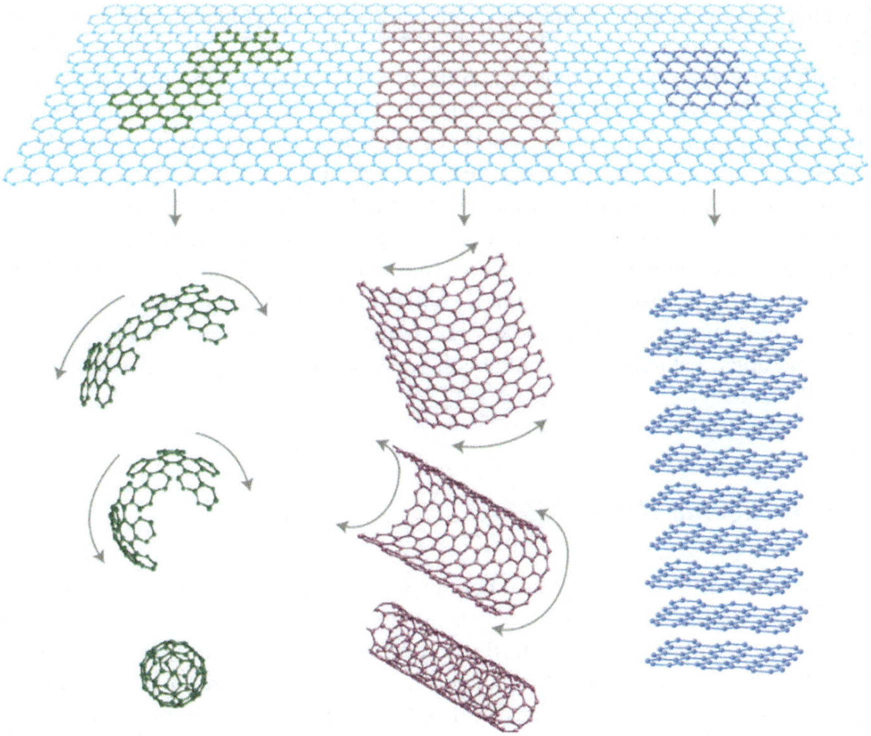

Fig. 4 Graphene, a monolayer of its corresponding bulk form graphite is a fundamental building material for forming structures with other dimensionalities (i.e. zero-dimensional buckyballs, one-dimensional nanotube, and three-dimensional graphite). Reprinted by permission from [57], © 2007 Nature Publishing Group (2007)

electric field effect wherein an exponential reduction in resistivity upon adding either electron or hole carriers was measured [5, 6, 29]. This translates to a high room-temperature mobility even at high carrier concentrations, indicative of ballistic transport on a sub-micrometer scale. The interaction of the surrounding electrons around the carbon atoms with the periodic potential of graphene's honeycomb lattice further leads to new quasiparticles known as massless Dirac fermions which can be precisely described by the (2+1)-dimensional Dirac equation with an effective speed of light $V_F \approx 10^6$ m^{-1} s^{-1} at low energies, contrary to the Schrodinger equation for condensed-matter physics [30–35]. Considering the conservation of chirality [36] and pseudospin properties [31–34, 37, 38] in graphene, many interesting quantum electrodynamics phenomena can be determined by assessing its corresponding electronic properties. This includes the chiral quantum Hall effects which highlight the presence of a minimum quantum conductivity ($\approx 4e^2/h$) in the limit of zero charges carriers [29], and a strong suppression of quantum interference effects at low temperatures (unlike typical metallic systems).

Consequently, some exceptional characteristics of graphene include room-temperature electron mobility [39] reaching 2.5×10^5 cm^2 V^{-1} s^{-1}, a high intrinsic strength of 130 GPa [40], Young's modulus of 1 TPa, excellent thermal conductivity above 3000 W mK^{-1} [41], high electrical stability [42], and complete impermeability to any gases [43] among others, easily surpassing reports on silicon-based materials. In addition, the graphene surface can be further modified using covalent and noncovalent techniques to obtain functionalized graphene to fulfill different requirements (for example insulating or high optical bandgaps). This however reduces the high conductivity properties exhibited by the intrinsic graphene material. Nonetheless, this creates application potential in fields such as polymer nanocomposites, super-capacitor devices, solar cells, water splitting, biosensing / bioimaging, biotherapeutics, catalytic, environmental, memory devices, and transistor devices [44–56].

3.2 *Hexagonal Boron Nitride*

Boron nitride (BN)is a heat- and chemically resistant refractory compound with alternatively linked boron and nitrogen atoms. Various crystalline forms of BN were reported: cubic BN, wurtzite BN, rhombohedral BN, and hexagonal BN (h-BN) [58]. Among which, h-BN has been actively studied due to several positive attributes: high thermal conductivity of 400 W mK^{-1} [59], Young's modulus of 0.8–1.3 TPa [60–62], high dielectric strength of ~12 MV cm^{-1} [63], excellent chemical stability (oxidation resistance up to 840 °C in air [64], and 1100 °C in oxygen [65]), and a large electrical bandgap of 5.5–6.0 eV making it practically insulating. In addition, h-BN benefits from an atomically smooth surface almost free of dangling bonds and charge traps [66], and a lattice structure very similar to graphene, making it one of the ideal growth substrates for obtaining graphene films with the highest electronic quality [67]. Figure 5 illustrates some structural models

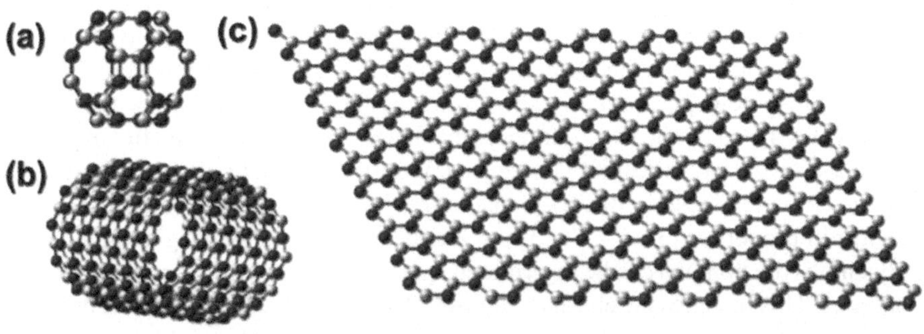

Fig. 5 Structural models of representative boron nitride (BN) nanomaterials: **a** 0D $B_{24}N_{24}$ fullerene; **b** 1D BN single-walled nanotube; **c** 2D mono-layered BN. Reprinted by permission from [58], Copyright © 2015 Published by Elsevier Ltd.(2015)

of representative boron nitride nanomaterials analogous to the graphene counterpart, which includes zero-dimensional fullerenes [8, 68], one-dimensional nanotubes [69], and two-dimensional monolayer BN [58] among other structures (nanomeshes, nanoparticles, nanowires, nanoribbons, nanoporous BN reported).

Despite the many similarities, the key difference between h-BN and graphene lies in the bonding configurations. In contrast to the pure covalent C-C bonds in graphene, h-BN is made up of heteroatom B-N bonds which are partially ionic in nature. Given the difference in the nature of the electron localizations around the boron and nitrogen atoms for different types of bonds (σ and π), its measured optical, electrical, and chemical properties are also different from the graphene counterparts. Nonetheless, this opens up new integration opportunities with other 2D materials as will be discussed in subsequent sections.

3.3 Molybdenum Disulfide (Transition Metal Dichalcogenides)

Molybdenum disulfide (MoS_2) belongs to a class of materials known as the transition metal dichalcogenides (TMD) [70] with the formula MX_2, where M is a transition metal element from group IV (Ti, Zr, Hf and so on), group V (for instance V, Nb or Ta) or group VI (Mo, W and so on), and X is a chalcogen (S, Se or Te), see Fig. 6. These materials form layered structures of form X–M–X, where a plane of transition metal atoms is separated by two hexagonal planes of chalcogen atoms. Figure 7 shows a structural schematic for the case of MoS_2. In this case, Molybdenum is the transition metal element, and Sulphur is the chalcogen element. It is to be noted that the schematic shows the 1T (tetragonal symmetry) which consists of one layer per repeat unit, with octahedral coordination, although there can also be other structural polytypes as well: 2H (hexagonal symmetry, two layers per repeat unit, trigonal prismatic coordination), and 3R (rhombohedral symmetry, three layers per repeat unit, trigonal prismatic coordination). As a result of these different polytypes, the lattice constants$a$ can range from 3.1 to 3.7 Å for different polytypes [71, 72]. The corresponding interlayer spacing for the 1 T case was also determined as ~6.5 Å [70]. Incorporating different transition metal elements into TMD leads to different exciting electrical and optical properties, with properties spanning from semiconducting to metallic, and some also exhibiting charge density wave and superconductivity characteristics (for example NbS_2, $NbSe_2$ [73] and TaS_2, $TaSe_2$ [74]).

In particular, $MoS_2$is a semiconducting material which exhibits a bulk indirect bandgap of 1.3 eV, which increases to a direct bandgap of 1.8–1.9 eV when reduced to a single-layer [75, 76], evident from the observation of strong photoluminescence signals from monolayer MoS_2. This is consistent with theoretical first-principles calculations [76, 77] and shown in Fig. 8. Density functional theory (DFT) calculations revealed that the conduction-band states at the K-point are mainly due to

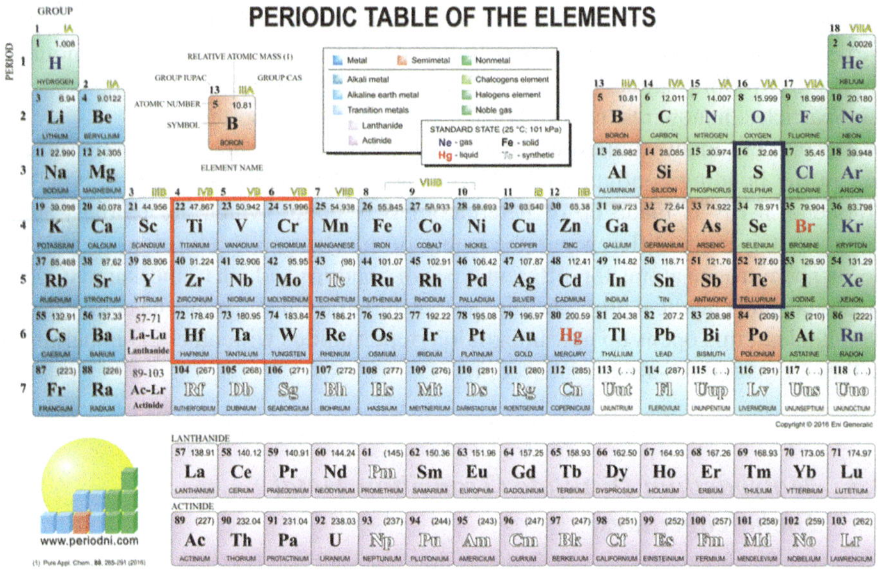

Fig. 6 Periodic table of the elements showing the key candidates for the transition metal dichalcogenides class of materials, comprising of a transition metal element from either group IV, V or VI, and a chalcogen (S, Se, Te) in a X-M-X configuration, where a plane of transition metal atoms is separated by two hexagonal planes of chalcogen atoms. Figure reproduced from Ref. [93]

Fig. 7 Structural model of 2D molybdenum disulfide (MoS_2) layers in 1T (tetragonal symmetry, one layer per repeat unit, octahedral coordination). The chalcogen atoms (S) are in yellow and the metal atoms (Mo) are in grey. Reprinted by permission from [70], © 2011 Macmillan Publishers Limited.(2011)

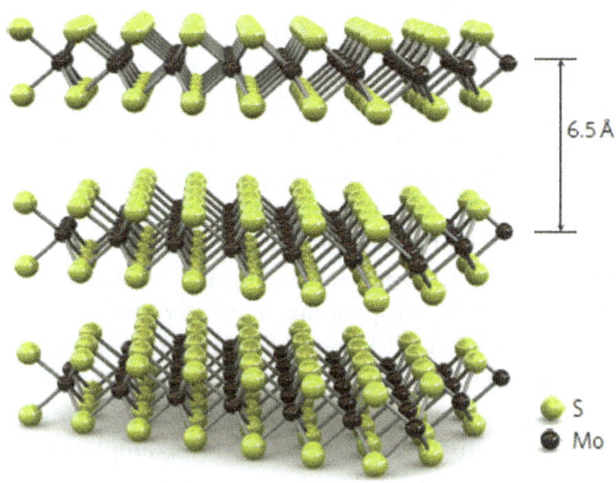

localized *d*orbitals on the Mo atoms in the layered structure and relatively unaffected by interlayer coupling. However, there is a strong interlayer coupling for the states near the Γ-point due to a combination of the *d*orbitals on Mo atoms and the antibondingp_z-orbitals on the S atoms, and shows a thickness dependence [78]. As the number of MoS_2 layers reduces, the transition at the Γ-point shift significantly

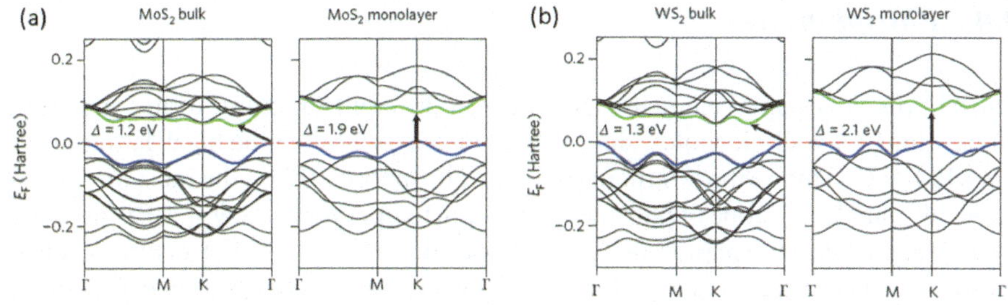

Fig. 8 Band structures calculated from first-principles density functional theory (DFT) for bulk and monolayer**a** $MoS_2$and **b** WS_2. The horizontal dashed lines indicate the Fermi level. The arrows indicate the fundamental bandgap (direct or indirect). Reprinted with permission from [76]. ©2011 American Physical Society

from an indirect one to a larger, direct one for a monolayer MoS_2 (i.e. transformation from indirect bandgap to direct bandgap). This prediction also applies to all other $MoX_2$and WX_2compounds with decreasing layer numbers, covering the bandgap energy range 1.1–1.9 eV [72, 76, 79–82]. With a transition to direct bandgap, and increase in bandgap energy, there is a corresponding change in the absorption spectra, photoconductivity, and photoluminescence [75, 76, 78] for a monolayer MoS_2 as compared to the bulk form. The photoluminescence quantum yield for a direct bandgap monolayer MoS_2 was however observed to be significantly lower (4×10^{-3}) [75] than unity, and can be attributed to the considerable defect density (such as sulphur vacancies) leading to increased non-radiative recombination. Recent efforts to address this issue by utilizing an air-stable solution-based chemical treatment by a non-oxidising organic superacidbis(trifluoromethane) sulfonimide to eliminate these defects had been highly successful, resulting in a final photoluminescence quantum yield in excess of 0.95 [83]. The ability to obtain monolayer TMDs with near-perfect properties creates opportunities for the development of highly efficient 2D optoelectronics devices such as light-emitting diodes, lasers, and solar cells. The electronic structure of monolayer MoS_2 also enables valley polarization, which is not seen in bilayer MoS_2 [84–86]. It has also been recently demonstrated that by doping MoS_2 with vanadium, the modified MoS_2 is capable of maintaining permanent valley polarization and hence is a potential candidate for valleytronics [87]. Monolayer MoS_2 also exhibits strong spin-orbit splitting [88–90], making them promising candidates for spintronic devices as well.

To aid in the identification of monolayer MoS_2, it is common to utilize the Raman spectroscopy technique [91], where the shifts in the main Raman peaks corresponding to the in-plane E^1_{2g} and E_{1u} phonon modes, and the out-of-plane E_{1g} mode facilitates this purpose. The Raman shifts had been attributed to the influence of neighbouring layers on the effective restoring forces on atoms and the increase of dielectric screening of long-range Coulomb interactions [92].

3.4 Phosphorene

Phosphorus, a Group V element can appear in different structural forms, including cubic white, monoclinic violet, amorphous red, and orthorhombic black lattice structures [94–97]. Among those structures, black phosphorus(BP) is the most thermodynamically stable allotrope of phosphorus at room temperature [98, 99]. BP is obtained by heating white phosphorus under high pressures. The crystal structure of black phosphorus is illustrated in Fig. 9, in which the crystal structure was experimentally determined as orthorhombic, with its unit cell containing eight atoms, and the lattice constants determined as $a = 4.38$ Å, $b = 3.31$ Å, and $c = 10.5$ Å [99]. The BP atoms form crystalline 2D puckered sheets [100], with an interlayer separation distance of ~5.3 Å [101, 102]. Similar to other 2D materials, the in-planecovalent bonding between the P-atoms is stronger as compared to the weak Van der Waals interactions between layers [103], allowing the ease of separation by the popular micromechanical exfoliation approach or other approaches mentioned earlier. A monolayer of black phosphorus layer is also commonly known as phosphorene.

One of the key advantageous properties of BP is its direct bandgap (E_g) regardless of the number of layers. As compared to the bulk BP with a E_g of ~ 0.3 eV, this value increases as the number of layers reduce and reaching ~2.05 eV for a monolayer [104, 105], see Fig. 10. With the wide range of tunable bandgap spanning visible to mid-infrared regions, this opens up new potential applications (for example: in the field of communication) for wavelength regions not covered by semiconducting TMD materials, see Fig. 11 [106]. Another advantage is its relatively high charge carrier mobility (300–1000 cm^2 V^{-1} s^{-1}) [107], as compared to monolayer MoS_2 TMDs (2 orders lower) [108], and comparable to the existing silicon technologies. Another positive trait of phosphorene is in its intrinsic ambipolar behavior (it can be a p- or n-type semiconductor) in contrast to TMDs, which show only unipolar (n-type) behavior [109]. Hence, BP could be a potential candidate for integration

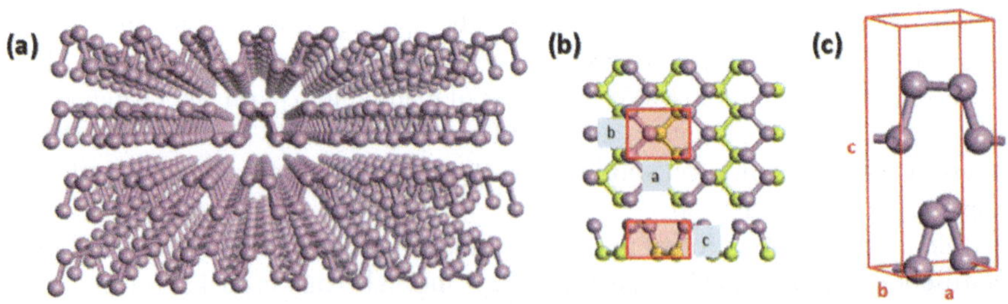

Fig. 9 **a** Layered crystal structure of black phosphorous. (b) The two-dimensional top and side view of phosphorene (a monolayer of black phosphorus). The atoms in different colors represent atoms in different planes. The lattice constants (**a–c**) of the unit cell of black phosphorous-containing four P-atoms is outlined by the red line (**c**). A three-dimensional zoom-in view of the atomic structure of phosphorene, showing eight P-atoms, and the lattice constants. Reprinted by permission from [122], © 2016 Taylor & Francis Group, LLC. (2016)

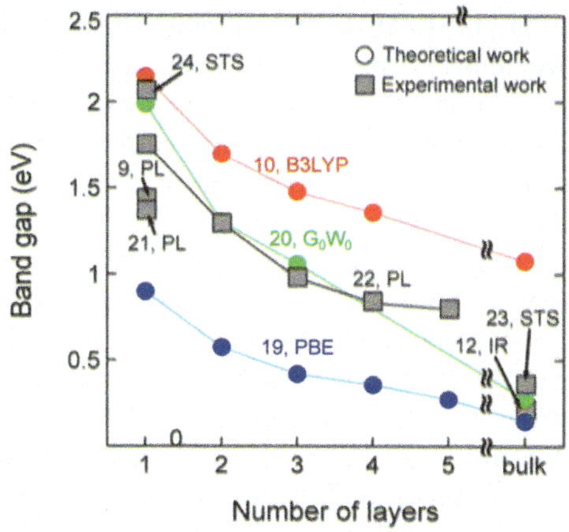

Fig. 10 Thickness dependence of the black phosphorus bandgap, including experimentally measured values, and calculated data using different approaches. PBE, G_0W_0, B3LYP are ab initio approaches for band structure calculation, while photoluminescence (PL), scanning tunneling spectroscopy (STS), infrared spectroscopy (IR) are experimental measurement methods. The reader is referred back to the original publication [105] for the references shown in this figure. Reprinted (adapted) with permission from Ref. [105]. Copyright (2019) American Chemical Society

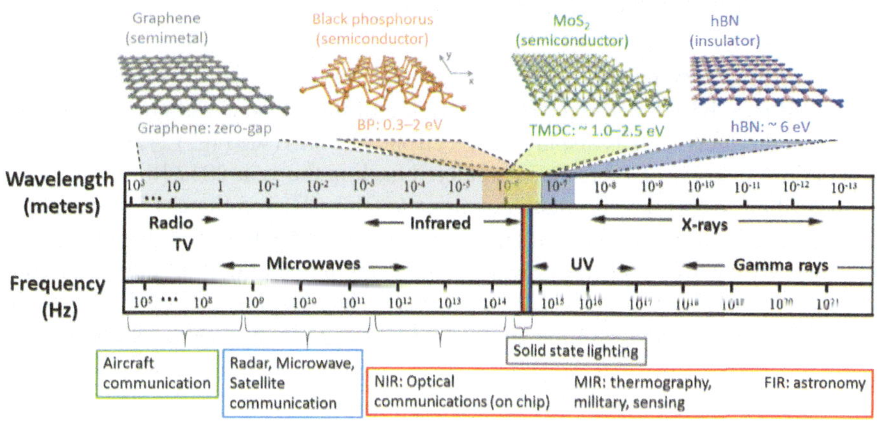

Fig. 11 A schematic of some of the actively researched two-dimensional materials (such as graphene, phosphorene, molybdenum disulfide, and boron nitride) exhibiting unique electrical and optical properties, and their potential real-world applications in the field of communications. Reprinted, with permission, from [106]. © 2015 Copyright IEEE. (2015)

into CMOS related technologies. BP is also found to be suitable for thermoelectric applications due to its high Seebeck coefficient (dimensionless thermoelectric figure of merit ZT = 2.5) [110, 111], while their thermal conductivities are comparable to TMDs ($\lambda \approx 50$ Wm^{-1} K^{-1}), while showing a strong in-plane anisotropy in which the thermal conductivity along the zigzag direction is ≈2.5 times higher than that of the armchair direction [112]. Also, phosphorene can withstand a high degree of elasticity and flexibility (tensile strain up to 30%) [113, 114] as compared to existing silicon material (tensile strain of 2%) [115], making it a potential candidate for flexible electronics. In addition, these properties exhibited by phosphorene can be further changed or enhanced by the number of layers, strain, physical and chemical functionalization, and defect engineering [116–118].

However, there are also some challenges encountered by BP that needs to be considered. Firstly, BP was found to be unstable under ambient conditions [119, 120] due to its hydrophilicity which resulted in the etching of its layers and an increase in volume. Devices made of few-layer phosphorene films were found to degrade when no additional passivation was applied on the exposed BP layers. Secondly, similar to other promising 2D candidates, the next challenge is to scale up the deposition of phosphorene films to reach commercialization levels, especially when precise control and preservation of its puckered lattice orientation in large scale deposition is required [121].

4 2D Materials Applications in Various Fields

4.1 Graphene-Based Applications

Since graphene exhibits a plethora of unique and attractive properties not seen in the existing silicon-based materials and integration of this 2D material into mainstream consumer products such as display and electronic devices (see Fig. 12) and photonics applications (see Fig. 13) has been predicted back in the year 2012 [8].

According to Ref. [123], many countries (such as United States, Europe, South Korea, Japan, and China) had started placing emphasis on the research and development and the eventual transfer of technology from the laboratory to manufacturing. Among which, China had invested considerably (hundreds of millions) through the China Innovation Alliance of the Graphene Industry (CGIA) established in the year

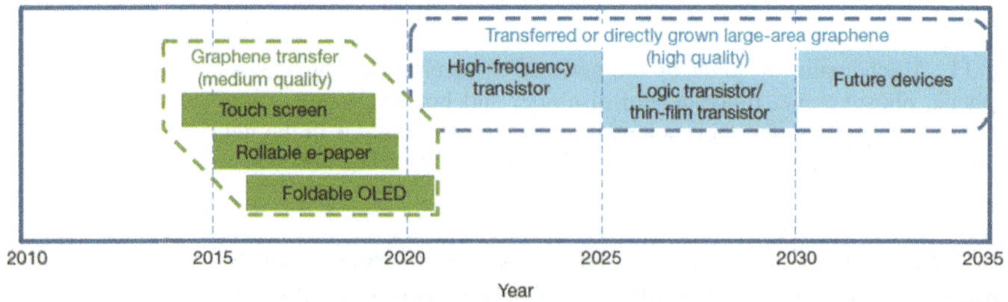

Fig. 12 A technology roadmap projection of Graphene-based display and electronic devices. Reprinted by permission from [8]. ©2012 Macmillan Publishers Limited (2012)

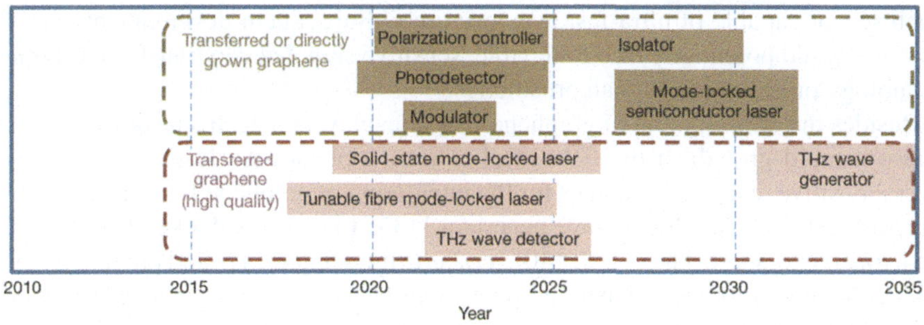

Fig. 13 A technology roadmap projection of Graphene-based photonic applications. Reprinted by permission from [8]. ©2012 Macmillan Publishers Limited (2012)

2013 to focus on the commercialization of graphene within the country, via graphene industrial parks and international cooperation.

As a result of this active interest, initial prototypes of flexible display and electronic devices integrating graphene had been reported [124–126] as can be seen in Fig. 14. With the incorporation of graphene-related technologies into consumer products (such as mobile devices), and drawing on the unique advantages of this material such as (1) high charge mobility which can enable ultra-fast internet surfing and data transfer [127], (2) high film transparency which can lead to brighter screens and concurrently act as transparent electrodes[128], (3) high intrinsic strength which improves product durability, and (4) excellent thermal conductivity which enables good heat dissipation, graphene appears to be one of the highly attractive candidates to continue this technology trend.

With further innovations in graphene growth technologies, such as the resistive-heating cold-wall chemical vapor deposition technique [129], the same high-quality graphene can be obtained at a much higher throughput rate (100 times faster) and lower cost (99% cheaper) than the conventional CVD approach. The researchers involved in this work had also demonstrated a working flexible and transparent

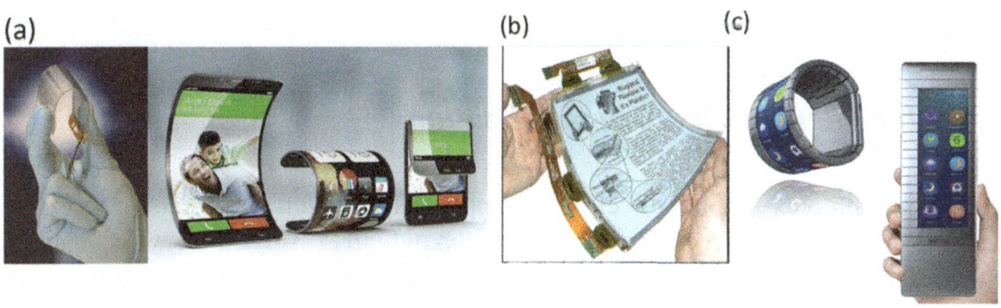

Fig. 14 Demonstration of flexible displays utilizing graphene technology [124, 126] in (**a**) and (**b**), while commercially available products such as the bendable smartphones produced by Moxi, a company based in China are now available [125] (**c**). Reprinted by permission from [125]. © 2014 Macmillan Publishers Limited (2014)

touch sensor capable of ultra-fast response times. Subsequent developments in this direction could potentially lead to flexible, sensitive skin that can transform robotics technology or applied in human prosthetics.

Besides the development of graphene-based display and electronic devices, there is also a rapid growth in the field of photonics applications. There are currently several positive research outcomes for the integration of graphene for terahertz wave detector[130–133], terahertz wave generator [133, 134], mode-locked lasers [135–138], tunable terahertz laser [139, 140], terahertz modulator[141] and photodetectors [142–145] among others. Hence, graphene can be viewed as an enabler or extension for future technologies requiring terahertz imaging and sensing in the fields of security, medical, industrial, environmental, biological, and optical communications among others.

4.2 *Hexagonal Boron Nitride Applications*

Arising from the unique advantages of h-BN, various applications utilizing this material either in the powder form or hot-pressed shapes have been reported [146], and that include the various fields of chemistry, metallurgy, high-temperature technology, electrotechnology, and electronics [147] as shown in Table 1.

From the 2D perspective, when atomically thin high-quality h-BN layers are considered, the reported direct bandgap of 5.9 eV [148] opens up new application potentials in the field of ultraviolet (UV) lasing, far-UV light emitting diodes (LEDs) [149, 150], and could promote the epitaxial growth of high-quality graphene due to the close matching lattice parameters of h-BN and graphene. Its electrically insulating properties could also act as an appropriate gate dielectric for transistor applications, or encapsulant for flexible or high-temperature electronics devices [151, 152].

Furthermore, with the demonstrated ability to deposit atomically thin h-BN films on a large scale using techniques such as chemical vapor deposition [153–159], atomic layer deposition [160, 161]sputtering [162] and pulsed laser deposition [163–165], h-BN stands a good chance to be integrated into future technologies.

4.3 *Molybdenum Disulfide Applications*

MoS_2 complements graphene well when applied as field-effect transistor in digital electronics, benefitting from the presence of a bandgap (1.3–1.9 eV), which enables an excellent on/off current ratio in the range of ~10^8 [70, 166], not achievable by the graphene material. Given the continuous scaling trend of transistors to ever-shrinking dimensions, the issues with statistical and quantum effects, heat dissipation, and short channel effects encountered by the existing silicon devices may be better addressed by the 2D semiconductor materials such as MoS_2. To further improve conductance, doping can be utilized but this has to be a trade-off with decreased mobility due to

Table 1 The typical applications of boron nitride. The legend for the property desired are as follows: *R*: High-temperature refractoriness; *T*: thermal conductivity; *E*: electrical insulator; *I*: chemical inertness, nonwetness; *L*: lubricity; *M*: machining ease. Reprinted with permission from [147]. © 1989 Elsevier Science Publishers Ltd

Application	Property			Desired		
	R	*T*	*E*	*I*	*L*	*M*
Hexagonal boron nitride powder						
• Solid lubricant for high bear temperature	x			x		
• Mould release agent for die casting of glass and metals	x		x	x		
• Active filler for rubber, resins and plastics	x		x	x	x	
• Additive to oil and high temperature grease				x	x	
• Ultrahigh-pressure transmitting agent	x			x	x	
• Coating for evaporation plants	x			x		
• Coating for graphite hot pressing moulds	x			x	x	
• Embedding medium for heating wires	x		x	x		
• Boron source preparation of c-BN and ceramic composites	x					
Hexagonal boron nitride hot pressed shapes						
• Crucibles for melting glass and metals	x	x		x		x
• Break rings for horizontal continuous castings	x			x	x	x
• Compounds for high temperature electric furnaces	x	x	x	x		x
• Structural parts for magnetohydrodynamic devices	x		x	x		x
• Dielectric for radar antennas and windows			x			x
• Insulator for low- and high-frequency equipment	x		x			x
• Insulator for plasma-jet furnaces, arc pulse generators and ion engines	x		x			x
• Holders, mounting plates, substrates and heat sinks in valve and transistor circuits		x	x			x
• Pump components, pipes and nozzles for handling liquid metals	x			x		x
• Protective tubes and insulating sleeves for thermocouples	x	x	x	x		x
• Protective sleeves for elctrodes in automatic welding	x		x	x		x
• Wafers for boron-doping of semiconductors	x					x
• Moulds for hot pressing of ceramics like B_6O or $B_{6.5}C$	x	x		x	x	x
• Nutron absorbers and shields for nuclear reactors	x	x		x		x

scattering effects [167, 168]. Figure 15a shows the schematic of a top-gated transistor based on a monolayer MoS_2 film [70], and subsequently setting the foundations for more advanced integrated circuits such as logic gates (see Fig. 15b–d) [169] static random access memory, and five-stage ring oscillator [170] among others.

Since MoS_2in its monolayer form exhibits a direct semiconducting bandgap, this material has a great potential to be applied in the field of optoelectronics, whereby the ability to generate, detect, interact with or control light is required. When this

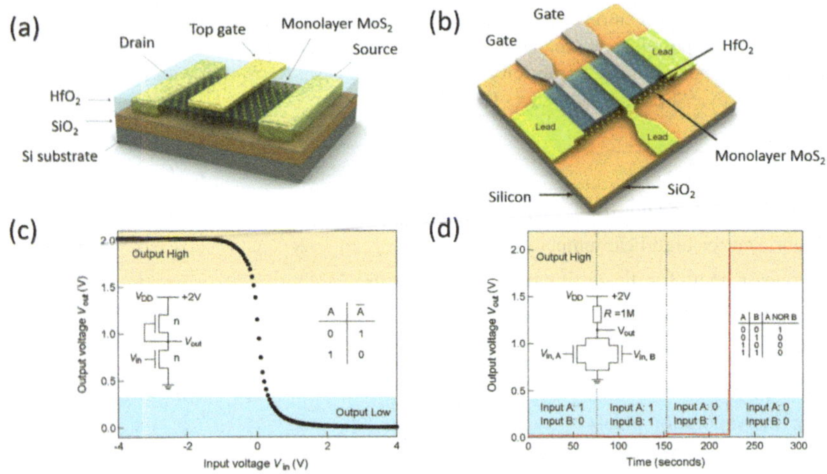

Fig. 15 **a** Schematic of a top-gated monolayer MoS_2 field-effect transistor device with hafnium oxide top-gate dielectric, **b** Schematic of an integrated circuit comprising of two transistors built on a monolayer MoS_2 which can function either as a logical inverter (**c**) or as a logical NOR gate (**d**), which is one of the universal gates that can be combined to form all other logic operations [171]. Figure (**a**) is reproduced from Ref. [70], and figures (**b**)–(**d**) are reproduced from Ref [169]. Reprinted (adapted) with permission from Ref [169]. Copyright (2019) American Chemical Society

is coupled with the high mechanical strength of this material (Young's modulus comparable to steel) [172, 173] could potentially lead to new opportunities for flexible and transparent optoelectronics. Till date, successful research scale demonstration of atomically thin MoS_2 material for optoelectronics had been reported which includes photodetectors/ phototransistors [24, 174–184], light emitting diodes [185, 186], photovoltaics [186–190], sensing [191–196], and waveguides [197, 198] among others. Figure 16 shows an example of a MoS_2 based multiple quantum well light emitting diode reported in Ref. [185]. Other applications recently explored for MoS2 material also includes electrodes for energy storage [199, 200].

These unique properties of atomically thin MoS_2 films (one individual candidate from the 2D transition metal dichalcogenides (TMD) class of materials) coupled with the demonstrated ability to obtain precise layer thickness control on different large-area (several centimeters square) deposition techniques such as chemical vapor deposition [201–206], atomic layer deposition [207, 208], pulsed laser deposition [209, 210], and sputtering [24, 211, 212] indicates the high commercialization potential of this material for future electronic or optoelectronic devices. With plenty of other candidates in this 2D TMD class of materials, future research works in this direction could be both exciting and rewarding.

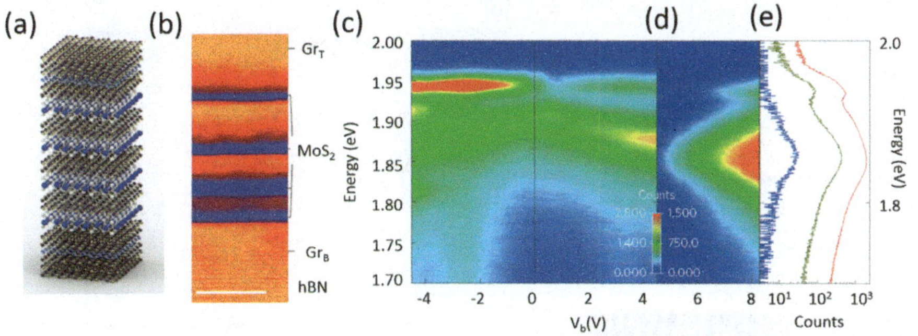

Fig. 16 **a** Schematic and **b** scanning tunneling electron microscopy image of the multiple quantum well (MQW) heterostructure [$hBN/GrB/2hBN/MoS_2/2hBN/MoS_2/2hBN/MoS_2/2hBN/MoS_2/2hBN/GrT/hBN$], consisting of multiple stacks of hexagonal boron nitride and MoS_2, and contacted by graphene top and bottom electrodes. **c** Maps of photoluminescence and (d) electroluminescence spectra for the same device. $E_{Light} = 2.33$ eV. (e) Individual electroluminescence spectra plotted on a logarithmic scale as a function of layer thickness. Reprinted by permission from [185].© 2015 Macmillan Publishers Limited (2015)

4.4 Black Phosphorus Applications

Black phosphorus or few-layers phosphorene has been evaluated for various electronics applications, which includes field-effect transistors [107, 120, 213–225], molecular sensors [226–232], solar cells [233–235]photocatalytic water splitting [236–238]batteries [239–243], thermoelectrics [110, 111, 244, 245], and biomedical applications [246].

Given the direct bandgap nature of this material regardless of layer thickness (as compared to TMDs which exhibits direct bandgap only when reduced to a monolayer) varying from 0.3 eV for bulk BP and increasing to ~2 eV for a monolayer), there is keen interest in integrating this material for optoelectronics applications spanning near and mid-infrared wavelengths to visible wavelengths such as detectors [247–250], switches [251–254], imaging [255, 256]light emitting diodes and lasers [257, 258]. The high mechanical flexibility of phosphorene also creates opportunities for wearable devices. The wide range of tunable bandgap for few-layer phosphorene also makes it a potential candidate for high-speed optical communication and nanoscale optical circuits [259]. By combining phosphorene with existing technologies such as silicon photonics and metallic nano-plasmonics structures (see Fig. 17), Chen et al. [260] has successfully demonstrated an operating 3D hybrid integrated photonic structures capable of detecting telecom-band, near-infrared light with high intrinsic responsivity (10 A/W).

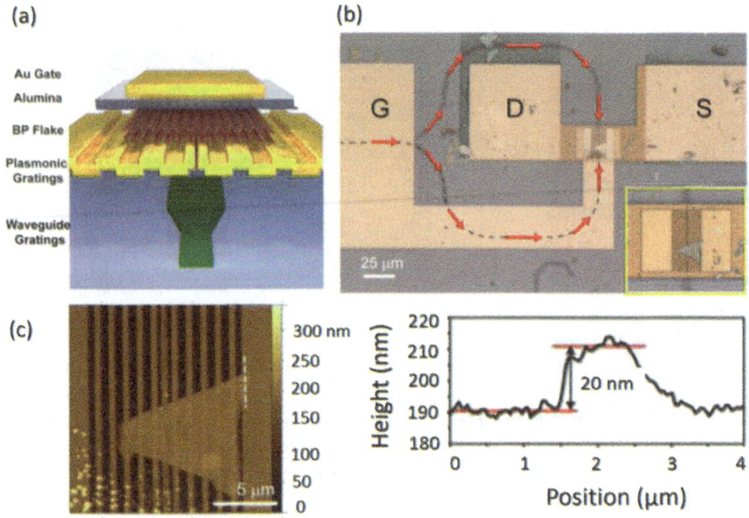

Fig. 17 **a** Schematic of a 3D hybrid integrated photonic structure comprising of silicon photonics, nano-plasmonics, and black phosphorus. **b** Optical microscope image of the completed 3D hybrid photonic device and **c** Atomic force microscope (AFM) image of the black phosphorus–nanogap region. The topographic profile (lower panel) shows that the black phosphorus flake is 20 nm thick. "Reprinted (adapted) with permission from Ref. [260]. Copyright (2019) American Chemical Society"

4.5 Flexible Next-Generation Device Integration—A Complete 2D Approach

Given the high mechanical flexibility of the investigated 2D materials and its unique thickness-dependent electronic and optoelectronic properties, these are highly attractive candidates to be applied for next-generation flexible electronics and optoelectronics applications, as compared to the more traditional brittle materials such as silicon. As an example, Kelly et al. [261] have successfully demonstrated a simple low-cost approach to fabricate an array of thin-film transistors made up entirely of atomically thin 2D materials (WSe_2 as the channel, h-BN as the separator/top-gate dielectric, and graphene as source, drain, and top-gate electrodes). By additionally depositing an ionic liquid (1-ethyl-3-methylimidazolium bis(trifluoromethylsulfonyl) imide (EMIm TFSI) on the entire structure, it enhances electrolytic gating, leading to improved device performance, with higher operating currents than comparable organic thin-film transistors (see Fig. 18).

Table 2 summarizes some of the demonstrated applications of 2D Van der Waals heterostructures till date comprising of a combination of semi-metal, semiconductor, and insulating candidates [262]. Some of the more commonly investigated

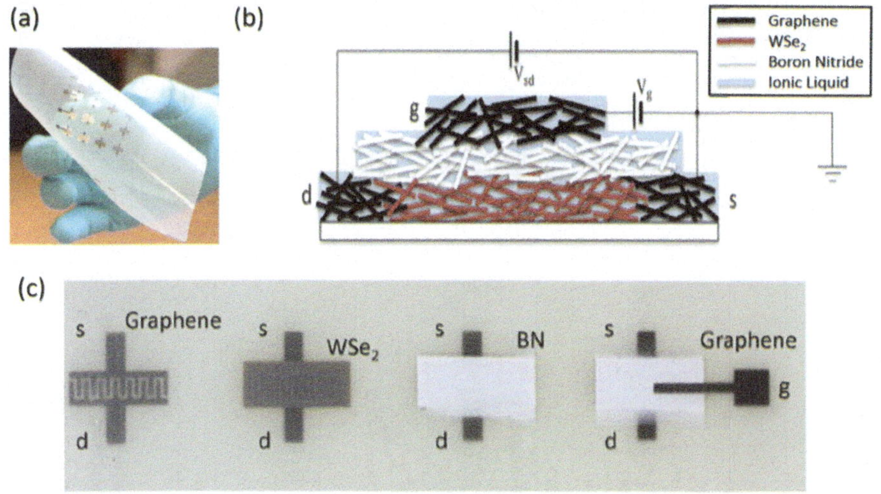

Fig. 18 **a** Demonstration of a flexible array of all-printed thin-film transistors on a flexible alumina-coated polyethylene-terephthalate (PET) substrate. **b** Schematic of the thin-film top-gated transistor structure featuring an inkjet-printed network of WSe_2 nanosheets as a channel material, inkjet-printed network of graphene nanosheets for source, drain, and gate electrodes, spray-deposited boron nitride as the gate dielectric/separator, and an ionic liquid to facilitate electrolytic gating. **c** Sequential photographs of the involved printing steps. "Reprinted (adapted) with permission from Ref. [261]. Copyright (2019) American Chemical Society"

applications are in the fields of energy harvesting (solar cells), electronics (field-effect transistors), light detection (photodetectors), and light generation (electroluminescence). There are also increasing research efforts in exotic directions such as quantum-dot qubits, single-photon emitters, superconducting qubits and topological quantum computing elements [263]. With the active interest in material screening and evaluation of diverse device architectures, evident by the rapid increase in related publications, we are now ever closer to the stage of commercialization, to deliver better-performing, or interesting novel applications.

5 Progress Towards Commercialization

To reiterate, atomically thin two-dimensional materials had been proposed to be one of the approaches to fulfill the ever-shrinking device dimension targets outlined by the international technology roadmap for semiconductors (ITRS). To meet this requirement economically, the industry players need to adopt a large scale deposition approach which can deliver atomically thin 2D layers to meet several key objectives: (1) low-cost, (2) high reproducibility, (3) precise thickness control, and (4) excellent film qualities, on par with the micromechanically exfoliated approaches on pristine bulk crystals.

Table 2 Varied 2D heterostructures and their applications. Reprinted with permission from [262] © 1989 Elsevier Science Publishers Ltd.

Heterojunction type	Layer structure	Application	Refs.
Semi-metal/	Graphene/MoS_2	DNA biosensor	[264]
semiconductor	Graphene/WS_2	Solar cells	[265]
	Graphene/WS_2/Graphene/h-BN	Vertical FETs	[266]
	Graphene/MoS_2	Vertical FETs	[267, 268]
	Graphene/h-BN/MoS_2/Graphene	FETs	[269]
	Graphene/MoS_2/Graphene	Photodetector	[270]
	Graphene/WS_2/Graphene	Photodetector	[271]
	Graphene/MoS_2	Photodetector	[183, 272]
	Graphene/h-BN/MoS_2(WS_2)/h-BN/Graphene	Electroluminescence	[185]
Semi-metal/	Graphene/h-BN	FETs	[67]
insulator	h-BN/Graphene/h-BN/Graphene/h-BN	Vertical FETs	[273]
	h-BN/Graphene/MoS_2/Graphene/h-BN	Vertical FETs	
	Graphene/h-BN/Graphene	Thermoelectrical power	[274]
Semiconductor/	WSe_2/MoS_2	Solar cells	[275]
semiconductor	p-WSe_2/n-WSe_2	CMOS	[267]
	p-MoS_2/n-MoS_2	Solar cells	[276]
	MoS_2/p-Si	Solar cells	[188]
	WSe_2/MoS_2	Solar cells	[277]
	Graphene/WSe_2/MoS_2/Graphene	Solar cells	
	WSe_2/MoS_2	p–n diode, photodetector, electroluminescence	[278]
	Black phosphorus/MoS_2	p–n diode, photodetector	[279]

(continued)

Table 2 (continued)

Heterojunction type	Layer structure	Application	Refs.
	MoS_2/WS_2	Charge transfer	[280]
	WS_2/MoS_2	Solar cells	[281]
	WSe_2/WS_2	Solar cells	[282]
	WSe_2/MoS_2	Solar cells	[283]
	$WSe_2/MoSe_2$	Solar cells	[284]
	MoS_2/p-Ge	Band-to-band tunneling FET	[285]
	MoS_2/p-Si	Solar cells; electroluminescence	[186]
Semiconductor/ insulator	MoS_2/h-BN/graphene	FETs	[286]

One successful demonstration of a large-area deposition tool to obtain high-quality graphene layers has been reported by Bae et al. [287], in which monolayer graphene films with a large 30-inch diagonal width can be obtained by a roll-to-roll production process which performs chemical vapor deposition onto flexible copper substrates, followed by a series of steps leading to the dry transfer-printing on a target substrate (see Fig. 19). Using a similar setup, the graphene layers can also be functionalized by wet-chemical doping. Figure 20 shows that using such an approach, a uniform and highly transparent (97.4% optical transmittance) large-area graphene layer on a PET film is achievable, and which exhibits excellent mechanical flexibility and suitability

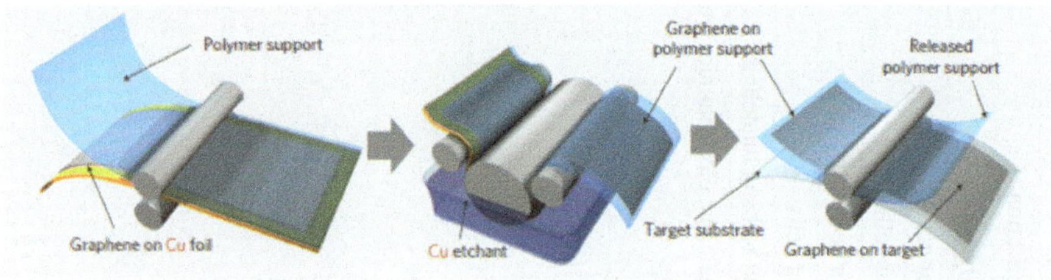

Fig. 19 Schematic of the roll-based production of graphene films grown on a copper foil. The process includes adhesion of polymer supports, copper etching (rinsing), and dry transfer-printing on a target substrate. A wet-chemical doping can be carried out using a setup similar to that used for etching. Reprinted by permission from [287]. © 2010 Macmillan Publishers Limited (2010)

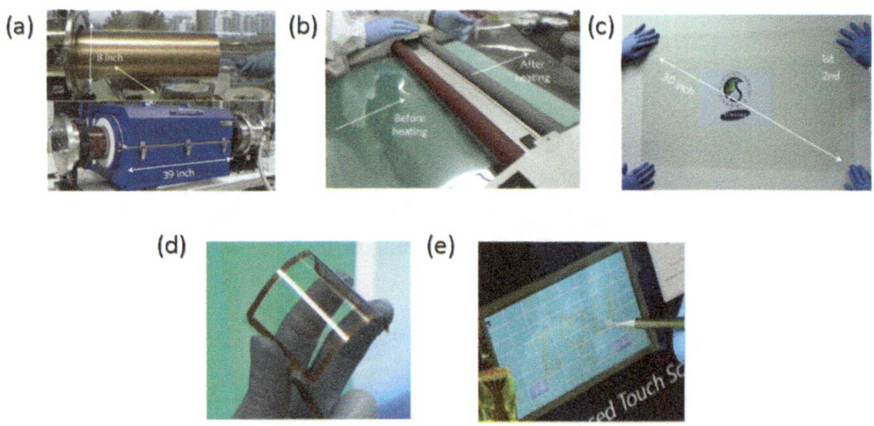

Fig. 20 Photographs of the roll-based production of graphene films. **a** Copper foil wrapping around a 7.5-i. quartz tube to be inserted into an 8-inch quartz reactor. The lower image shows the stage in which the copper foil reacts with CH_4 and H_2 gases at high temperatures. **b** Roll-to-roll transfer of graphene films from a thermal release tape to a PET film at 120 °C. **c** A transparent ultralarge-area graphene film was transferred on a 35-in. PET sheet. **d** An assembled graphene/PET touch panel showing outstanding flexibility. **e** A graphene-based touch-screen panel connected to a computer with control software. Reprinted by permission from [287].© 2010 Macmillan Publishers Limited (2010)

for touch-screen panel applications. With multiple stacked doped graphene layers, very low sheet resistance (~30 $\Omega/\square$) at ~90% transparency has been demonstrated, which is superior to commercial transparent electrodes such as indium tin oxides. This opens up new potential applications in the fields of transparent conductive layers, bio-applications, photonics, nanoelectronics, sensors, high-frequency transistors, and novel electronic devices such as Van der Waals heterostructures [68] as discussed earlier.

With this major leap in large-area graphene processing capability, it is then of keen interest to see similar technologies developed for other 2D materials discussed in this chapter and beyond. The ability to perform large-area deposition for the prospective 2D materials with different properties (semi-metal, semiconducting, insulating) will be the crucial bridge between academic pursuits and commercialization. Consequently, when silicon-based devices have reached its fundamental scaling limits, the timely and successful demonstration of large-area deposition of various 2D materialscould be the solution to meet the ITRS roadmap.

6 Conclusions

Atomically thin two-dimensional materials such as graphene, hexagonal boron nitride, molybdenum disulfide, and black phosphorus exhibit unique material, electrical and optical properties from their corresponding bulk counterparts, sparking an immediate interest from academics and industry partners alike. Many approaches of obtaining these monolayers or few-layers had been reported to vary success, with the characterization results indicating that such 2D layers exhibit a high potential to complement/replace the existing technologies based on conventional materials such as silicon. The promising results demonstrated at the research level reveals new opportunities for nanoelectronics and optoelectronics applications, enabling technologies for many fields such as imaging and sensing, medical, industrial, environmental, biological, field-effect transistors, solar cells, batteries, thermoelectrics, light emitting diodes, lasers and ultra-fast optical communications. When combining different 2D materials in Van der Waals heterostructures, more potential applications are possible, especially in the area of flexible electronics and optoelectronics, in which these 2D materials exhibit high mechanical flexibility not reported by the existing materials such as silicon. Undoubtedly, these 2D materials are a relatively new and exciting area to explore, and the materials combination appears to be only limited by our imagination. With the experience gained from graphene in terms of material processing, characterization, and device fabrications, researchers applying these skills to other unique 2D materials are sure to find unexpected and exciting potential applications. For a long-term perspective, to progress towards the goal of commercialization and integration of these novel 2D materials in nanoelectronics and optoelectronics, low-cost large-area uniform deposition approaches with precise control of layer thickness and film quality will be a prerequisite. This will also enable researchers to better understand the physical and chemical properties of each potential 2D material candidate. New opportunities with each newly discovered 2D material awaits our further exploration.

References

1. P. Gargini, *The International Technology Roadmap for Semiconductors*. https://www.itrs2.net/. Accessed 1 Feb 2019
2. P.R. Wallace, Phys. Rev. **71**, 622 (1947)
3. J.W. McClure, Phys. Rev. **104**, 666 (1956)
4. J.C. Slonczewski, P.R. Weiss, Phys. Rev. **109**, 272 (1958)
5. K.S. Novoselov, A.K. Geim, S.V. Morozov, D. Jiang, Y. Zhang, S.V. Dubonos, I.V. Grigorieva, A.A. Firsov, Science **306**, 666 (2004)
6. K.S. Novoselov, D. Jiang, F. Schedin, T.J. Booth, V.V. Khotkevich, S.V. Morozov, A.K. Geim, Proc. Natl. Acad. Sci. U.S.A. **102**, 10451 (2005)
7. C.-G. Andres, B. Michele, M. Rianda, S. Vibhor, J. Laurens, S.J.v.d.Z. Herre, A.S. Gary, 2D Materials **1**, 011002 (2014)
8. K.S. Novoselov, V.I. Falko, L. Colombo, P.R. Gellert, M.G. Schwab, K. Kim, Nature **490**, 192 (2012)
9. A. Castellanos-Gomez, M. Barkelid, A.M. Goossens, V.E. Calado, H.S.J. van der Zant, G.A. Steele, Nano Lett. **12**, 3187 (2012)
10. C. Jin, F. Lin, K. Suenaga, S. Iijima, Phys. Rev. Lett. **102**, 195505 (2009)
11. J.C. Meyer, A. Chuvilin, G. Algara-Siller, J. Biskupek, U. Kaiser, Nano Lett. **9**, 2683 (2009)
12. J.N. Coleman, M. Lotya, A. O'Neill, S.D. Bergin, P.J. King, U. Khan, K. Young, A. Gaucher, S. De, R.J. Smith, I.V. Shvets, S.K. Arora, G. Stanton, H.-Y. Kim, K. Lee, G.T. Kim, G.S. Duesberg, T. Hallam, J.J. Boland, J.J. Wang, J.F. Donegan, J.C. Grunlan, G. Moriarty, A. Shmeliov, R.J. Nicholls, J.M. Perkins, E.M. Grieveson, K. Theuwissen, D.W. McComb, P.D. Nellist, V. Nicolosi, Science **331**, 568 (2011)
13. C. Zhi, Y. Bando, C. Tang, H. Kuwahara, D. Golberg, Adv. Mater. **21**, 2889 (2009)
14. Y.-H. Lee, X.-Q. Zhang, W. Zhang, M.-T. Chang, C.-T. Lin, K.-D. Chang, Y.-C. Yu, J.T.-W. Wang, C.-S. Chang, L.-J. Li, T.-W. Lin, Adv. Mater. **24**, 2320 (2012)
15. Y. Zhan, Z. Liu, S. Najmaei, P.M. Ajayan, J. Lou, Small **8**, 966 (2012)
16. K.-K. Liu, W. Zhang, Y.-H. Lee, Y.-C. Lin, M.-T. Chang, C.-Y. Su, C.-S. Chang, H. Li, Y. Shi, H. Zhang, C.-S. Lai, L.-J. Li, Nano Lett. **12**, 1538 (2012)
17. A. Nagashima, N. Tejima, Y. Gamou, T. Kawai, C. Oshima, Phys. Rev. Lett. **75**, 3918 (1995)
18. M. Corso, W. Auwärter, M. Muntwiler, A. Tamai, T. Greber, J. Osterwalder, Science **303**, 217 (2004)
19. A. Pakdel, X. Wang, C. Zhi, Y. Bando, K. Watanabe, T. Sekiguchi, T. Nakayama, D. Golberg, J. Mater. Chem. **22**, 4818 (2012)
20. J. Yu, L. Qin, Y. Hao, S. Kuang, X. Bai, Y.-M. Chong, W. Zhang, E. Wang, ACS Nano **4**, 414 (2010)
21. K. Kang, S. Xie, L. Huang, Y. Han, P.Y. Huang, K.F. Mak, C.-J. Kim, D. Muller, J. Park, Nature **520**, 656 (2015)
22. J.J. Pyeon, S.H. Kim, D.S. Jeong, S.-H. Baek, C.-Y. Kang, J.-S. Kim, S.K. Kim, Nanoscale **8**, 10792 (2016)
23. J. Tao, J. Chai, X. Lu, L.M. Wong, T.I. Wong, J. Pan, Q. Xiong, D. Chi, S. Wang, Nanoscale **7**, 2497 (2015)
24. Z.P. Ling, R. Yang, J.W. Chai, S.J. Wang, W.S. Leong, Y. Tong, D. Lei, Q. Zhou, X. Gong, D.Z. Chi, K.W. Ang, Opt. Express **23**, 13580 (2015)
25. Graphene Supermarket, https://graphene-supermarket.com. Accessed 2 Feb 2019
26. Manchester Nanomaterials, https://mos2crystals.com. Accessed 2 Feb 2019
27. 2Dsemiconductors USA, https://www.2dsemiconductors.com. Accessed 2 Feb 2019

28. HQ Graphene, https://www.hqgraphene.com. Accessed 2 Feb 2019
29. K.S. Novoselov, A.K. Geim, S.V. Morozov, D. Jiang, M.I. Katsnelson, I.V. Grigorieva, S.V. Dubonos, A.A. Firsov, Nature **438**, 197 (2005)
30. G.W. Semenoff, Phys. Rev. Lett. **53**, 2449 (1984)
31. A.M.J. Schakel, Phys. Rev. D **43**, 1428 (1991)
32. J. González, F. Guinea, M.A.H. Vozmediano, Phys. Rev. Lett. **77**, 3589 (1996)
33. M.I. Katsnelson, Eur. Phys. J. B. Condens Matter Complex Syst. **51**, 157 (2006)
34. M.I. Katsnelson, K.S. Novoselov, A.K. Geim, Nat Phys **2**, 620 (2006)
35. G. Rut, A. Rycerz, Philos. Mag. **95**, 599 (2015)
36. F.D.M. Haldane, Phys. Rev. Lett. **61**, 2015 (1988)
37. E.V. Gorbar, V.P. Gusynin, V.A. Miransky, I.A. Shovkovy, Phys. Rev. B **66**, 045108 (2002)
38. J. Tworzydło, B. Trauzettel, M. Titov, A. Rycerz, C.W.J. Beenakker, Phys. Rev. Lett. **96**, 246802 (2006)
39. A.S. Mayorov, R.V. Gorbachev, S.V. Morozov, L. Britnell, R. Jalil, L.A. Ponomarenko, P. Blake, K.S. Novoselov, K. Watanabe, T. Taniguchi, A.K. Geim, Nano Lett. **11**, 2396 (2011)
40. C. Lee, X. Wei, J.W. Kysar, J. Hone, Science **321**, 385 (2008)
41. A.A. Balandin, Nat. Mater. **10**, 569 (2011)
42. J. Moser, A. Barreiro, A. Bachtold, Appl. Phys. Lett. **91**, 163513 (2007)
43. J.S. Bunch, S.S. Verbridge, J.S. Alden, A.M. van der Zande, J.M. Parpia, H.G. Craighead, P.L. McEuen, Nano Lett. **8**, 2458 (2008)
44. D.C. Elias, R.R. Nair, T.M.G. Mohiuddin, S.V. Morozov, P. Blake, M.P. Halsall, A.C. Ferrari, D.W. Boukhvalov, M.I. Katsnelson, A.K. Geim, K.S. Novoselov, Science **323**, 610 (2009)
45. K.P. Loh, Q. Bao, P.K. Ang, J. Yang, J. Mater. Chem. **20**, 2277 (2010)
46. R.R. Nair, W. Ren, R. Jalil, I. Riaz, V.G. Kravets, L. Britnell, P. Blake, F. Schedin, A.S. Mayorov, S. Yuan, M.I. Katsnelson, H.-M. Cheng, W. Strupinski, L.G. Bulusheva, A.V. Okotrub, I.V. Grigorieva, A.N. Grigorenko, K.S. Novoselov, A.K. Geim, Small **6**, 2877 (2010)
47. A. Vacchi Isabella, C. Ménard-Moyon, A. Bianco, Chemical functionalization of graphene family members, in *Physical Sciences Reviews*. https://doi.org/10.1515/psr-2016-0103
48. G. Gao, D. Liu, S. Tang, C. Huang, M. He, Y. Guo, X. Sun, B. Gao, Sci. Rep. **6**, 20034 (2016)
49. J.M. Englert, C. Dotzer, G. Yang, M. Schmid, C. Papp, J.M. Gottfried, H.-P. Steinrück, E. Spiecker, F. Hauke, A. Hirsch, Nat. Chem. **3**, 279 (2011)
50. M.S. Chang, Y.S. Kim, J.H. Kang, J. Park, S.J. Sung, S.H. So, K.T. Park, S.J. Yang, T. Kim, C.R. Park, Chem. Mater. **29**, 307 (2017)
51. T. Kuila, S. Bose, A.K. Mishra, P. Khanra, N.H. Kim, J.H. Lee, Prog. Mater. Sci. **57**, 1061 (2012)
52. V. Georgakilas, J.N. Tiwari, K.C. Kemp, J.A. Perman, A.B. Bourlinos, K.S. Kim, R. Zboril, Chem. Rev. **116**, 5464 (2016)
53. I. Ali, A.A. Basheer, X.Y. Mbianda, A. Burakov, E. Galunin, I. Burakova, E. Mkrtchyan, A. Tkachev, V. Grachev, Environ. Int. **127**, 160 (2019)
54. J. Liu, Q. Ma, Z. Huang, G. Liu, H. Zhang, Adv. Mater. **31**, 1800696 (2019)
55. Q. Ma, C.H. Lui, J.C.W. Song, Y. Lin, J.F. Kong, Y. Cao, T.H. Dinh, N.L. Nair, W. Fang, K. Watanabe, T. Taniguchi, S.-Y. Xu, J. Kong, T. Palacios, N. Gedik, N.M. Gabor, P. Jarillo-Herrero, Nat. Nanotechnol. **14**, 145 (2019)
56. J. Yin, H. Peng, Nat. Nanotechnol. **14**, 105 (2019)
57. A.K. Geim, K.S. Novoselov, Nat. Mater. **6**, 183 (2007)
58. X.-F. Jiang, Q. Weng, X.-B. Wang, X. Li, J. Zhang, D. Golberg, Y. Bando, J. Mater. Sci. Technol. **31**, 589 (2015)
59. I. Jo, M.T. Pettes, J. Kim, K. Watanabe, T. Taniguchi, Z. Yao, L. Shi, Nano. Lett. **13**, 550 (2013)
60. L. Chun, B. Yoshio, Z. Chunyi, H. Yang, G. Dmitri, Nanotechnology **20**, 385707 (2009)
61. A. Bosak, J. Serrano, M. Krisch, K. Watanabe, T. Taniguchi, H. Kanda, Phys. Rev. B **73**, 041402 (2006)
62. X. Wei, M.-S. Wang, Y. Bando, D. Golberg, Adv. Mater. **22**, 4895 (2010)
63. Y. Hattori, T. Taniguchi, K. Watanabe, K. Nagashio, ACS Nano **9**, 916 (2015)

64. L.H. Li, J. Cervenka, K. Watanabe, T. Taniguchi, Y. Chen, ACS Nano **8**, 1457 (2014)
65. Z. Liu, Y. Gong, W. Zhou, L. Ma, J. Yu, J.C. Idrobo, J. Jung, A.H. MacDonald, R. Vajtai, J. Lou, P.M. Ajayan, Nat. Commun. **4**, 2541 (2013)
66. M.S. Bresnehan, M.J. Hollander, M. Wetherington, M. LaBella, K.A. Trumbull, R. Cavalero, D.W. Snyder, J.A. Robinson, ACS Nano **6**, 5234 (2012)
67. C.R. Dean, A.F. Young, I. Meric, C. LeeC, L. Wang, S. Sorgenfrei, K. Watanabe, T. Taniguchi, P. Kim, K.L. Shepard, J. Hone, Nat. Nano **5**, 722 (2010)
68. A.K. Geim, I.V. Grigorieva, Nature **499**, 419 (2013)
69. T. Xu, Y. Zhou, X. Tan, K. Yin, L. He, F. Banhart, L. Sun, Adv. Funct. Mater. n/a-n/a
70. B. Radenovic, A. Radisavljevic, J. Brivio, V. Giacometti, A. Kis, Nat. Nano. **6**, 147 (2011)
71. P.A. Young, J. Phys. D: Appl. Phys **1**, 936 (1968)
72. Y. Ding, Y. Wang, J. Ni, L. Shi, S. Shi, W. Tang, Phys. B **406**, 2254 (2011)
73. A.R. Beal, H.P. Hughes, W.Y. Liang, J. Phys. C: Solid State Physics **8**, 4236 (1975)
74. J.A. Wilson, F.J. Di Salvo, S. Mahajan, Adv. Phys. **24**, 117 (1975)
75. K.F. Mak, C. Lee, J. Hone, J. Shan, T.F. Heinz, Phys. Rev. Lett. **105**, 136805 (2010)
76. A. Kuc, N. Zibouche, T. Heine, Phys. Rev. B **83**, 245213 (2011)
77. Y. Cheng, U. Schwingenschlögl, in *MoS_2: Materials, Physics, and Devices: Materials, Physics, and Devices*. ed. by Z.M. Wang (Springer, Cham, 2014), p. 103
78. A. Splendiani, L. Sun, Y. Zhang, T. Li, J. Kim, C.-Y. Chim, G. Galli, F. Wang, Nano. Lett. **10**, 1271 (2010)
79. L. Liu, S.B. Kumar, Y. Ouyang, J. Guo, IEEE Trans. Electron. Devices **58**, 3042 (2011)
80. T. Li, G. Galli, J. Phys. Chem. C **111**, 16192 (2007)
81. C. Ataca, H. Şahin, S. Ciraci, J. Phys. Chem. C **116**, 8983 (2012)
82. T. Böker, R. Severin, A. Müller, C. Janowitz, R. Manzke, D. Voß, P. Krüger, A. Mazur, J. Pollmann, Phys. Rev. B **64**, 235305 (2001)
83. M. Amani, D.-H. Lien, D. Kiriya, J. Xiao, A. Azcatl, J. Noh, S.R. Madhvapathy, R. Addou, S. Kc, M. Dubey, K. Cho, R.M. Wallace, S.-C. Lee, J.-H. He, J.W. Ager, X. Zhang, E. Yablonovitch, A. Javey, Science **350**, 1065 (2015)
84. T. Cao, G. Wang, W. Han, H. Ye, C. Zhu, J. Shi, Q. Niu, P. Tan, E. Wang, B. Liu, J. Feng, Nat. Commun. **3**, 887 (2012)
85. H. Zeng, J. Dai, W. Yao, D. Xiao, X. Cui, Nat. Nano. **7**, 490 (2012)
86. K.F. Mak, K. He, J. Shan, T.F. Heinz, Nat. Nano. **7**, 494 (2012)
87. N. Singh, U. Schwingenschlögl, Adv. Mater.**29**, 1600970-n/a (2017)
88. D. Xiao, G.-B. Liu, W. Feng, X. Xu, W. Yao, Phys. Rev. Lett. **108**, 196802 (2012)
89. Z.Y. Zhu, Y.C. Cheng, U. Schwingenschlögl, Phys. Rev. B **84**, 153402 (2011)
90. J. Klinovaja, D. Loss, Phys. Rev. B **88**, 075404 (2013)
91. C. Lee, H. Yan, L.E. Brus, T.F. Heinz, J. Hone, S. Ryu, ACS Nano **4**, 2695 (2010)
92. A. Molina-Sánchez, L. Wirtz, Phys. Rev. B **84**, 155413 (2011)
93. E. Generalic, *Periodic Table of the Elements*. https://www.periodni.com/images.html. Accessed 5 Feb 2019
94. H. Thurn, H. Kerbs, Angew. Chem. **5**, 1047 (1966)
95. J.M. Zaug, A.K. Soper, S.M. Clark, Nat. Mater. **7**, 890 (2008)
96. H. Liu, Y. Du, Y. Deng, P.D. Ye, Chem. Soc. Rev. **44**, 2732 (2015)
97. D.E.C. Corbridge, E.J. Lowe, Nature **170**, 629 (1952)
98. P.A.G. O'Hare, B.M. Lewis, I. Shirotani, Thermochim. Acta. **129**, 57 (1988)
99. R. Hultgren, N.S. Gingrich, B.E. Warren, J. Chem. Phys. **3**, 351 (1935)
100. H.O.H. Churchill, P. Jarillo-Herrero, Nat. Nano. **9**, 330 (2014)
101. D. Warschauer, J. Appl. Phys. **34**, 1853 (1963)
102. A. Morita, Appl. Phys. A **39**, 227 (1986)
103. L. Shulenburger, A.D. Baczewski, Z. Zhu, J. Guan, D. Tománek, Nano. Lett. **15**, 8170 (2015)
104. L. Liang, J. Wang, W. Lin, B.G. Sumpter, V. Meunier, M. Pan, Nano. Lett. **14**, 6400 (2014)
105. A. Castellanos-Gomez, J. Phys. Chem. Lett. **6**, 4280 (2015)
106. Z.P. Ling, R. Yang, J.W. Chai, S.J. Wang, Y. Tong, Q. Zhou, X. Gong, D.Z. Chi, K. W. Ang, in *2015 IEEE International Wireless Symposium (IWS 2015)*, Shenzhen, China, 30 Mar 2015–1 Apr 2015

107. L. Li, Y. Yu, G.J. Ye, Q. Ge, X. Ou, H. Wu, D. Feng, X.H. Chen, Y. Zhang, Nat. Nano. **9**, 372 (2014)
108. V. Sorkin, H. Pan, H. Shi, S.Y. Quek, Y.W. Zhang, Crit. Rev. Solid State **39**, 319 (2014)
109. H. Du, X. Lin, Z. Xu, D. Chu, J. Mate. Chem. C **3**, 8760 (2015)
110. R. Fei, A. Faghaninia, R. Soklaski, J.-A. Yan, C. Lo, L. Yang, Nano. Lett. **14**, 6393 (2014)
111. J. Zhang, H.J. Liu, L. Cheng, J. Wei, J.H. Liang, D.D. Fan, J. Shi, X.F. Tang, Q.J. Zhang, Sci. Rep. **4**, 6452 (2014)
112. H. Jang, J.D. Wood, C.R. Ryder, M.C. Hersam, D.G. Cahill, Adv. Mater. **27**, 8017 (2015)
113. Q. Wei, X. Peng, Appl. Phys. Lett. **104**, 251915 (2014)
114. J.-W. Jiang, H.S. Park, Nat. Commun. **5**, 4727 (2014)
115. K.E. Petersen, Proc. IEEE **70**, 420 (1982)
116. L. Kou, C. Chen, S.C. Smith, J. Phys. Chem. Lett. **6**, 2794 (2015)
117. M.Z. Rahman, C.W. Kwong, K. Davey, S.Z. Qiao, Energy Environ. Sci. **9**, 709 (2016)
118. Z.-P. Ling, K.-W. Ang, APL Mater. **3**, 126104 (2015)
119. O.I. Joshua, A.S. Gary, S.J.v.d.Z. Herre, C.-G. Andres, 2d Mater. **2**, 011002 (2015)
120. C.-G. Andres, V. Leonardo, P. Elsa, O.I. Joshua, K.L. Narasimha-Acharya, I.B. Sofya, J.G. Dirk, B. Michele, A.S. Gary, J.V. Alvarez, W.Z. Henny, J.J. Palacios, S.J.v.d.Z. Herre, 2d Mater. **1**, 025001 (2014)
121. P.R. Sajanlal, T.S. Sreeprasad, A.K. Samal, T. Pradeep, Nano. Rev. **2**, 5883 (2011)
122. V. Sorkin, Y. Cai, Z. Ong, G. Zhang, Y.W. Zhang, Crit. Rev. Solid State **42**, 1 (2017)
123. X. Xiao, Y. Li, Z. Liu, Nat. Mater **15**, 697 (2016)
124. C. Ayinon, *Robots with Electronic Skin.* https://www.yalescientific.org/2015/11/robots-with-electronic-skin/. Accessed 1 Mar 2019
125. *Graphene Bendable Phones.* https://www.cqmxi.com/product/detail-88.html. Accessed 1 Mar 2019
126. J.-H. Ahn, B.H. Hong, Nat. Nano. **9**, 737 (2014)
127. T.J. Echtermeyer, L. Britnell, P.K. Jasnos, A. Lombardo, R.V. Gorbachev, A.N. Grigorenko, A.K. Geim, A.C. Ferrari, K.S. Novoselov, Nat. Commun. **2**, 458 (2011)
128. E. Torres Alonso, G. Karkera, G.F. Jones, M.F. Craciun, S. Russo, ACS Appl. Mater. Interfaces **8**, 16541 (2016)
129. T.H. Bointon, M.D. Barnes, S. Russo, M.F. Craciun, Adv. Mater. **27**, 4200 (2015)
130. A. Tredicucci, M.S. Vitiello, IEEE J. Sel. Top. Quantum Electron. **20**, 130 (2014)
131. X. Cai, A.B. Sushkov, R.J. Suess, M.M. Jadidi, G.S. Jenkins, L.O. Nyakiti, R.L. Myers-Ward, S. Li, J. Yan, D.K. Gaskill, T.E. Murphy, H.D. Drew, M.S. Fuhrer, Nat. Nano. **9**, 814 (2014)
132. L. Vicarelli, M.S. Vitiello, D. Coquillat, A. Lombardo, A.C. Ferrari, W. Knap, M. Polini, V. Pellegrini, A. Tredicucci, Nat. Mater. **11**, 865 (2012)
133. Y. Deepika, T. Stephane Boubanga, W. Takayuki, A. Stevanus, R. Victor, O. Taiichi, 2d Mater.**3**, 045009(2016)
134. R. Asadi, Z. Ouyang, M.M. Mohammd, Nanoscale **7**, 11379 (2015)
135. Z. Sun, T. Hasan, A.C. Ferrari, Physica E **44**, 1082 (2012)
136. R. Mary, G. Brown, S.J. Beecher, F. Torrisi, S. Milana, D. Popa, T. Hasan, Z. Sun, E. Lidorikis, S. Ohara, A.C. Ferrari, A.K. Kar, Opt. Express **21**, 7943 (2013)
137. A.A. Lagatsky, Z. Sun, T.S. Kulmala, R.S. Sundaram, S. Milana, F. Torrisi, O.L. Antipov, Y. Lee, J.H. Ahn, C.T.A. Brown, W. Sibbett, A.C. Ferrari, Appl. Phys. Lett. **102**, 013113 (2013)
138. T. Hasan, Z. Sun, F. Wang, F. Bonaccorso, P.H. Tan, A.G. Rozhin, A.C. Ferrari, Adv. Mater. **21**, 3874 (2009)
139. S. Chakraborty, O.P. Marshall, T.G. Folland, Y.J. Kim, A.N. Grigorenko, K.S. Novoselov, Science **351**, 246 (2016)
140. M.M. Jadidi, A.B. Sushkov, R.L. Myers-Ward, A.K. Boyd, K.M. Daniels, D.K. Gaskill, M.S. Fuhrer, H.D. Drew, T.E. Murphy, Nano. Lett. **15**, 7099 (2015)
141. Q.-Y. Wen, W. Tian, Q. Mao, Z. Chen, W.-W. Liu, Q.-H. Yang, M. Sanderson, H.-W. Zhang, Sci. Rep. **4**, 7409 (2014)
142. N. Guo, W. Hu, T. Jiang, F. Gong, W. Luo, W. Qiu, P. Wang, L. Liu, S. Wu, L. Liao, X. Chen, W. Lu, Nanoscale **8**, 16065 (2016)

143. C. Cheng, B. Huang, X. Mao, Z. Zhang, Z. Zhang, Z. Geng, P. Xue, H. Chen, Nanoscale **9**, 4082 (2017)
144. Y. Liu, Q. Xia, J. He, Z. Liu, Nanoscale Res. Lett. **12**, 93 (2017)
145. C.-H. Liu, Y.-C. Chang, T.B. Norris, Z. Zhong, Nat. Nano. **9**, 273 (2014)
146. B. Ertug, in *D.B. Ertug*. ed. by S. Applications (Intech Rijeka, Croatia, 2013), p. 33
147. A. Lipp, K.A. Schwetz, K. Hunold, J. Eur. Ceram. Soc. **5**, 3 (1989)
148. K. Watanabe, T. Taniguchi, H. Kanda, Nat. Mater. **3**, 404 (2004)
149. Y. Kubota, K. Watanabe, O. Tsuda, T. Taniguchi, Science **317**, 932 (2007)
150. K. Watanabe, T. Taniguchi, T. Niiyama, K. Miya, M. Taniguchi, Nat. Photon. **3**, 591 (2009)
151. M. Šiškins, C. Mullan, S.-K. Son, J. Yin, K. Watanabe, T. Taniguchi, D. Ghazaryan, K.S. Novoselov, A. Mishchenko, Appl. Phys. Lett. **114**, 123104 (2019)
152. L. Wang, S. Zihlmann, A. Baumgartner, J. Overbeck, K. Watanabe, T. Taniguchi, P. Makk, C. Schönenberger, Nano. Lett. **19**, 4097 (2019)
153. L. Song, L. Ci, H. Lu, P.B. Sorokin, C. Jin, J. Ni, A.G. Kvashnin, D.G. Kvashnin, J. Lou, B.I. Yakobson, P.M. Ajayan, Nano. Lett. **10**, 3209 (2010)
154. Y. Shi, C. Hamsen, X. Jia, K.K. Kim, A. Reina, M. Hofmann, A.L. Hsu, K. Zhang, H. Li, Z.-Y. Juang, M.S. Dresselhaus, L.-J. Li, J. Kong, Nano. Lett. **10**, 4134 (2010)
155. S. Sushant, D. Andrei, L. Ning, C. Chris, J. K. Moon, T. Emanuel, K.B. Sanjay, C. Luigi, 2d Mater. **4**, 025052 (2017)
156. G. Lu, T. Wu, Q. Yuan, H. Wang, H. Wang, F. Ding, X. Xie, M. Jiang, Nat. Commun. **6**, 6160 (2015)
157. K.H. Lee, H.-J. Shin, J. Lee, I.-Y. Lee, G.-H. Kim, J.-Y. Choi, S.-W. Kim, Nano. Lett. **12**, 714 (2012)
158. X. Song, J. Gao, Y. Nie, T. Gao, J. Sun, D. Ma, Q. Li, Y. Chen, C. Jin, A. Bachmatiuk, M.H. Rümmeli, F. Ding, Y. Zhang, Z. Liu, Nano. Res. **8**, 3164 (2015)
159. A. Hemmi, C. Bernard, H. Cun, S. Roth, M. Klöckner, T. Kälin, M. Weinl, S. Gsell, M. Schreck, J. Osterwalder, T. Greber, Rev. Sci. Instrum. **85**, 035101 (2014)
160. H. Park, T.K. Kim, S.W. Cho, H.S. Jang, S.I. Lee, S.-Y. Choi, Sci. Rep. **7**, 40091 (2017)
161. A. Haider, C. Ozgit-Akgun, E. Goldenberg, A.K. Okyay, N. Biyikli, J. Am. Ceram. Soc. **97**, 4052 (2014)
162. C. Hu, S. Kotake, Y. Suzuki, M. Senoo, Vacuum **59**, 748 (2000)
163. D. Velázquez, R. Seibert, H. Man, L. Spentzouris, J. Terry, J. Appl. Phys. **119**, 095306 (2016)
164. N.R. Glavin, M.L. Jespersen, M.H. Check, J. Hu, A.M. Hilton, T.S. Fisher, A.A. Voevodin, Thin Solid Films **572**, 245 (2014)
165. G. Reisse, S. Weissmantel, Appl. Phys. A **69**, S749 (1999)
166. C. Zhou, X. Wang, S. Raju, Z. Lin, D. Villaroman, B. Huang, H.L.-W. Chan, M. Chan, Y. Chai, Nanoscale **7**, 8695 (2015)
167. W. Lu, C.M. Lieber, Nat. Mater. **6**, 841 (2007)
168. Y. Yoon, K. Ganapathi, S. Salahuddin, Nano. Lett. **11**, 3768 (2011)
169. B. Radisavljevic, M.B. Whitwick, A. Kis, ACS Nano **5**, 9934–9938 (2011)
170. H. Wang, L. Yu, Y.-H. Lee, Y. Shi, A. Hsu, M.L. Chin, L.-J. Li, M. Dubey, J. Kong, T. Palacios, Nano. Lett. **12**, 4674 (2012)
171. S. Brown, Z. Vranesic, *Fundamentals of Digital Logic with VHDL Design*, 3rd edn. (McGraw-Hill, New York, 2008).
172. Q. Peng, S. De, Phys. Chem. Chem. Phys. **15**, 19427 (2013)
173. S. Bertolazzi, J. Brivio, A. Kis, ACS Nano **5**, 9703 (2011)
174. Y. Li, C.-Y. Xu, J.-Y. Wang, L. Zhen, Sci. Rep. **4**, 7186 (2014)
175. O. Lopez-Sanchez, D. Lembke, M. Kayci, A. Radenovic, A. Kis, Nat. Nano. **8**, 497(2013)
176. L. Yang, X. Cui, J. Zhang, K. Wang, M. Shen, S. Zeng, S.A. Dayeh, L. Feng, B. Xiang, Sci. Rep. **4**, 5649 (2014)
177. D.-S. Tsai, K.-K. Liu, D.-H. Lien, M.-L. Tsai, C.-F. Kang, C.-A. Lin, L.-J. Li, J.-H. He, ACS Nano **7**, 3905 (2013)
178. H.S. Lee, S.-W. Min, Y.-G. Chang, M.K. Park, T. Nam, H. Kim, J.H. Kim, S. Ryu, S. Im, Nano. Lett. **12**, 3695 (2012)

179. W. Wu, L. Wang, R. Yu, Y. Liu, S.-H. Wei, J. Hone, Z.L. Wang, Adv. Mater **28**, 8463 (2016)
180. Z. Yin, H. Li, H. Li, L. Jiang, Y. Shi, Y. Sun, G. Lu, Q. Zhang, X. Chen, H. Zhang, ACS Nano **6**, 74 (2012)
181. S.H. Yu, Y. Lee, S.K. Jang, J. Kang, J. Jeon, C. Lee, J.Y. Lee, H. Kim, E. Hwang, S. Lee, J.H. Cho, ACS Nano **8**, 8285 (2014)
182. G. Yoo, S.L. Choi, S.J. Park, K.-T. Lee, S. Lee, M.S. Oh, J. Heo, H.J. Park, Sci. Rep. **7**, 40945 (2017)
183. W. Zhang, C.-P. Chuu, J.-K. Huang, C.-H. Chen, M.-L. Tsai, Y.-H. Chang, C.-T. Liang, Y.-Z. Chen, Y.-L. Chueh, J.-H. He, M.-Y. Chou, L.-J. Li, Sci. Rep. **4**, 3826 (2014)
184. Y. Chen, X. Wang, P. Wang, H. Huang, G. Wu, B. Tian, Z. Hong, Y. Wang, S. Sun, H. Shen, J. Wang, W. Hu, J. Sun, X. Meng, J. Chu, ACS Appl. Mater. Interfaces **8**, 32083 (2016)
185. F. Withers, O. Del Pozo-Zamudio, A. Mishchenko, A.P. Rooney, A. Gholinia, K. Watanabe, T. Taniguchi, S.J. Haigh, A.K. Geim, A.I. Tartakovskii, K.S. Novoselov, Nat. Mater. **14**, 301 (2015)
186. O. Lopez-Sanchez, E. Alarcon Llado, V. Koman, A. Fontcuberta i Morral, A. Radenovic, A. Kis, ACS Nano **8**, 3042(2014)
187. L. Hao, Y. Liu, W. Gao, Z. Han, Q. Xue, H. Zeng, Z. Wu, J. Zhu, W. Zhang, J. Appl. Phys. **117**, 114502 (2015)
188. M.-L. Tsai, S.-H. Su, J.-K. Chang, D.-S. Tsai, C.-H. Chen, C.-I. Wu, L.-J. Li, L.-J. Chen, J.-H. He, ACS Nano **8**, 8317 (2014)
189. T.A. Shastry, I. Balla, H. Bergeron, S.H. Amsterdam, T.J. Marks, M.C. Hersam, ACS Nano **10**, 10573 (2016)
190. E. Fortin, W.M. Sears, J. Phys. Chem. Solids **43**, 881 (1982)
191. A. Shokri, N. Salami, Sens. Actuator B-Chem. **236**, 378 (2016)
192. K. Jin, L. Xie, Y. Tian, D. Liu, J. Phys. Chem. C **120**, 11204 (2016)
193. K. Kalantar-zadeh, J.Z. Ou, ACS Sens. **1**, 5 (2016)
194. W. Zhang, P. Zhang, Z. Su, G. Wei, Nanoscale **7**, 18364 (2015)
195. B. Cho, M.G. Hahm, M. Choi, J. Yoon, A.R. Kim, Y.-J. Lee, S.-G. Park, J.-D. Kwon, C.S. Kim, M. Song, Y. Jeong, K.-S. Nam, S. Lee, T.J. Yoo, C.G. Kang, B.H. Lee, H.C. Ko, P.M. Ajayan, D.-H. Kim, Sci. Rep. **5**, 8052 (2015)
196. M. Park, Y.J. Park, X. Chen, Y.-K. Park, M.-S. Kim, J.-H. Ahn, Adv. Mate. **28**, 2556 (2016)
197. L. Liu, K. Xu, X. Wan, J. Xu, C.Y. Wong, H.K. Tsang, Photonics Res. **3**, 206 (2015)
198. C. Cheng, H. Liu, Z. Shang, W. Nie, Y. Tan, B.d.R. Rabes, J.R. Vázquez de Aldana, D. Jaque, F. Chen, Opt. Mater. Express **6**, 367 (2016)
199. A.L. Santhosha, P.K. Nayak, K. Pollok, F. Langenhorst, P. Adelhelm, J. Phys. Chem. C **123**, 12126 (2019)
200. D. Sarkar, D. Das, S. Das, A. Kumar, S. Patil, K.K. Nanda, D.D. Sarma, A. Shukla, ACS Energy Lett. **4**, 1602 (2019)
201. H. Nguyen, C.-F. Huang, W. Luo, G. Xia, Z. Chen, Z. Li, C. Raymond, D. Doyle, F. Zhao, Mate. Lett. **168**, 1 (2016)
202. H. Wang, L. Yu, Y. H. Lee, W. Fang, A. Hsu, P. Herring, M. Chin, M. Dubey, L.J. Li, J. Kong, T. Palacios, in *2012 International Electron Devices Meeting*, San Francisco, California, USA, 10–13 Dec 2012
203. C. Ahn, J. Lee, H.-U. Kim, H. Bark, M. Jeon, G.H. Ryu, Z. Lee, G.Y. Yeom, K. Kim, J. Jung, Y. Kim, C. Lee, T. Kim, Adv. Mater. **27**, 5223 (2015)
204. S.H. Baek, Y. Choi, W. Choi, Nanoscale Res. Lett. **10**, 388 (2015)
205. M. Amani, R.A. Burke, X. Ji, P. Zhao, D.-H. Lien, P. Taheri, G.H. Ahn, D. Kirya, J.W. Ager, E. Yablonovitch, J. Kong, M. Dubey, A. Javey, ACS Nano **10**, 6535 (2016)
206. S. Li, Y.-C. Lin, X.-Y. Liu, Z. Hu, J. Wu, H. Nakajima, S. Liu, T. Okazaki, W. Chen, T. Minari, Y. Sakuma, K. Tsukagoshi, K. Suenaga, T. Taniguchi, M. Osada, Nanoscale **11**, 16122 (2019)
207. M. Mattinen, T. Hatanpää, T. Sarnet, K. Mizohata, K. Meinander, P.J. King, L. Khriachtchev, J. Räisänen, M. Ritala, M. Leskelä, Adv. Mater. Interfaces 1700123-n/a (2017)
208. T. Jurca, M.J. Moody, A. Henning, J.D. Emery, B. Wang, J.M. Tan, T.L. Lohr, L.J. Lauhon, T.J. Marks, Angew. Chem. Int. Ed. **56**, 4991 (2017)

209. M.I. Serna, S.H. Yoo, S. Moreno, Y. Xi, J.P. Oviedo, H. Choi, H.N. Alshareef, M.J. Kim, M. Minary-Jolandan, M.A. Quevedo-Lopez, ACS Nano **10**, 6054 (2016)
210. G. Siegel, Y.P.V. Subbaiah, M.C. Prestgard, A. Tiwari, APL Mater. **3**, 056103 (2015)
211. S. Ishihara, Y. Hibino, N. Sawamoto, T. Ohashi, K. Matsuura, H. Machida, M. Ishikawa, H. Wakabayashi, A. Ogura, ECS J. Solid. State. Sci. Technol. **5**, Q3012 (2016)
212. S. Hussain, J. Singh, D. Vikraman, A.K. Singh, M.Z. Iqbal, M.F. Khan, P. Kumar, D.-C. Choi, W. Song, K.-S. An, J. Eom, W.-G. Lee, J. Jung, Sci. Rep. **6**, 30791 (2016)
213. S.P. Koenig, R.A. Doganov, H. Schmidt, A.H. Castro Neto, B. Özyilmaz, Appl. Phys. Lett. **104**, 103106 (2014)
214. W. Zhu, M.N. Yogeesh, S. Yang, S.H. Aldave, J.-S. Kim, S. Sonde, L. Tao, N. Lu, D. Akinwande, Nano. Lett. **15**, 1883 (2015)
215. B. Yang, B. Wan, Q. Zhou, Y. Wang, W. Hu, W. Lv, Q. Chen, Z. Zeng, F. Wen, J. Xiang, S. Yuan, J. Wang, B. Zhang, W. Wang, J. Zhang, B. Xu, Z. Zhao, Y. Tian, Z. Liu, Adv. Mater. **28**, 9408 (2016)
216. Q. Wei, J. He, S. Yang, H. Jia, Y. Liu, W. Liu, Y. Liu, T. Li, Opt. Quant. Electron. **48**, 344 (2016)
217. Y.Y. Illarionov, M. Waltl, G. Rzepa, J.-S. Kim, S. Kim, A. Dodabalapur, D. Akinwande, T. Grasser, ACS Nano **10**, 9543 (2016)
218. Y. Du, H. Liu, Y. Deng, P.D. Ye, ACS Nano **8**, 10035 (2014)
219. J. Xu, J. Jia, S. Lai, J. Ju, S. Lee, Appl. Phys. Lett. **110**, 033103 (2017)
220. Y. Du, L. Yang, H. Zhou, P.D. Ye, I.E.E.E. Electron, Device. Lett. **37**, 429 (2016)
221. Z.P. Ling, X. Feng, H. Jiang, Z. He, X. Liu, K.W. Ang, Black phosphorus transistors with enhanced hole transport and subthreshold swing using ultra-thin HfO_2 high-k gate dielectric, in Paper presented at the 2016 IEEE silicon nanoelectronics workshop (SNW), Honolulu, Hawaii, USA, 12–13 June 2016
222. Z.P. Ling, K. Majumdar, S. Sakar, S. Mathew, J.T. Zhu, K. Gopinadhan, T. Venkatesan, K.W. Ang, Nickel-phosphide contact for effective Schottky barrier modulation in black phosphorus p-channel transistors. Paper presented at the 2016 international symposium on VLSI technology, systems and application (VLSI-TSA), Hsinchu, Taiwan, 25–27April 2016
223. Z.-P. Ling, S. Sakar, S. Mathew, J.-T. Zhu, K. Gopinadhan, T. Venkatesan, K.-W. Ang, Sci. Rep. **5**, 18000 (2015)
224. Z.-P. Ling, J.-T. Zhu, X. Liu, K.-W. Ang, Sci. Rep. **6**, 26609 (2016)
225. K.W. Ang, Z.P. Ling, J. Zhu, in *2015 IEEE International Conference on Digital Signal Processing (DSP)*, Singapore, 21–24 July 2015
226. S. Cui, H. Pu, S.A. Wells, Z. Wen, S. Mao, J. Chang, M.C. Hersam, J. Chen, Nat. Commun. **6**, 8632 (2015)
227. S. Varghese, S. Varghese, S. Swaminathan, K. Singh, V. Mittal, Electronics **4**, 651 (2015)
228. A.N. Abbas, B. Liu, L. Chen, Y. Ma, S. Cong, N. Aroonyadet, M. Köpf, T. Nilges, C. Zhou, ACS Nano **9**, 5618 (2015)
229. P. Li, D. Zhang, J. Liu, H. Chang, Y.e. Sun, N. Yin, ACS Appl. Mater. Interfaces **7**, 24396 (2015)
230. P. Yasaei, A. Behranginia, T. Foroozan, M. Asadi, K. Kim, F. Khalili-Araghi, A. Salehi-Khojin, ACS Nano **9**, 9898 (2015)
231. G. Korotcenkov, Sensors (Basel) **19**, 1010 (2019)
232. R. Zhou, J. Peng, S. Yang, D. Liu, Y. Xiao, G. Cao, Nanoscale **10**, 18878 (2018)
233. Y. Yang, J. Gao, Z. Zhang, S. Xiao, H.-H. Xie, Z.-B. Sun, J.-H. Wang, C.-H. Zhou, Y.-W. Wang, X.-Y. Guo, P.K. Chu, X.-F. Yu, Adv. Mater. **28**, 8937 (2016)
234. L. Bai, L. Sun, Y. Wang, Z. Liu, Q. Gao, H. Xiang, H. Xie, Y. Zhao, J. Mater. Chem. **A5**, 8280 (2017)
235. W. Chen, K. Li, Y. Wang, X. Feng, Z. Liao, Q. Su, X. Lin, Z. He, J. Phys. Chem. Lett. **8**, 591 (2017)
236. H. Uk Lee, S.C. Lee, J. Won, B.-C. Son, S. Choi, Y. Kim, S.Y. Park, H.-S. Kim, Y.-C. Lee, J. Lee, Sci. Rep. **5**, 8691 (2015)

237. X. Zhu, T. Zhang, Z. Sun, H. Chen, J. Guan, X. Chen, H. Ji, P. Du, S. Yang, Adv. Mater. **29**, 1605776-n/a (2017)
238. H. Wang, X. Yang, W. Shao, S. Chen, J. Xie, X. Zhang, J. Wang, Y. Xie, J. Am. Chem. Soc. **137**, 11376 (2015)
239. W. Li, Y. Yang, G. Zhang, Y.-W. Zhang, Nano. Lett. **15**, 1691 (2015)
240. Q. Yao, C. Huang, Y. Yuan, Y. Liu, S. Liu, K. Deng, E. Kan, J. Phys. Chem. C **119**, 6923 (2015)
241. J. Sun, H.-W. Lee, M. Pasta, H. Yuan, G. Zheng, Y. Sun, Y. Li, Y. Cui, Nat. Nano. **10**, 980 (2015)
242. S. Banerjee, S.K. Pati, ChemComm. **52**, 8381 (2016)
243. M. Dahbi, N. Yabuuchi, M. Fukunishi, K. Kubota, K. Chihara, K. Tokiwa, X.-F. Yu, H. Ushiyama, K. Yamashita, J.-Y. Son, Y.-T. Cui, H. Oji, S. Komaba, Chem. Mater. **28**, 1625 (2016)
244. Y. Saito, T. Iizuka, T. Koretsune, R. Arita, S. Shimizu, Y. Iwasa, Nano. Lett. **16**, 4819 (2016)
245. E. Flores, J.R. Ares, A. Castellanos-Gomez, M. Barawi, I.J. Ferrer, C. Sánchez, Appl. Phys. Lett. **106**, 022102 (2015)
246. S. Anju, J. Ashtami, P.V. Mohanan, Mater. Sci. Eng. C **97**, 978 (2019)
247. L. Viti, J. Hu, D. Coquillat, W. Knap, A. Tredicucci, A. Politano, M.S. Vitiello, Adv. Mater. **27**, 5567 (2015)
248. M. Huang, M. Wang, C. Chen, Z. Ma, X. Li, J. Han, Y. Wu, Adv. Mater. **28**, 3481 (2016)
249. Q. Guo, A. Pospischil, M. Bhuiyan, H. Jiang, H. Tian, D. Farmer, B. Deng, C. Li, S.-J. Han, H. Wang, Q. Xia, T.-P. Ma, T. Mueller, F. Xia, Nano. Lett. **16**, 4648 (2016)
250. C. Chen, N. Youngblood, D. Mohr, D. Yoo, T. Johnson, R. Peng, S.-H. Oh, M. Li, in *Conference on Lasers and Electro-Optics*, San Jose, California, 5–10 June 2016
251. M.A. Huber, F. Mooshammer, M. Plankl, L. Viti, F. Sandner, L.Z. Kastner, T. Frank, J. Fabian, M.S. Vitiello, T.L. Cocker, R. Huber, Nat. Nano. **12**, 207 (2017)
252. S. Uddin, P.C. Debnath, K. Park, Y.-W. Song, Sci. Rep. **7**, 43371 (2017)
253. F. Liu, C. Zhu, L. You, S.-J. Liang, S. Zheng, J. Zhou, Q. Fu, Y. He, Q. Zeng, H.J. Fan, L.K. Ang, J. Wang, Z. Liu, Adv. Mater. **28**, 7768 (2016)
254. A. Penillard, C. Tripon-Canseliet, M. Rosticher, I. Maksimovic, Z. Liu, E. Géron, J. Appl. Phys. **119**, 024506 (2016)
255. M. Engel, M. Steiner, P. Avouris, Nano. Lett. **14**, 6414 (2014)
256. Z. Sun, Y. Zhao, Z. Li, H. Cui, Y. Zhou, W. Li, W. Tao, H. Zhang, H. Wang, P.K. Chu, X.-F. Yu, Small **13**, n/a-n/a (2017)
257. J. Sotor, G. Sobon, W. Macherzynski, P. Paletko, K.M. Abramski, Appl. Phys. Lett. **107**, 051108 (2015)
258. I. Kriegel, S. Toffanin, F. Scotognella, Appl. Opt. **55**, 9288 (2016)
259. N. Youngblood, C. Chen, S.J. Koester, M. Li, Nat. Photon **9**, 247 (2015)
260. C. Chen, N. Youngblood, R. Peng, D. Yoo, D.A. Mohr, T.W. Johnson, S.-H. Oh, M. Li, Nano. Lett. **17**, 985 (2017)
261. A.G. Kelly, T. Hallam, C. Backes, A. Harvey, A.S. Esmaeily, I. Godwin, J. Coelho, V. Nicolosi, J. Lauth, A. Kulkarni, S. Kinge, L.D.A. Siebbeles, G.S. Duesberg, J.N. Coleman, Science **356**, 69 (2017)
262. M.-Y. Li, C.-H. Chen, Y. Shi, L.-J. Li, Mater. Today **19**, 322 (2016)
263. X. Liu, M.C. Hersam, Nat. Rev. Mater. **4**, 669 (2019)
264. P.T.K. Loan, W. Zhang, C.-T. Lin, K.-H. Wei, L.-J. Li, C.-H. Chen, Adv. Mater. **26**, 4838 (2014)
265. M. Shanmugam, R. Jacobs-Gedrim, E.S. Song, B. Yu, Nanoscale **6**, 12682 (2014)
266. T. Georgiou, R. Jalil, B.D. Belle, L. Britnell, R.V. Gorbachev, S.V. Morozov, Y.-J. Kim, A. Gholinia, S.J. Haigh, O. Makarovsky, L. Eaves, L.A. Ponomarenko, A.K. Geim, K.S. Novoselov, A. Mishchenko, Nat. Nano. **8**, 100 (2013)
267. W.J. Yu, Z. Li, H. Zhou, Y. Chen, Y. Wang, Y. Huang, X. Duan, Nat. Mater. **12**, 246 (2013)
268. R. Moriya, T. Yamaguchi, Y. Inoue, S. Morikawa, Y. Sata, S. Masubuchi, T. Machida, Appl. Phys. Lett. **105**, 083119 (2014)

269. T. Roy, M. Tosun, J.S. Kang, A.B. Sachid, S.B. Desai, M. Hettick, C.C. Hu, A. Javey, ACS Nano **8**, 6259 (2014)
270. W.J. Yu, Y. Liu, H. Zhou, A. Yin, Z. Li, Y. Huang, X. Duan, Nat. Nano. **8**, 952 (2013)
271. L. Britnell, R.M. Ribeiro, A. Eckmann, R. Jalil, B.D. Belle, A. Mishchenko, Y.J. Kim, R.V. Gorbachev, T. Georgiou, S.V. Morozov, A.N. Grigorenko, A.K. Geim, C. Casiraghi, A.H.C. Neto, K.S. Novoselov, Science **340**, 1311 (2013)
272. K. Roy, M. Padmanabhan, S. Goswami, T.P. Sai, G. Ramalingam, S. Raghavan, A. Ghosh, Nat Nano. **8**, 826 (2013)
273. L. Britnell, R.V. Gorbachev, R. Jalil, B.D. Belle, F. Schedin, A. Mishchenko, T. Georgiou, M.I. Katsnelson, L. Eaves, S.V. Morozov, N.M.R. Peres, J. Leist, A.K. Geim, K.S. Novoselov, L.A. Ponomarenko, Science **335**, 947 (2012)
274. C.-C. Chen, Z. Li, L. Shi, S.B. Cronin, Nano Res. **8**, 666 (2015)
275. M.M. Furchi, A. Pospischil, F. Libisch, J. Burgdörfer, T. Mueller, Nano Lett. **14**, 4785 (2014)
276. S. Wi, H. Kim, M. Chen, H. Nam, L.J. Guo, E. Meyhofer, X. Liang, ACS Nano **8**, 5270 (2014)
277. C.-H. Lee, G.-H. Lee, A.M. van der Zande, W. Chen, Y. Li, M. Han, X. Cui, G. Arefe, C. Nuckolls, T.F. Heinz, J. Guo, J. Hone, P. Kim, Nat. Nano. **9**, 676 (2014)
278. R. Cheng, D. Li, H. Zhou, C. Wang, A. Yin, S. Jiang, Y. Liu, Y. Chen, Y. Huang, X. Duan, Nano. Lett. **14**, 5590 (2014)
279. Y. Deng, Z. Luo, N.J. Conrad, H. Liu, Y. Gong, S. Najmaei, P.M. Ajayan, J. Lou, X. Xu, P.D. Ye, ACS Nano **8**, 8292 (2014)
280. X. Hong, J. Kim, S.-F. Shi, Y. Zhang, C. Jin, Y. Sun, S. Tongay, J. Wu, Y. Zhang, F. Wang, Nat. Nano. **9**, 682 (2014)
281. Y. Gong, J. Lin, X. Wang, G. Shi, S. Lei, Z. Lin, X. Zou, G. Ye, R. Vajtai, B. I. Yakobson, H. Terrones, M. Terrones, B.K. Tay, J. Lou, S.T. Pantelides, Z. Liu, W. Zhou, P. M. Ajayan, Nat. Mater. **13**, 1135 (2014)
282. X. Duan, C. Wang, J.C. Shaw, R. Cheng, Y. Chen, H. Li, X. Wu, Y. Tang, Q. Zhang, A. Pan, J. Jiang, R. Yu, Y. Huang, X. Duan, Nat. Nano. **9**, 1024 (2014)
283. M.-Y. Li, Y. Shi, C.-C. Cheng, L.-S. Lu, Y.-C. Lin, H.-L. Tang, M.-L. Tsai, C.-W. Chu, K.-H. Wei, J.-H. He, W.-H. Chang, K. Suenaga, L.-J. Li, Science **349**, 524 (2015)
284. Y. Gong, S. Lei, G. Ye, B. Li, Y. He, K. Keyshar, X. Zhang, Q. Wang, J. Lou, Z. Liu, R. Vajtai, W. Zhou, P.M. Ajayan, Nano. Lett. **15**, 6135 (2015)
285. D. Sarkar, X. Xie, W. Liu, W. Cao, J. Kang, Y. Gong, S. Kraemer, P.M. Ajayan, K. Banerjee, Nature **526**, 91 (2015)
286. G.-H. Lee, Y.-J. Yu, X. Cui, N. Petrone, C.-H. Lee, M.S. Choi, D.-Y. Lee, C. Lee, W.J. Yoo, K. Watanabe, T. Taniguchi, C. Nuckolls, P. Kim, J. Hone, ACS Nano **7**, 7931 (2013)
287. S. Bae, H. Kim, Y. Lee, X. Xu, J.-S. Park, Y. Zheng, J. Balakrishnan, T. Lei, H. Ri Kim, Y.I. Song, Y.-J. Kim, K.S. Kim, B. Ozyilmaz, J.-H. Ahn, B.H. Hong, S. Iijima, Nat. Nano **5**, 574 (2010)

Enhancement of the NO Gas Sensitivity by UV Light in Sn-Doped ZnO Nanostructures

Irmak Karaduman Er, Ali Orkun Çağırtekin, Tuğba Çorlu, Memet Ali Yıldırım, Aytunç Ateş, and Selim Acar

Abstract ZnO has a long history as a material for gas sensing, varistors, piezoelectric transducers, as a light-transmitting electrode in optoelectronic devices, electro-optic modulators and as a sunscreen. As a gas sensor, ZnO and doped ZnO materials have been invensively used. This chapter looks at the importance of chemical solution-based technologies and the SILAR method on the gas-sensing properties in details. We believe that a combined understanding of theoretical and experimental aspects will open new fields for metal oxide gas sensor and beyond.

Keywords Sn doped ZnO · UV light · NO gas sensing

1 Introduction

The amount of toxic, explosives, flammable and hazardous gases has been significantly increased because of industrialization and rapid economic growth in the world [1, 2]. This stimulates the scientists to work in the field of different sensors for the purpose of environmental and human safety. In the last decade, the specific demand

I. K. Er
Department of Medical Services and Techniques, Eldivan Medical Services Vocational School, Çankırı Karatekin University, Çankırı, Turkey

A. O. Çağırtekin · S. Acar (✉)
Department of Physics, Science Faculty, Gazi University, Ankara, Turkey
e-mail: sacar@gazi.edu.tr

T. Çorlu
Innovative Technologies Application and Research Center, Suleyman Demirel University, Isparta, Turkey

M. A. Yıldırım
Department of Electric and Electronics Engineering, Engineering Faculty, Erzincan University, Erzincan, Turkey

A. Ateş
Department of Material Engineering, Engineering and Natural Sciences Faculty, Yıldırım Beyazıt University, Ankara, Turkey

S. J. Ikhmayies et al. (eds.), *Advances in Optoelectronic Materials*, Advances in Material Research and Technology,
https://doi.org/10.1007/978-3-030-57737-7_4

for gas detection and monitoring has emerged particularly as the awareness of the need to protect the environment has grown [3]. Therefore, designing and developing highly efficient sensing materials is significant in meeting people's demand for working and living requirements. Metal oxides are well known to be effective in detecting various gases because they have a high area, low response times and high sensitivity. The great interest comes from their numerous advantages, like small sizes, high sensitivities in detecting very low concentrations (at the level of part per million or even part per billion) of a wide range of gaseous chemical compounds, the possibility of online operation and, due to possible bench production, low cost [2, 3].

Among various gas-sensing materials studied so far, ZnO has a long history as a material for gas sensing, varistors and piezoelectric transducers, as a light-transmitting electrode in optoelectronic devices, electro-optic modulators and as a sunscreen. ZnO is an n-type wide bandgap semiconductor with hexagonal wurtzite structure, having a resistivity in the range of 10^{-4}–10^{12} Ω-cm. ZnO usually crystallizes in the form of wurtzite (as in the case of zincite) or zinc blende structure. The wurtzite structure is commonly found for ZnO due to the stability in ambient conditions [4, 5].

ZnO has attracted keen research interest in recent years due to its promising properties for optoelectronics, biomedical, gas sensing and many other applications. It generally dominated the market for semiconductor gas sensors due to its low cost, moderate conductivity, high stability and diverse response to reducing as well as oxidizing gases [5].

2 Different Producing Methods of ZnO Nanostructure Thin Films

Conventional physical routes (vacuum methods such as sputtering and evaporation) render better control over stoichiometry producing uniform, compact films and generally produces good quality transparent ZnO thin films [6]. They are generally safe (no toxic gas emissions) and perform high deposition rate at room temperature. However, they require expensive capital instruments. Accordingly, they are difficult to expand on the other hand are cost-effective compared with vapor-phase methods and simple. Thus, they offer the desirable cheapness and possibility of scaling up to industrial level [7].

Accordingly, chemical methods have come out to be a good alternative for material preparation during the past few decades. In chemical deposition methods, a liquid precursor undergoes a chemical change at a solid surface, leaving a solid layer. ZnO thin films from chemical deposition methods tend to be conformal, rather than directional. The different category of chemical deposition methods includes spray pyrolysis [8], sol–gel [9], chemical bath deposition (CBD) [10], electroless [11], anodization [12], laser ablation [13], hydrothermal method [14] and SILAR [15] etc.

Spray pyrolysis is a method of depositing films having thicknesses in the region of thin film and thick film [8].

Electroless is a process typically used to obtain thick (micrometers) metal structures on metallic or nonmetallic substrates [11]. The process offers simplicity and cheapness.

In chemical bath deposition (CBD) method, deposition of thin films occurs due to substrate maintained in contact with a dilute chemical bath containing cationic and anionic solutions. Consequently, the film attains a terminal thickness [10].

In anodization method,the metal to be anodized is made an anode and immersed in an oxygen-containing electrolyte [12].

In laser ablation method, it is a rapid, efficient and cost-effective liquid-phase technique for producing various kinds of nanoscale materials. Some of the remarkable features of this method are the simplicity of the experimental setup and the ability to control the size of synthesized material by changing different parameters such as laser wavelength, pulse laser duration, the solution temperature, laser fluency, the pH of the solution and inexpensive equipment to control the ablation atmosphere and so on [13].

In hydrothermal method, formation of hydrothermal method morphologies can be achieved with its advantages such as low cost, environmental, potential controllability in shape and size [14]. Hydrothermal method is generally described as a fast, clean and economical method except for autoclave cost. It is a chemical reaction in the presence of a solvent above room temperature and at pressure greater than 1 atm in a closed system. The slight changes in experimental parameters like precursors, reaction temperature and time pH value strongly affect the precise morphology [15]. Shape, size and morphology significantly change the properties of nanomaterials. Three-dimensional (3D) creation of metallic or semiconductor materials has been reported to be important advantages for advanced nanoelectronic and nanooptoelectronic applications. ZnO nanorod, nanoflowers and nanoneedles-like structures, three-dimensional structure, wide and direct band spacing, large surface volume ratio and new surface chemical properties are specified as highly attractive structures for nanoelectronics, nanooptoelectronics and nanosensors. Different precursors cause different gas-sensing properties [16, 17]. The potential application of nanomaterials depends on the shape of nanostructure, and the controlled morphology can modulate the properties of nanomaterials with hydrothermal method. Therefore, ZnO nanostructures fabricating with hydrothermal method hierarchical in a controllable manner are required to improve the performance and broaden the application range of ZnO [18].

The successive ionic layer adsorption and reaction (SILAR) method is the most used solution technique and also one of the oldest methods for thin film growth. In SILAR, thin films are obtained by immersing the substrate into separately placed cationic and anionic precursors for reaction at chosen temperatures. Between every immersion, the substrate is rinsed in distilled water or ion exchanged water to avoid homogeneous precipitation in the solution [19]. Sequential reaction on the substrate surface under optimized conditions of concentration and pH of the reacting solutions results in the formation of the film. Thus, precipitation formation, i.e., wastage of material, is avoided in SILAR method [19].

It does not require high-quality substrates and can operate at room temperature without the need for vacuum. In addition, SILAR method provides optimal fabrication conditions to obtain high-quality thin films. Stoichiometric precipitation can easily be obtained. Since the basic building materials are ions instead of atoms, the preparation parameters easily controllable, best orientation and particle structure can be obtained at room temperature [20, 21].

In spite of its simplicity, SILAR has number of advantages: (i) Unlike vapor deposition method, SILAR does not require vacuum at any stage; (ii) the deposition can be carried out at or close to room temperature; (iii) unlike high power methods such as radio frequency magnetron sputtering, it does not cause local overheating that can be detrimental for materials to be deposited; and (iv) there are virtually no restrictions on substrate material, dimensions or its surface profile [20, 21]. Thus, any insoluble surface to which the solution has free access will be a suitable substrate for the deposition making the technique convenient for large area deposition [21].

Gao et al. reported that they produced ZnO nanocluster thin films modified SILAR method and investigated their structural, morphological and optical properties [22]. Vargas-Hernandez et al. studied that the comparison between ZnO thin films deposited by three processes was studied by means of X-ray diffraction (XRD), optical absorption and micro Raman [23]. Gao et al. reported that nanoporous ZnO film was prepared by the stepwise solution route: the ultrasonic irradiation mediated successive ionic layer adsorption and reaction (SILAR) method. They studied the morphology and microstructure of ZnO thin films [24]. Suresh Kumar et al. reported that they produced ZnO nanostructure thin films by a simple SILAR process at relatively low temperature for its self-cleaning application [25]. Ravichandran et al. studied that ZnO thin films were deposited using SILAR method from three different precursors. They annealed all the films in air and vacuum ambiences at 350 °C and characterized for their structural, optical, photoluminescence, electrical and surface morphological properties [26].

2.1 Dopants in ZnO Nanostructured Thin Films

In recent years, the important and effective method to improve sensing properties of ZnO, doping method has drawn many researchers' attention. In fact, doping with noble metals or other additives has been reported to enhance the sensitivity and selectivity of ZnO gas sensors [27]. The changes of the electronic band and morphological structure due to doping with other metal or additives lead to enhancing the gas-sensing properties with increasing surface-to-volume ratio and creating more centers for gas interaction on the ZnO metal oxide semiconductor surface [28]. It is very important to optimize the process parameters of film growth and doping levels to have a desired enhanced device performance. Tin (Sn) is well known to improve the electrical conductivity since the ionic radius of Sn^{4+} ion (0.69 Å) is smaller than that of Zn^{2+} ion (0.74 Å). When Sn was added into ZnO for doping, Sn^{4+} substitutes Zn^{2+} site in the ZnO crystal structure resulting in two more free electrons to contribute to the electric conduction [29]. The ionic radius of Sn^{4+} is smaller than Zn^{2+}; therefore, Sn^{4+} ions can replace Zn^{2+} ions in substitution sites.

3 Gas-Sensing Applications Using ZnO Nanostructure Thin Films

During the last few decades, undoped and doped ZnO nanostructure thin films are the promising candidate among the earliest discovered metal oxide materials. As the toxic gases affect the human health and life even at ppm level, hence many efforts have done to improve the gas-sensing properties of ZnO-based gas sensors. Gupta et al. studied different ZnO nanostructures and showed that the response of ZnO sensors to H_2S arisen from grain boundary only whereas for NO gas both grain boundaries and intragrain resistances contribute toward the response. At the same time, oxygen vacancies in the lattice also improve the sensor response [30]. Badhe et al. studied In doped ZnO thin films for H_2S gas sensors and observed that 3 at.% In doping exhibited high gas sensitivity with fast response and recovery time which depend on the surface morphology and indium concentration [31]. Sahay et al. observed that performance of liquefied petroleum gas (LPG) gas was strongly dependent on crystallite size of Al-doped ZnO thin films showing maximum response for 0.5 at% doping concentration at 325 °C operating temperature [32]. Choi et al. studied the H_2 sensing response of SnO_2 thin films with different crystal orientations. They suggested that the (101) orientated SnO_2 films exhibited higher gas response than (002) and (101) orientated ones [33]. Gong et al. investigated the role of Cu doping in ZnO film in enhancing the capability of the films to adsorb CO molecules. They suggested that the Cu site in ZnO films played an important role in adsorbing CO molecules at both low and high temperatures [34]. Navale et al. reported the selective NO_X sensing characteristics of Al-doped ZnO synthesized in the form of porous pellets. They compared its gas-sensing performance in presence of different gases such as SO_X, HCl, liquefied petroleum gas (LPG), H_2S, H_2, NH_3, alcohol and acetone and found it selectively detects NO_X due to Al-doping [35]. Al-Hadeethi et al. reported that 2D Sn-doped ZnO ultra thin nanosheet networks synthesized by a simple and facile hydrothermal process and investigated their gas-sensing properties. Sn doping resulted in distortions and lattice defects, particularly O-vacancies in the ZnO nanostructures which were the most susceptible adsorption sites for generation of oxygenated anionic species from molecular O_2 from atmosphere [36]. Luo et al. studied that the gas-sensing properties of pure ZnO and Sn-doped thick films to ethanol were investigated. The sensitivity of Sn-doped ZnO was about fourfold than that of pure ZnO [37]. Li et al. reported that nanostructured pure ZnO (PZO) and

Sn-doped ZnO (SZO) were deposited on glass templates by chemical deposition method. However, it remarkably shortened the response time and recovery time of the sensor. Both PZO and SZO showed higher response to acetone than to ethanol, while the response time and recovery time of PZO and SZO to ethanol were shorter than those to acetone, which indicates that the selectivity of ZnO nanostructures to ethanol and acetone were different [38]. Literature review shows that the gas response depends on various factors, such as morphology, dopants and their concentrations, thickness of thin films and operating temperatures. This chapter deals with study of the gas-sensing properties of undoped and doped ZnO nanostructure thin films on these various parameters.

3.1 The Effect of Operating Temperature

It is well known that the operating temperature is important in evaluating the performance of metal oxide semiconductor-based gas sensors due to its considerable influence on the surface state of sensing materials and on the reaction during the gas-sensing process. The operating temperature is important in evaluating the performance of metal oxide semiconductor-based gas sensors due to its considerable influence on the surface state of sensing materials and on the reaction during the gas-sensing process [39]. When a gas sensor is exposed to air, oxygen molecules are adsorbed on the surface of the sensitive material. This material traps electrons from the conduction band and produces negatively charged chemisorbed oxygen species $O^{\delta-}$ such as O_2^-, O^- and O^{2-}. Hence, the concentration of holes in valence band increases and the resistance (of the sensitive material) decreases due to increased concentration of available carriers [40]. The type of such chemisorbed oxygen species at the surface depends on the operating temperature. There is significant influence of operating temperature on the sensitivity of gas sensors, and all sensors must be tested at different temperatures to find the optimum operating conditions for gas sensing [41].

3.2 The Effect of Surface Morphology

In materials science, morphology is the study of shape, size, texture, spatial organization and phase distribution of particles in a material [42]. Different nanostructures can be obtained depending on their material composition, crystal structure and manufacturing method [42]. Each synthesis process offers a diversity of parameters such as temperature, pressure, reagent concentration, treatment time and pH that when controlled they allow obtaining different morphologies, compositions and crystallinity of the products [43–46]. Besides, a lot of three-dimentional tridimensional shapes can be synthesized such as spheres, rods, tubes, needles, cubes and octahedrons. Morphology variation allows to control its functionality in an effective way, since surface is directly related to interface that such material will have to outside [46]. Such interface is related to the large number of atoms that determine their physicochemical properties. Moreover, most nanomaterials are thermodynamically unstable, which make their morphologies unbalanced, that they are different from the shape of monocrystals of a given material [47].

3.3 *The Effect of Annealing Process*

In such a case, annealing in N_2 was found as the most useful regime for thin films which is most probably due to oxygen loss annealed in N_2. It can be induced that the increase in response in N_2 was attributed to the increase of oxygen vacancies on the sensing surface. However, oxygen is more prominent at the surface annealed in O_2 environment, because of this the response of thin films begins to decrease [48, 49]. This is probably correlated with the changes brought by the defects states in the structure of the film related to oxygen deficiencies. The response is also related to the oxidizing ability and the adsorbing ability of detected gas on surface of the material [50].

Annealing environments affect the electrical and surface properties of metal oxide semiconductors. Jlassi et al. produced NiO and ZnO thin films and reported that the roughness and grain size increased as the annealing atmosphere changed from air to nitrogen [51]. Arfaoui et al. investigated the effect of annealing environment on morphological and electrical properties as well as the sensing performance of Mo_xO_y samples exposed to ethanol vapor. They reported the response of Mo_xO_y samples change with changing annealing atmosphere [52]. Sadr et al. reported that annealing process rapidly increased the gas-sensing properties of porous silicon [53]. Moreover, during the process of annealing, the oxygen vacancies on the surfaces of the sensing films begin to reduce which is electrically and chemically active [54, 55]. These surface defects also affect the adsorption behaviors of gas molecules on metal oxide surfaces [53].

4 Experimental Details

$Zn_{1-x}Sn_xO$ ($x = 0, 0.05, 0.10, 0.15, 0.20$) nanostructure thin films were grown on the interdigitated Au electrodes at room temperature by SILAR method. To grow $Zn_{1-x}Sn_xO$ nanostructure thin films, aqueous zinc-ammonia complex ions ($[Zn(NH_3)_4]^{2+}$) and aqueous tin-ammonia complex ions ($[Sn(NH_3)_4]^{4+}$) were chosen for the cation precursors, in which trace metals basis of $ZnCl_2$ (99.9%, Sigma–Aldrich) of 0.1 M, $SnCl_4$ (99.9%, Sigma–Aldrich) of 0.1 M as sources for Zn, Sn and aqueous ammonia solution (NH_3-28%, Sigma–Aldrich) were used. Deionized water was used as a solvent. The obtained $[Zn(NH_3)_4]^{2+}$ and $[Sn(NH_3)_4]^{4+}$

complexes were mixed in appropriate proportions according to the composition for ZnO, $Zn_{0.95}Sn_{0.05}O$, $Zn_{0.90}Sn_{0.10}O$, $Zn_{0.85}Sn_{0.15}O$ and $Zn_{0.80}Sn_{0.20}O$ nanostructure thin films. All the growth process parameters of $Zn_{1-x}Sn_xO$ nanostructure thin films by SILAR method were described in detail in our previous studies [54–57]. $Zn_{1-x}Sn_xO$ nanostructure thin films were grown by repeating 40 SILAR cycles. The nanostructure thin films were annealed in a nitrogen atmosphere (300 °C, 13 min), and characteristic parameters of the films were investigated.

The gas-sensing performance of the sensors was tested with a gas-sensing measurement system. The gas-sensing measurement system working principle is given in our previous work [20]. The gas-sensing measurements were carried out for NO gas at different concentrations and different operating temperatures by monitoring the resistance changes. Dry air was used as the carrier gas which is % 99.999 purity (dry air was the 'blank' gas used to purge the sensor.). The flow rate of the dry air undergoing testing was fixed at 500 cm^3/min during the measurements. Air flow rate, under the same conditions in order to observe the behavior of different concentrations, must be kept always constant. To ensure stable zero-level resistance in ambient air prior to exposure to gas, the stabilization of the thin film resistance is important because it ensures stable zero level for gas-sensing applications. For the corresponding operating temperature of the gas chamber, it is the prime requisite to stabilize the resistance in air atmosphere before ejecting the gas into the chamber. It indicates the resistance of thin film in air. Relative humidity was kept constant (about 25%) for all measurements, monitored by a Honeywell HIH-4000 humidity sensor. The NO concentration and dry airflow rates were controlled by computer-controlled mass flow controllers (MKS Series). A LakeShore 325 temperature controller with platinum RTDs was used to maintain a constant temperature. The sensor resistance was continuously monitored with a computer-controlled system by using a Keithley 2400 source meter, and data was collected in real time using a computer with corresponding data acquisition hardware and software. Before the measurements, sensors under test were fixed into the chamber, and the samples are kept constant at 130 °C for 20 min in order to enhance the adhesion.

To investigate structural, morphological, compositional and optical properties of the $Zn_{1-x}Sn_xO$ nanostructure thin films, XRD, SEM, EDAX and optical absorption measurements were used. For structural, morphological, compositional and optical studies, the Panalytical Empyrean X-ray diffractometer (using Cu Kα $\lambda = 1.5405$ Å radiation), the FEI Quanta FEG 450 model scanning electron microscope (SEM) with energy dispersive X-ray analysis (EDAX) attachment and the Perkin-Elmer UV/VS Lambda 2S Spectrometer were used, respectively.

5 Results and Discussions

5.1 Structural Analysis

Figure 1 shows the XRD patterns of the $Zn_{1-x}Sn_xO$ nanostructure thin films. As seen in Fig. 1, all the nanostructure thin films have polycrystalline nature and well-defined peaks belonging to ZnO and SnO_2. The pure ZnO film ($x = 0$) has hexagonal wurtzite phase with (100), (002), (101), (102), (110) and (103) diffraction peaks (JPCDS Card No: 36–1451) [54, 56, 58]. The intensities and full width at half maximum (FWHM) values of the characteristic peaks of ZnO film change and some peaks disappear with increasing Sn concentration (x), whereas the characteristic peaks belonging to the tetragonal phase (JPCDS Card No: 41-1445) of SnO_2 film begin to appear (in $x = 0.10, 0.15, 0.20$ values) and the intensity of the peaks increases with increasing Sn concentration (x). It has been observed that the dominant crystal phase in all the XRD patterns of the nanostructure thin films is the hexagonal wurtzite phase of ZnO. The crystal quality of the nanostructure thin films deteriorated with increasing Sn concentration (x). This deterioration may be attributed to changing in the atomic environment due to the extrinsic doping (Sn) of ZnO nanostructure thin film. The changing in the atomic environment should be differences in ionic radii of Zn and Sn elements [59–61]. Also, as seen in Fig. 1, the peak positions with Sn concentration (x) have been determined to shift with slightly larger diffraction angles. It has been related to the unit cell size changes in nanostructures with the substitution of Sn^{4+} for Zn^{2+}, because the ionic radius of Zn^{2+} (0.74 Å) and Sn^{4+}(0.71 Å) are different values [59, 62, 63]. The grain size values of ZnO, $Zn_{0.95}Sn_{0.05}O$, $Zn_{0.90}Sn_{0.10}O$, $Zn_{0.85}Sn_{0.15}O$ and $Zn_{0.80}Sn_{0.20}O$ nanostructure thin films calculated using the Scherrer formula [54, 57] were 42.12 nm, 35.79 nm, 33.00 nm, 29.55 nm and 25.61 nm, respectively.

5.2 Surface Morphological Analysis

The surface morphology of the nanostructure thin films as sensor materials influences their electrical and gas-sensing properties, which are important factors in applications of devices. Thus, it is very important to characterize the surface morphology of the films. Figure 2 shows the SEM images of $Zn_{1-x}Sn_xO$ nanostructure thin films. As seen in Fig. 2, all the nanostructure thin films have a compact, dense surface morphology and cover the substrate very well. As clearly seen in Fig. 2a, ZnO film ($x = 0$) has nanorod structure and localized clusters on the surface morphology. $Zn_{0.95}Sn_{0.05}O$ film has a porous sphere and localized flowerlike structures (in Fig. 2b). $Zn_{0.90}Sn_{0.10}O$ film has a uniform and well-grown porous sphere structure (in Fig. 2c). The SEM image of $Zn_{0.85}Sn_{0.15}O$ film is composed of a large number of uniform nanoballs (in Fig. 2d). It is clearly seen that the uniformity and smoothness of $Zn_{0.80}Sn_{0.20}O$ film have decreased compared to the others (in Fig. 2e). Consequently, the diversity

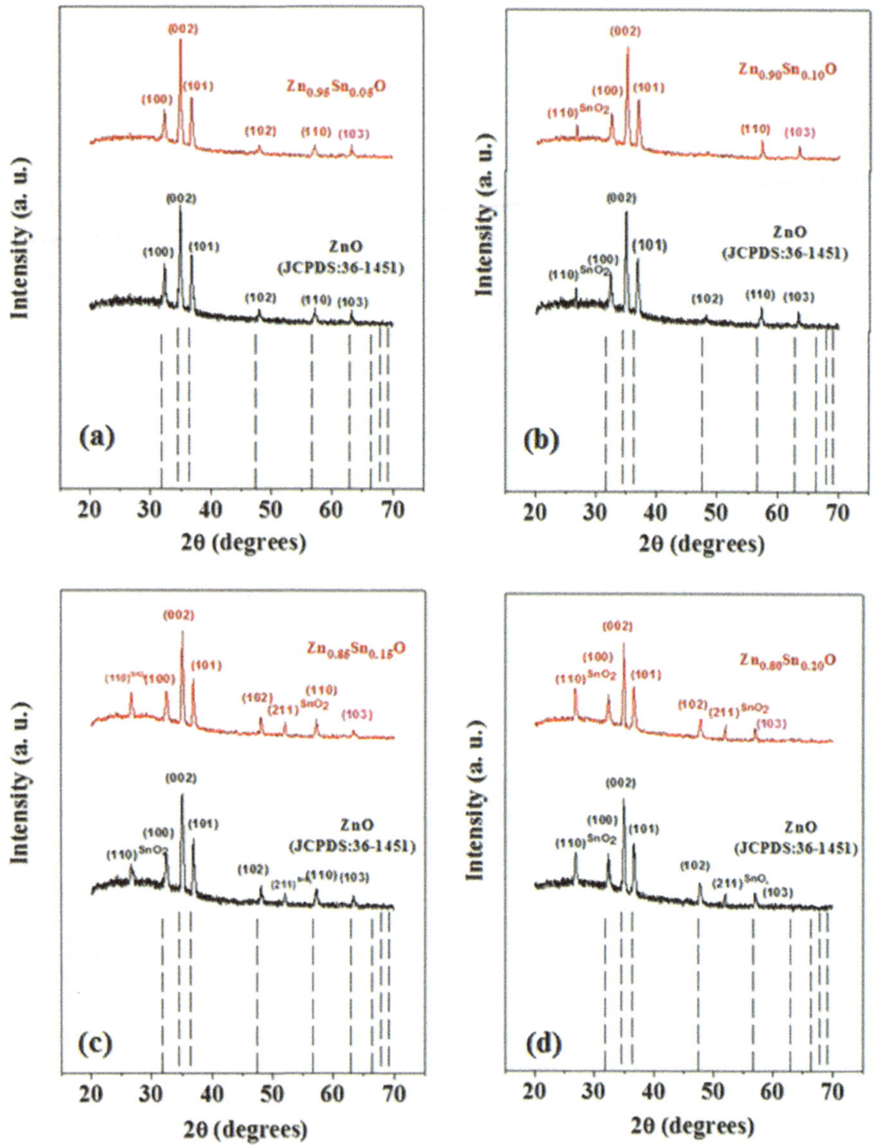

Fig. 1 XRD patterns of $Zn_{1-x}Sn_xO$ nanostructure thin films [64]

in surface morphology of Sn-doped ZnO materials is a key factor in the sensing performance of the gas sensors.

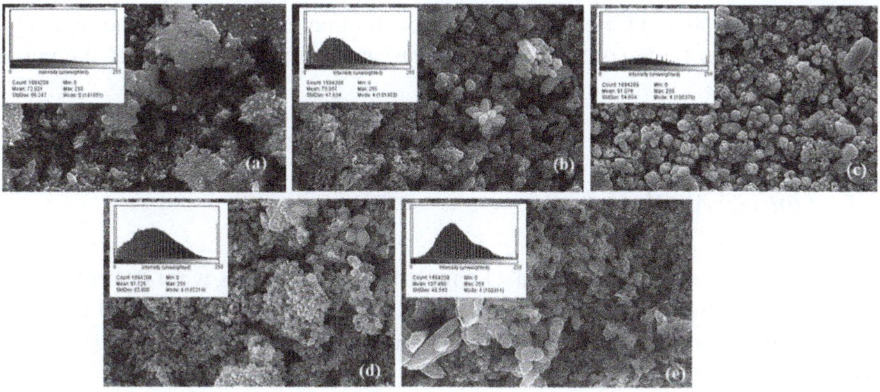

Fig. 2 SEM images of ZnO (**a**), $Zn_{0.95}Sn_{0.05}O$ (**b**), $Zn_{0.90}Sn_{0.10}O$ (**c**), $Zn_{0.85}Sn_{0.15}O$ (**d**) and $Zn_{0.80}Sn_{0.20}O$ (**e**) nanostructure thin films [64]

5.3 Compositional Analysis

Figure 3 shows EDAX spectra of $Zn_{1-x}Sn_xO$ nanostructure thin films. EDAX analysis shows the presence of Zn, Sn and O elements in the films (in Fig. 3). The atomic percent values of these elements in the films are given in the inset of EDAX spectra.

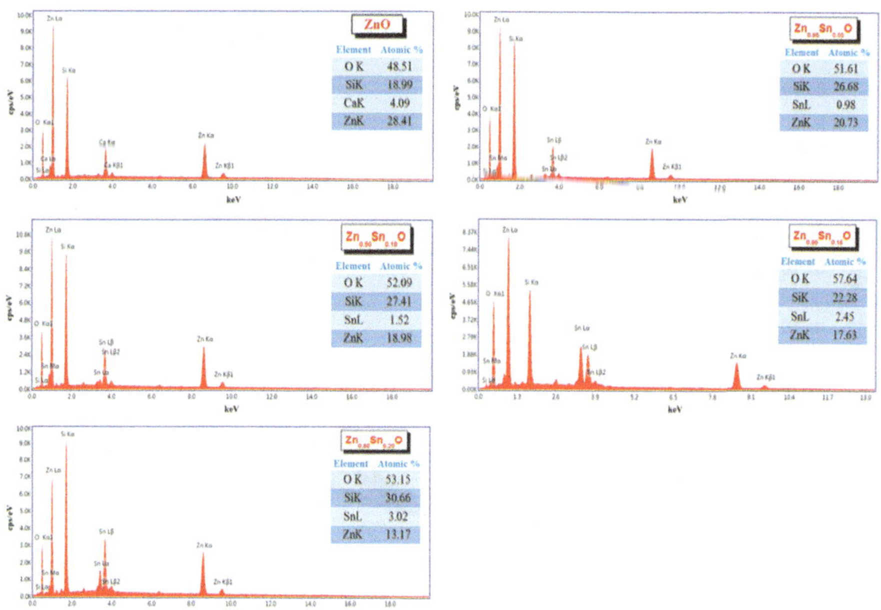

Fig. 3 EDAX analysis of $Zn_{1-x}Sn_x$Onanostructure thin films [64]

The presence of Si and Ca elements in the spectra may originate from the substrate. The atomic ratios (at.%) of Sn in ZnO film are approximately as same as the initial ratio.

5.4 Optical Analysis

The optical properties of $Zn_{1-x}Sn_xO$ nanostructure thin films were determined by using the optical absorption measurements. The optical direct bandgap of the films can be determined by extrapolation of the linear region of $(\alpha h\nu)^2$ versus $(h\nu)$ plots using Tauc equation [54, 65],

$$\alpha = \frac{A}{h\nu}\left(h\nu - E_g\right)^{1/2} \quad (1)$$

where A is a constant, α is the absorption coefficient, E_g is the optical bandgap, and $h\nu$ is a photon energy. Figure 4 shows $(\alpha h\nu)^2$ versus $(h\nu)$ plots of $Zn_{1-x}Sn_xO$ nanostructure thin films. As seen in Fig. 4, the bandgap values increased from 3.01 to 3.20 eV with increasing Sn concentration (x). The increasing in bandgap with Sn concentration (x) can be attributed to a real bandgap change between ZnO and SnO_2 because of the wider bandgap of SnO_2 compared to the ZnO, morphological

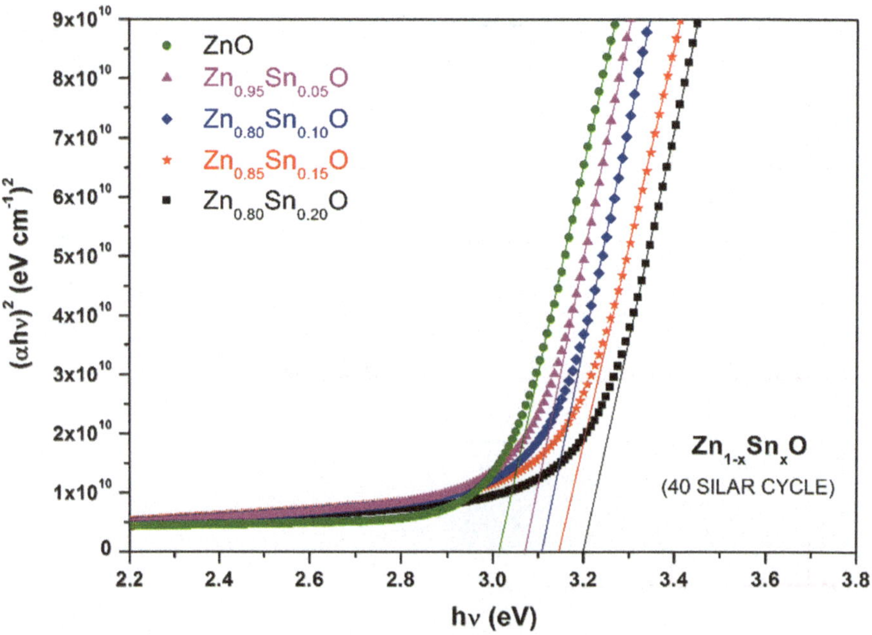

Fig. 4 Plot of $(\alpha h\nu)^2$ versus $h\nu$ for $Zn_{1-}xSn_xO$ nanostructure thin films [64]

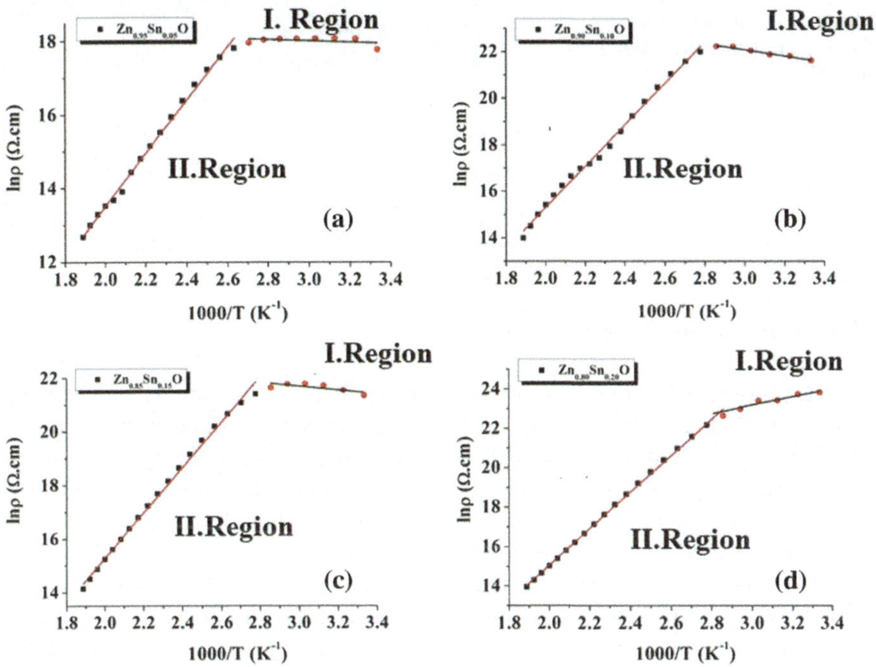

Fig. 5 Graphs of ln ρ $-1000/T$. **a** $Zn_{0.95}Sn_{0.05}O$, **b** $Zn_{0.90}Sn_{0.10}O$, **c** $Zn_{0.85}Sn_{0.15}O$, **d** $Zn_{0.80}Sn_{0.10}O$

changes, changes of grain size and unit cell size with the substitution of Sn^{4+} for Zn^{2+} because of the ionic radius of Zn^{2+} (0.74 Å) and Sn^{4+} (0.69 Å) [54, 57, 59, 66, 67].

The plots of ln ρ versus 1000/T are shown in Fig. 5. The measurements were carried out in the temperature range of 300–525 K. Different regions appearing on figures show that the films exhibit the nature of semiconductor. Resistivity decreases as temperature increases, and this situation is seen as an essential condition for the temperature dependence of the resistivity changes for semiconductors. Activation energies of films are calculated from Arrhenius plot [68]. The obtained E_I and E_{II} values are calculated in the low and high temperature regions. The activation energies are given in Table 1.

Since the ionic diameter of Sn^{4+} is smaller than the ionic diameter of Zn^{2+}, the tin additive made to pure ZnO is deposited in Zn^{2+} ion regions in the ZnO cage and may cause an increase in activation energy [69]. However, as seen from the resistance graphs, the potential energy to ionize tin is lower than that of zinc and it can easily ionize. Thus, activation can lead to an increase in the energies, even though it causes the resistance to fall.

Table 1 Activation energies of $Zn_{1-x}Sn_xO$ nanostructure thin films

	E_I (eV) (I. Region)	E_{II} (eV) (II. Region)
ZnO		
$Zn_{0.95}Sn_{0.05}O$	0.0102	0.230
$Zn_{0.90}Sn_{0.10}O$	0.034	0.245
$Zn_{0.85}Sn_{0.15}O$	0.057	0.33
$Zn_{0.80}Sn_{0.20}O$	0.069	0.39

5.4.1 Gas-Sensing Measurements

The operating temperature is an important factor that can essentially influence the catalytic properties of the sensing material, especially in the case of noble metals modified semiconducting oxides [70]. It affects the electron mobility and the electrical conductivity of the metal oxide material. The operating temperature influences the chemical dynamics at the gas-solid interface and thus determines the important sensing properties such as response, selectivity, stability, response and recovery times [71]. There usually exists a temperature region in which the sensor reveals the highest sensitivity. However, high temperature results in high power consumption and therefore causes the deterioration of the sensing surface. Optimization of operating temperature was carried out by measuring gas response over the range of 30–135 °C. Figure 6 shows gas response at various operating temperatures for thin films at 25 ppm NO concentration. The maximum gas response was observed at 55 °C operating temperature. The operating temperature plays a crucial role for gas adsorption/ desorption as well as the chemical reaction between surface-adsorbed oxygen

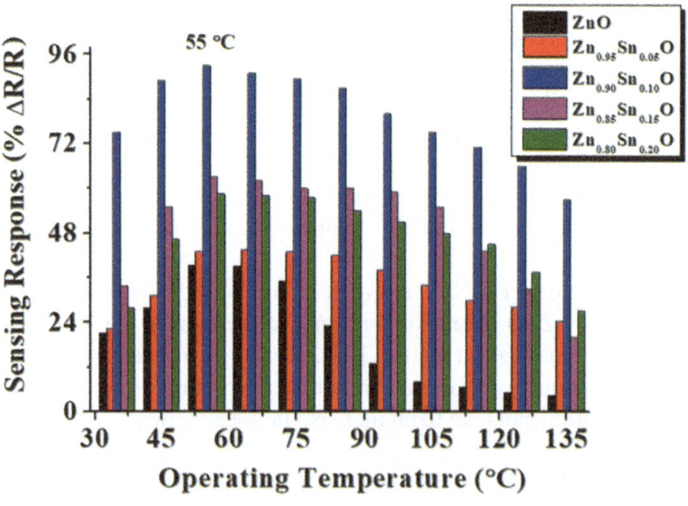

Fig. 6 Response of thin films for 25 ppm NO gas as a function of operating temperature [64]

and NO molecules. Sensing material exhibits a maximum response toward a target gas at an optimal temperature [71]. When the working temperature is up to 35 °C, the sensor begins to show response to NO and the response remarkably increases with the temperature up to 55 °C and then decreases with further increase in temperature. This behavior has been attributed to the decomposition and the adsorption of oxygen ions on the surface which depends on the temperature [69–72].

Responses are increased rapidly and reach to the maximum value at 55 °C and then decrease as the temperature is further increased. Because sufficient thermal energy is essential to overcome the activation energy barrier of oxygen chemisorption and surface reaction at temperature range from 35 to 55 °C, however, once the operating temperature is over 55 °C, the amount of desorbed gas will increase, leading to the reduction of response. Due to adsorption and desorption of gas molecules on the sensors' surface, the change of response can be observed. As seen, the maximum responses of all sensors are found at 55 °C. As a result, 55 °C operating temperature was preferred for all the gas-sensing measurements. Figure 7 depicts the response of thin films as a function of gas concentrations. It can be seen that responses increase with increasing gas concentration.

Figure 8 shows the responses of thin films toward 20 ppb NO gas concentration at room temperature. The responses of 20 ppb NO gas at room temperature were calculated 0.5%, 0.9%, 0.8% and 0.3% for $Zn_{0.95}Sn_{0.05}O$, $Zn_{0.90}Sn_{0.10}O$, $Zn_{0.85}Sn_{0.15}O$ and $Zn_{0.80}Sn_{0.20}O$, respectively.

The fast response and recovery time are critical for the application of gas sensors. The detection of the toxic and harmful gas filled in an environment in a shorter response time reduces the loss in lives. It is a very important parameter in industrial applications. The response or recovery time is defined as the time for reaching 90% of the full response change of sensor after testing gas is introduced. Figure 9 depicts the response and recovery times of $Zn_{1-x}Sn_xO$ thin films at 55 °C. The response times of 20 ppb NO gas were calculated 25 s, 5 s, 32 s and 6 s for (a) $Zn_{0.95}Sn_{0.05}O$, (b) $Zn_{0.90}Sn_{0.10}O$, (c) $Zn_{0.85}Sn_{0.15}O$, (d) $Zn_{0.80}Sn_{0.10}O$ (e), respectively. And also, the recovery times of 20 ppb NO gas were calculated 44 s, 11 s, 50 s and 11 s for (a) $Zn_{0.95}Sn_{0.05}O$, (b) $Zn_{0.90}Sn_{0.10}O$, (c) $Zn_{0.85}Sn_{0.15}O$, (d) $Zn_{0.80}Sn_{0.10}O$ (e), respectively. Sensors have rapid response and recovery times. Thus, the produced sensor works according to its own purpose.

The sensing mechanism of resistive gas sensors based on n-type MOS is essentially considered from the electrical resistance change owing to an interaction between the adsorbed oxygen ion and target gases. When the sensing surface exposed to the air, the adsorption of oxygen and formation of adsorbed molecular/atomic oxygen ions such as O_2^-, O^-, O^{2-} are the indispensable mechanism, which takes place on surfaces of MOS gas sensors [72]. Then, depletion layers are formed on the surface regions, causing an increase in resistance of the oxide by the following formulas [72]:

$$O_{2(gas)} \leftrightarrow O_{2(ads)} \quad (2)$$

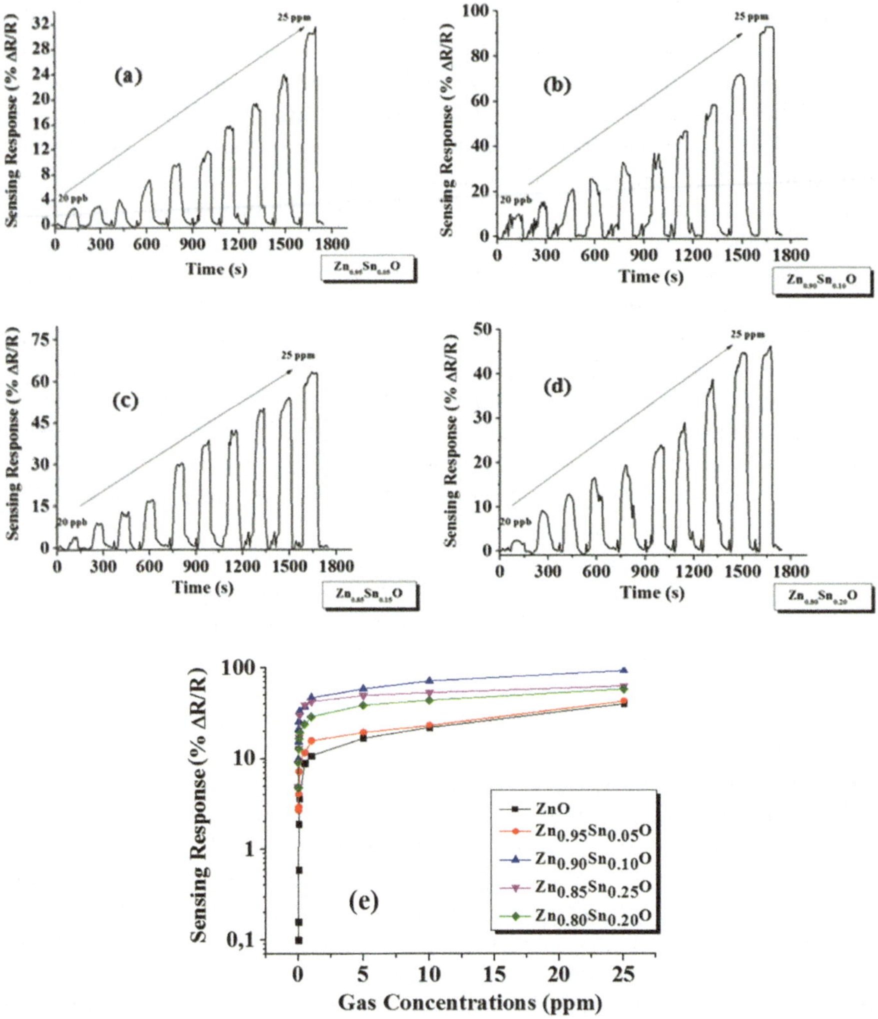

Fig. 7 Responses of thin films as functions of gas concentrations between 20 and 50 ppb. **a** $Zn_{0.95}Sn_{0.05}O$, **b** $Zn_{0.90}Sn_{0.10}O$, **c** $Zn_{0.85}Sn_{0.15}O$, **d** $Zn_{0.80}Sn_{0.20}O$ and **e** the sensing responses of all thin films at logarithmic scale at 55 °C

$$O_{2(ads)} + e^- \leftrightarrow O^-_{2(ads)} (T < 100\ ^\circ C) \tag{3}$$

$$O^-_2 + e^- \leftrightarrow 2O^- (100\ ^\circ C < T < 300\ ^\circ C) \tag{4}$$

The reaction between the electron and the chemisorbed oxygen results in a decrease in the electron concentration at the surface of the sensor. When the sensing surface exposed to NO gas, the adsorption of NO causes electron transportation from

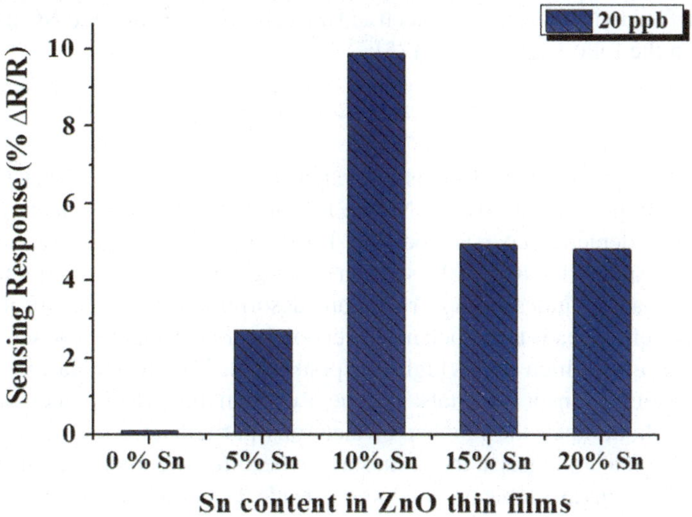

Fig. 8 Responses of 20 ppb NO gas at room temperature for $Zn_{1-x}Sn_xO$ thin films at 55 °C.

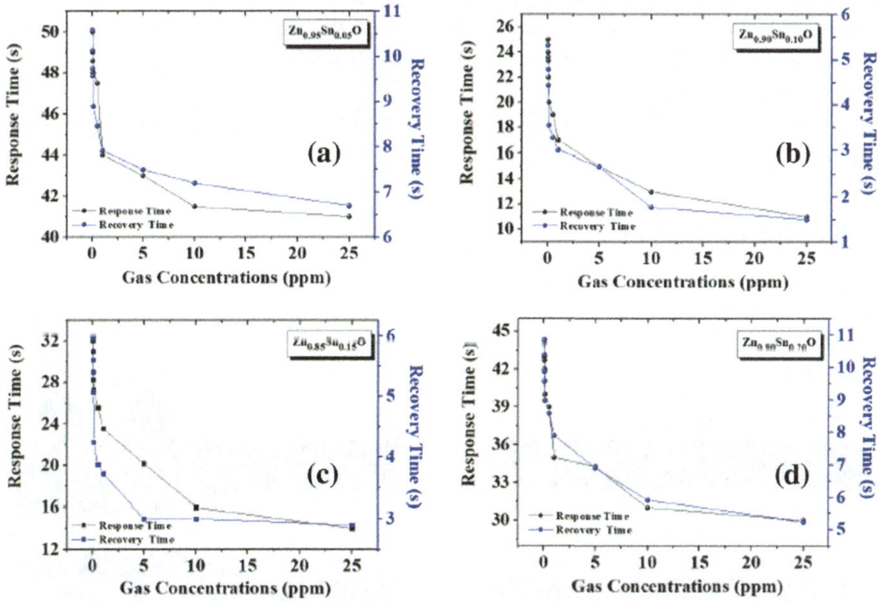

Fig. 9 Response and recovery times as functions of NO gas for $Zn_{1-x}Sn_xO$ thin films at 55 °C

sensing surface to NO due to high electrical attraction of the unpaired NO molecules, as shown in the following reaction [73];

$$2NO + 4e^- \rightarrow N_2 + 2O^{2-} \quad (5)$$

It is known that a depletion layer is generated in sensing material when it is exposed to an oxidizing gas such as O_2 and NO. In the air, the conduction channel becomes narrower and a depletion layer is produced. The depletion layer thickness increases as the sensor is exposed to NO and the sensor resistance increases. Furthermore, Sn acts as a catalyst that facilitates NO adsorption and desorption with the chemical reaction. The catalytic effect leads to a much narrower conduction channel in the sensing layer upon NO exposure which causes higher responses [72, 73]. The increase in a surface-to-volume ratio of sensor materials, i.e., the use of smaller metal oxide grains, is an effective method, since the number of electrons participating the reaction and/or interaction with a target gas at the grain surface becomes significant against the total number of electrons available for conduction inside the grain [74]. If the diameter (D) of each grain would be less than two times of the depth (L) of the space-charge region in air at certain conditions, the space-charge region occupies the entire grain and the sensor is in a very high resistance level. Under such conditions, i.e., a grain control region [75], the surface effect controls thoroughly the total resistance of each grain and therefore the whole sensor consisting of such fine grain, leading to amazingly high response to a target gas [74, 75].

Figure 10 depicts the schematic diagram of NO gas sensing for the $Zn_{1-x}Sn_xO$ thin films.

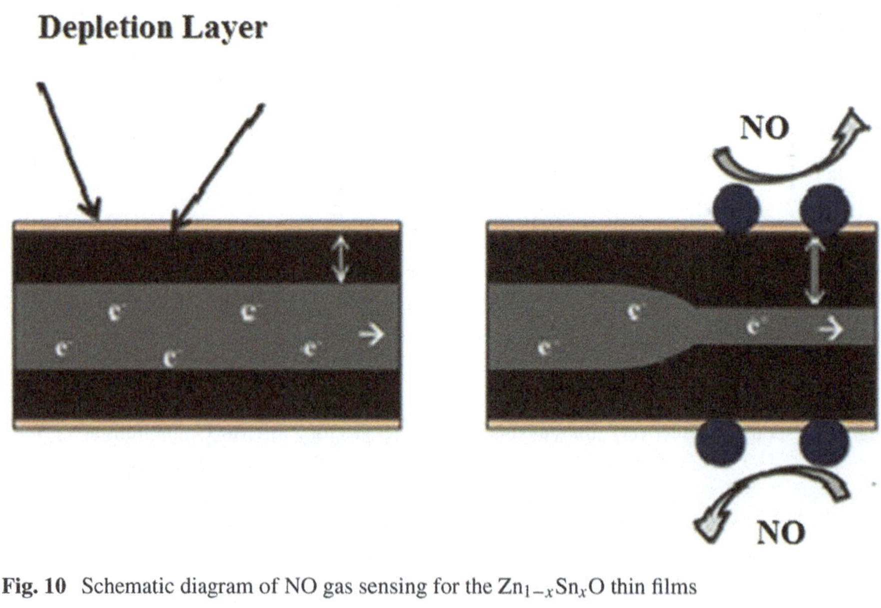

Fig. 10 Schematic diagram of NO gas sensing for the $Zn_{1-x}Sn_xO$ thin films

It can be clearly seen that the response of thin films varies depending on the annealing environment. Especially, thin films annealed in N_2 environment exhibited higher response at lower operating temperatures. In such a case, annealing in N_2 was found as the most useful regime for thin films which is most probably due to oxygen loss annealed in N_2. It can be induced that the increase in response in N_2 was attributed to the increase of oxygen vacancies on the sensing surface [76]. The response is also related to the oxidizing ability and the adsorbing ability of detected gas on the surface of the material [69]. Therefore, the introduction of more surface defects through annealing is able to significantly enhance the corresponding response values at lower operating temperatures. However, apart from this surface defects alteration, annealing also has aggravated the porosity of the ZnO nanowalls [77].

The sensor characteristics were further investigated by an impedance spectroscopy technique, whereby different contributions of the sensor elements could be distinguished [78]. By changing the frequency of the operating signal, the contribution of both grain and grain boundary to the total behavior of the sensor under different conditions could be detected [78]. Impedance measurements of the sample, with and without gas (25 ppm) exposure, give corresponding Nyquist plots consisting of two different semi-circles with a center nearly on the x-axis for two different impedance values. As the semi-circle in Nyquist plots corresponds to an equivalent parallel RC circuit, a parallel RC model fitting is done to determine the values of the impedance parameters from the curve obtained for both conditions. Figure 11 shows the Nyquist plots of thin films sample with and without the NO gas at 530 K.

When the Z and θ measurements at 530 K are performed, it is seen that when the Cole-Cole curves drawn by the equivalent circuit model are observed, the accumulation occurs in the high-frequency region and the received data decreases in the low-frequency region. From the Cole-Cole curves plotted in the presence and absence of 25 ppm NO gas, it is seen that each sensor has an increase in impedance values in the presence of gas and that the sensor of $Zn_{0,90}Sn_{0,10}O$ gives the highest change in the presence of gas.

6 UV Light Irradiation for Gas-Sensing Applications

It is well known that the working principle of the metal oxide gas sensor is based on the surface chemical reaction which perturbs the free carrier density within the metal oxide. It is the variation of the electrical properties of the sensor to gas that can be used for detecting the change in the surrounding atmosphere [79]. Metal oxide gas sensors are generally operated above 100 °C to overcome the energy limits of chemisorptions for reaching high sensitivity. However, when UV light is applied to the surface, high-energy electron-hole pairs are formed and enhance the concentration of surface [79]. These electrons in the electron-hole pairs provide higher sensitivities when they have higher energies than the bandgap energy of the surface. The holes react with the oxygen held on the surface to allow the oxygen vacancies in the surface to be

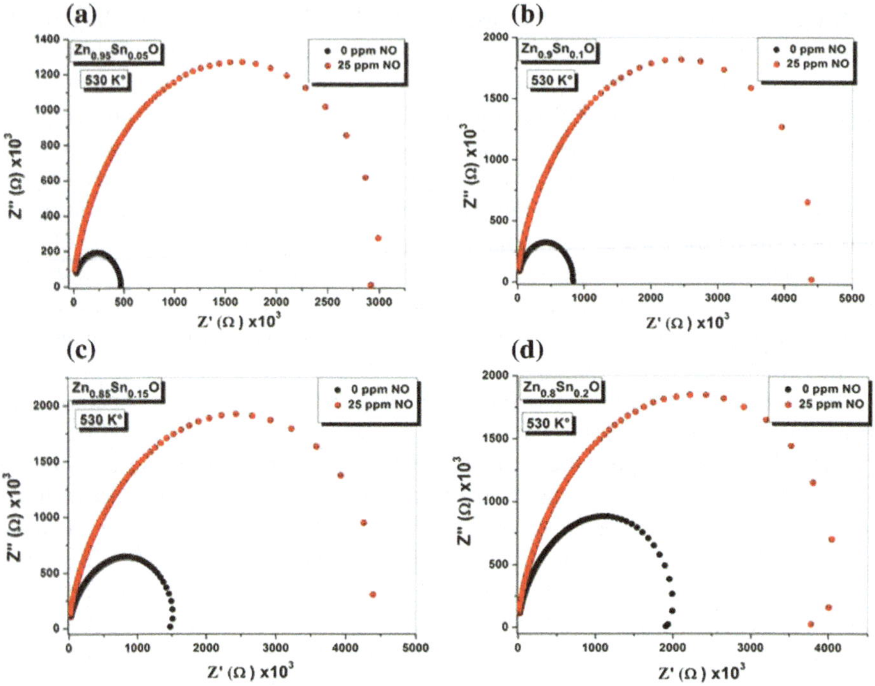

Fig. 11 Nyquist plots of **a** $Zn_{0.95}Sn_{0.05}O$, **b** $Zn_{0.90}Sn_{0.10}O$, **c** $Zn_{0.85}Sn_{0.15}O$ and **d** $Zn_{0.80}Sn_{0.20}O$ thin films at 530 K with and without 25 ppm NO gas

removed. The reaction of the holes with the oxygen can be shown by the following equations [80, 81];

$$h^{+}_{(h\nu)} + O^{2-}_{(ads)} \rightarrow O_{2(g)} \tag{6}$$

$$h^{+}_{(h\nu)} + O^{-}_{(ads)} \rightarrow \frac{1}{2}O_{2(h\nu)} \tag{7}$$

When the samples are stimulated by UV irradiation, the absorbed oxygen molecules on the surface of ZnO materials are ionized forming O^{2-} (hν) by capturing electrons from the ZnO conduction band. Oxygen adsorption and desorption reach an equilibrium state, consequently, the depletion layer is formed and caused a higher resistance.

When ZnO samples are exposed to an oxidizing gas such as NO, gas molecules will capture the electrons on the sensing surface, as shown in the following reaction [73, 79];

$$2NO + 4e^{-}_{(h\nu)} \rightarrow N_2 + 2O^{2-}_{(h)} \tag{8}$$

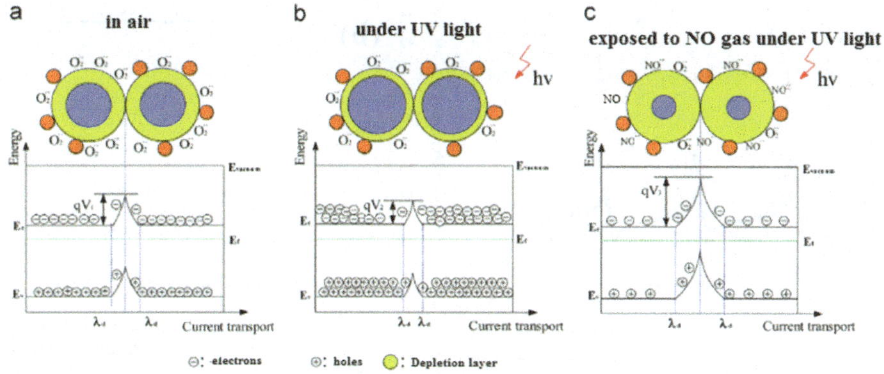

Fig. 12 Schematic diagram of effect of UV light irradiation exposed to NO gas

The NO reacts with the surface-adsorbed active oxygen species and is oxidized in the sensing reaction (the surface is reduced) [80]. Thus, the concentration of holes will increase, increasing the depletion layer widths in the adjacent grains and decreasing the conducting channel widths as a result of this resistance increase. NO can capture more electrons, which is attributed to the conducting channel width, and as a result, UV irradiation can improve the NO gas-sensing performance at room temperature [81]. Figure 12 shows the schematic diagram of effect of UV light irradiation exposed to NO gas.

Due to the strong influence of the structural formation and surface states of sensing layer materials, the governing mechanisms and sensing reactions could be subjected to alter as a result of a change in film properties such as in the presence of moisture or by the addition of any dopant to the structure [82]. For instance, it has been suggested that the possible mechanism in metal-doped semiconductor layers is totally different from that of pristine samples [83]. For metal-doped semiconductors, the mechanism can be explained based on the charge transfer transition of the metal ion by UV illumination which leads to creating additional holes that react with chemisorbed oxygen ions and desorb oxygen [83].

Exposing the sensors to UV irradiation led to a strong increase in the response at temperatures as low as room temperature. For this purpose, the 365-nm light was used as a UV source in all experiments. The relative humidity was kept constant for all measurements (25%). At low humidity, the conduction process is dominated mainly by electrical conduction whereas, at high humidity, proton conduction will play a role in enhancing the total conductivity and is dominated by the decomposition and the polarization of the absorbed water. In addition, the adsorbed water captures the electrons and the holes generated by UV irradiation at higher humidity [83, 84]. The sensors' responses to a constant 20-ppb NO concentration were investigated under UV irradiation, as shown in Fig. 13 (a). $Zn_{0.90}Sn_{0.10}O$ exhibited the highest response to 20-ppb NO gas. The response was calculated as 0.9% without UV light, whereas it was calculated as 43% under UV light for $Zn_{0.90}Sn_{0.10}O$ thin film. Figure 13b shows the responses as a function of doping concentration. The responses change with

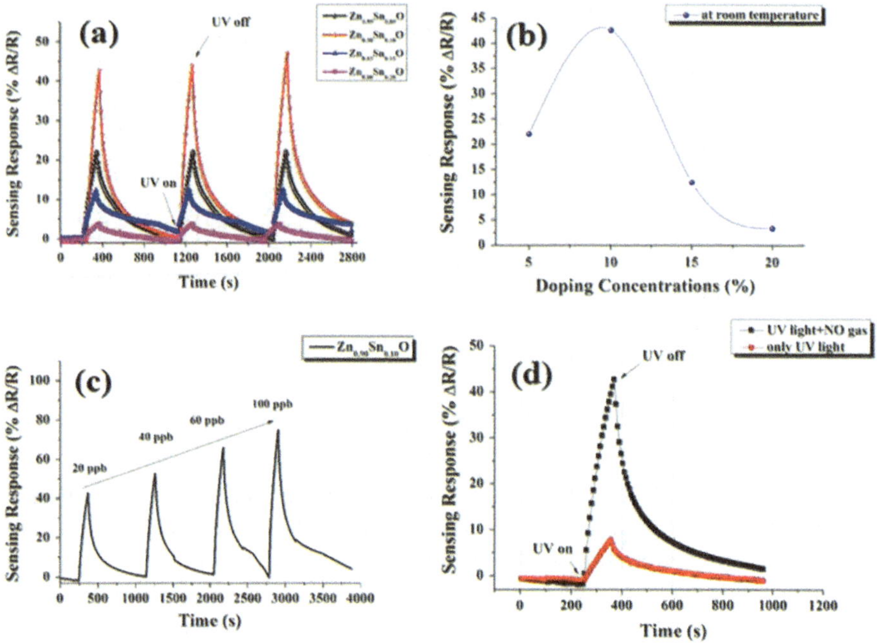

Fig. 13 Responses of all the $Zn_{1-x}Sn_xO$ sensors for 20 ppb NO gas at room temperature for three cycles (**a**) for different doping concentration (**b**), the dynamic gas-sensing measurements of $Zn_{0.90}Sn_{0.10}O$ sensor from 20 to 100 ppb (**c**) and the response with and without 20 ppb NO gas for $Zn_{0.90}Sn_{0.10}O$ sensor (**d**) under UV light irradiation [64]

doping concentration. Moreover, it is found that the response firstly increases when the content of Sn is 10% and afterward declines when the content of Sn reaches 15%. Figure 13c depicts the dynamic gas-sensing measurements for the $Zn_{0.90}Sn_{0.10}O$ thin film. The response increased with increasing gas concentrations. The response was calculated almost 80% for 100 ppb NO gas. Figure 13d shows the response of 20 ppb NO gas with and without UV light for $Zn_{0.90}Sn_{0.10}O$ thin film. The response was increased from 13 to 43% under UV light irradiation. It can be clearly seen that NO gas was affected the sensing surface and increased the gas response. Also, measurements cycles can depend on the sample properties. There is no need for long time measurement cycles in samples that can quickly detect the gas sent to the system and become stable. These long measurement cycles can lead to high power consumption [85, 86].

Figure 14 shows the stability of $Zn_{0.90}Sn_{0.10}O$ thin film with UV light irradiation. The thin films exhibited excellent stability to UV light for 5 cycles. When UV light turns off, the photoresponse returns back to initial value. The response time is much faster than the recovery time. The response time was calculated as 15 s, whereas the recovery time was calculated as 65 s for $Zn_{0.90}Sn_{0.10}O$ thin film.

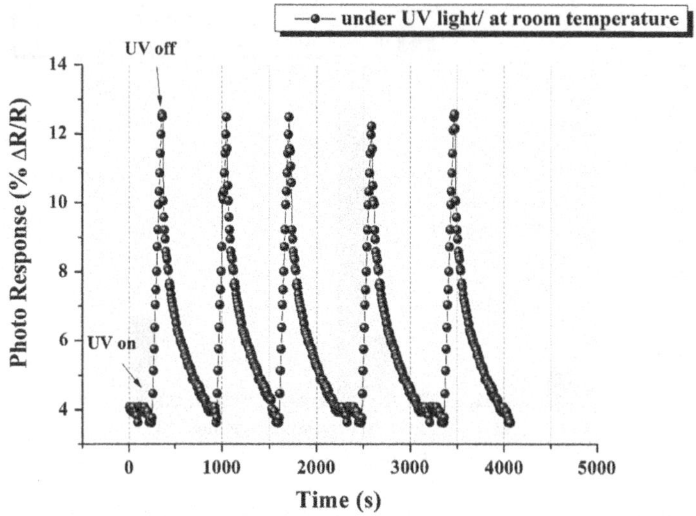

Fig. 14 Stability of $Zn_{0.90}Sn_{0.10}O$ thin film with UV light irradiation [64]

In this process, chemisorbed oxygen ions are photo desorbed from the surface, increasing the number of surface adsorption sites, and are replaced with photo-activated oxygen ions $O^{-2}(hv)$, which has much weaker attachment to the surface. Anionic oxygen species adsorb and desorb on the sensing surface related to UV light irradiation, and they will fill/empty pre-existing oxygen vacancies with electrons, respectively [87]. The oxygen vacancies are metastable and conducting and act as shallow donors. Moreover, the grains act as a reservoir of oxygen atoms that slowly diffuse to the reduced, surface-near regions to occupy the vacancies created by photoreduction. This leads to a gradual change in response after the UV light irradiation is turned off due to a slow oxygen-diffusion-based equilibration process [88].

Therefore, such significant reduction in power consumption could lie in that the UV light only provides the necessary activation energy required for the oxidation reactions at the surface. In contrast, for the thermal-activated response, much more redundant energy has to be consumed to keep the bulk of the oxide sensing materials at the high operating temperatures besides the surface regions [89, 90].

Figure 15 shows the selectivity of $Zn_{0.90}Sn_{0.10}O$ thin film with and without UV light irradiation. The responses increased with UV light irradiation. Especially, the response of NO was increased from 9.88 to 42.65. And, therefore, an increase was observed for CO and NH3 gas, too. But this increase was not noticeable.

The obtained gas-sensing results were encountered in the literature and tabulated in Table 2. It is confirmed that our sample has an acceptable response to very low NO gas concentration.

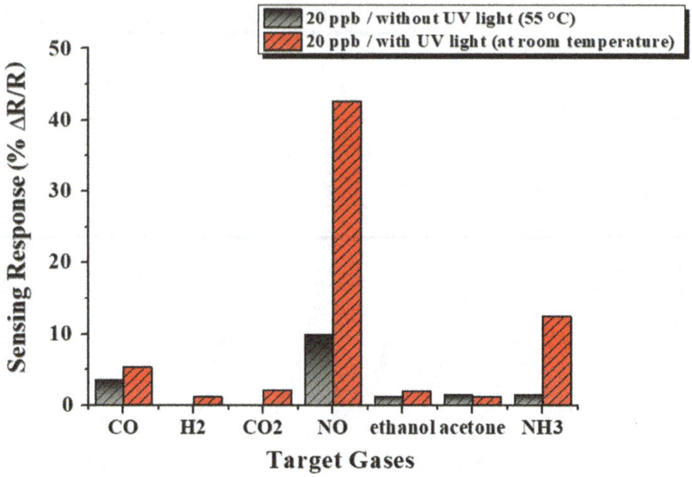

Fig. 15 Selectivity of $Zn_{0.90}Sn_{0.10}O$ thin film with and without UV light irradiation

7 Conclusions

In summary, we fabricated $Zn_{1-x}Sn_xO$ thin films by SILAR method and annealed at 300 °C, annealed thin films exhibited a good response to very low NO gas. The films showed an excellent response to low concentrations (20 ppb) of NO gas at room temperature with 365-nm illumination, while no response was observed without UV light. $Zn_{0.90}Sn_{0.10}O$ thin film exhibited the best gas-sensing properties compared to other samples. It is clear that the high responses obtained at room temperature were due to the surface reactions activated by UV irradiation. $Zn_{0.90}Sn_{0.10}O$ can thus be used to detect low concentrations of NO gas at room temperature.

Table 2 Gas-sensing measurements of different samples with and without UV light irradiation

Materials	Producing method	Gas concentration	Sensing response (without UV light)	Sensing response (with UV light)	Operating temperature	References
Ag-doped ZnO nanoflowers	Hydrothermal method	21.6 ppb NO	73.91	89.04	150 °C	[91]
In_2O_3 nanostructures	Arc-discharging a source	50 ppm NO	4.3	41.7	Room temperature	[92]
n-MoS2/p-GaN hybrid-structure	Layer-transfer technique	50 ppm NO	20	64.67	Room temperature	[93]
ZnSnO3 nanocube	Co-precipitation method	500 ppm ethanol	51	147	240 °C (without UV light) 220 °C (with UV light)	[94]
Au nanoparticles on ZnO nano-structured thin films	RF magnetron sputtering	1000 ppm H_2	80	172	250 °C	[95]
Na-doped ZnO nanowires	Hydrothermal method	5–60 ppm ethanol	–	29–40	Room temperature	[96]
g-C_3N_4/ZnO composites (8%)	İn situ precipitation method	40 ppm ethanol	–	121	Room temperature	[97]
ZnO/CdO n-n isotype heterostructure	Spray deposited	100 ppm acetone	20	540	Room temperature	[98]
$ZnSn_xO$ nanostructures	SILAR method	20 ppb NO	13	43	Room temperature	Our work

Acknowledgements This work was supported by TUBITAK with Project No: 115M658 and Gazi University Scientific Research Fund Project No: 05/2016-21.

References

1. H. Tian, H. Fan, J. Ma, Z. Liu, L. Ma, S. Lei, J. Fang, C. Long, J. Hazard. Mater. **341**, 102 (2018)
2. Y. Yin, Y. Shen, P. Zhou, R. Lu, A. Li, S. Zhao, W. Liu, D. Wei, K. Wei, Appl. Surf. Sci. **509**, 145335 (2020)
3. C. Dong, R. Zhao, L. Yao, Y. Ran, X. Zhang, Y. Wang, J. Alloys Compd. **820**, 153194 (2020)
4. R. Gao, X. Cheng, S. Gao, X. Zhang, Y. Xu, H. Zhao, L. Huo, Appl. Surf. Sci. **485**, 266 (2019)
5. I. Karaduman Er, T. Nurtayeva, M. Sbeta, A. O. Cagirtekin, S. Acar, A. Yildiz, J. Mater. Sci. Mater. El **30**, 10560 (2019)
6. E. Wongrat , N. Chanlek , C. Chueaiarroma, W. Thupthimchun , B. Samransuksamer , S. Choopun, Ceram. Int. **43**, S557 (2017)
7. S. Kanaparthi, S.G. Singh, Mater. Sci. Energy Technol. **3**, 91 (2020)
8.] M. Imai, M. Watanabe , A. Mochihara , H. Tominaga , K. Yoshino , Q. Shen, T. Toyoda, S. Hayase, J. Crystal Growth **468**, 473 (2017)
9. R. Peña-Garcia, Y. Guerra, R. Milani, D.M. Oliveira, A.R. Rodrigues, E. Padrón-Hernández, J. Magn. Magn. Mater. **498**, 166085 (2020)
10. J. Xu, Y. Yu, X. He, J. Sun, F. Liu, G. Lu, Mater. Lett. **81**, 145 (2012)
11. Z. Fu, Z. Pan, D. Sun, G. Zhan, H. Zhang, X. Zeng, G. Hu, C. Xiao, Z. Wei, Mater. Lett. **184**, 185 (2016)
12. M.-C. Huang, T. Hai Wang, B.-J. Wu, J.-C. Lin, C.-C. Wu, Appl. Surf. Sci. **360**, 442 (2016)
13. E. Vaghri, Z. Khalaj, D. Dorranian, Diam. Relat. Mater. **103**, 107697 (2020)
14. W. Guo, X. Li, H. Qin, Z. Wang, Physica E **73**, 163–168 (2015)
15. M. Jiao, N.V. Chien, N.V. Duy, N.D. Hoa, N.V. Hieu, K. Hjort, H. Nguyen, Mater. Lett. **169**, 231–235 (2016)
16. Y. Yang, X. Wang, G. Yi, H. Li, Z. Zhang, Matter. Lett. **2541**, 242–245 (2019)
17. S. Agarwal, P. Rai, E. Navarrete Gatell , E. Llobet, F. Güelle , M. Kumara , K. Awasthi, Sens. Actuators B, 2921, 24–31, (2019)
18. Y. Yang, X. Wang, G. Yi, H. Li, C. Shi, G. Sun, Z. Zhang, Nanomaterials **9**, 1599 (2019)
19. B. Soltabayev, M.A. Yıldırım, A. Ateş, S. Acar, Mater. Sci. Semıcon. Proc. **101**, 28 (2019)
20. T. Corlu, I. Karaduman, M.A. Yildirim, A. Ates, S. Acar, J. Electron. Mater **46**, 7 (2017)
21. M. A. Yıldırım, S. T Yıldırım, I. Cavanmirza, A. Ates¸ Chem. Phys. Lett. 647, 73 (2016).
22. X.D. Gao, X.M. Li, W.D. Yu, Appl. Surf. Sci. **229**, 275 (2004)
23. C. Vargas-Hernandez, F.N. Jimenez-Garcıa, J.F. Jurado, V. Henao Granada, Microelectron. J. **39**, 1349 (2008).
24. X.-D. Gao, X.-M. Li, W.-D. Yu, L. Li, F. Peng, C.-Y. Zhang, J. Cryst. Growth **291**, 175 (2006)
25. P. Suresh Kumar , A. Dhayal Raj, D. Mangalaraj, D. Nataraj, Thin Solid Films **518**, e183 (2010)
26. K. Ravichandran, P.V. Rajkumar, B. Sakthivel, K. Swaminathan, L. Chinnappa, Ceram. Inter. **40**, 12375 (2014)
27. P. Zhang, G. Pan, B. Zhang, J. Zhen, Y. Sun, Mater. Res. **17**(4), 817 (2014)
28. S. Luo, Y. Shen, Z. Wu, M. Cao, F. Gu, L. Wang, Mater. Sci. Semıcon. Proc. **41**, 535 (2016)
29. S.Y. Bae, C.W. Na, J.H. Kang, J. Park, J. Phys. Chem. B **109**, 2526 (2005)
30. S.K. Gupta, A. Joshı, M. Kaur, J. Chem. Sci. **122**(1), 57 (2010)
31. S.S. Badadhe, I.S. Mulla, Sens. Actuat. B **143**, 164 (2009)
32. P.P. Sahay, R.K. Nath, Sens. Actuat. B **133**, 222 (2008)
33. Y.-H. Choi, S.-H. Hong, Sens. Actuat. B **125**, 504 (2007)
34. H. Gong, J.Q. Hu, J.H. Wang, C.H. Ong, F.R. Zhu, Sens. Actuat. B **115**, 247 (2006)

35. S.C. Navale, V. Ravi, I.S. Mulla, S.W. Gosavi, S.K. Kulkarni, Sens. Actuat. B **126**, 382 (2007)
36. Y. Al-Hadeethi, A. Umar, S.H. Al-Heniti, R. Kumar, S.H. Kimc, X. Zhang, B.M. Raffah, Ceram. Int. **43**, 2418 (2017)
37. S. Luo, Y. Shen, Z. Wu, M. Cao, F. Gu, L. Wang, Mater. Sci. Semicon. Proc. **41**, 535 (2016)
38. X. Li, Y. Chang, Y. Long, Mater. Sci. Eng. C **32**, 817 (2012)
39. D. Han, Ceram. Int. **46**, 3304 (2020)
40. C. Balamurugan, D.-W. Lee, A. Subramania, Appl. Surf. Sci. **283**, 58 (2013)
41. D. Han, M. Zhao, J. Alloys Compd. **815**, 152406 (2020)
42. H. Gao, D. Wei, P. Lin, C. Liu, P. Sun, K. Shimanoe, N. Yamazoe, G. Lu, Sens. Actuat. B **253**, 1152 (2017)
43. S. Bhatia, N. Verma, R.K. Bedi, Appl. Surf. Sci. **407**, 495 (2017)
44. N. Zubaır, K. Akhtar, Trans. Nonferrous Met. Soc. China 29143 (2019)
45. S.C. Navale, I.S. , Mulla. Mater. Sci. Eng. C **29**, 1317 (2009)
46. P.S. Shewale, Y.S. Yu, J. Alloys Compd. **684**, 428 (2016)
47. L. Ma, S.Y. Ma, H. Kang, X.F. Shen, T.T. Wang, X.H. Jiang, Q. Chen, Mater. Lett. **209**, 188 (2017)
48. S. Ozturk, N. Kilinc, Z.Z. Ozturk, Procedia Eng. **47**, 434–437 (2012)
49. J.-J. Zhang, E.-J. Guo, L.-P. Wang, H.-Y. Yue, G.-J. Cao, L. Song, Trans. Nonferrous Meter. Soc. China **24**, 736 (2014)
50. A. Arfaoui, B. Ouni, S. Touihri, A. Mhamdi, A. Labidi, T. Manoubi, Opt. Mater **45**, 109 (2015)
51. M. Jlassi, I. Sta , M. Hajji, B. Ben Haoua, H. Ezzaouia, Mater. Sci. Semıcon. Proc. **26**, 395 (2014)
52. S. Sadr, M.E. Azim-Eroğlu, M. Rahimi, R.S. Dariani, Indian J. Pura Appl. Phys. **51**, 860 (2013)
53. C. Liu, L. Zhao, B. Wang, P. Sun, Q. Wang, Y. Ga, X. Liang, T. Zhang, G. Lu, J. Colloid Interface Sci. **495**, 207 (2017)
54. M.A. Yıldırım, S. Tuna Yıldırım, A. Ateş, J. Alloys Compd. **701**, 37 (2017).
55. T. Çorlu, I. Karaduman, M.A. Yıldırım, A. Ateş, S. Acar, High Temper.-High Pressures **46**, 155 (2016)
56. I. Karaduman, T. Corlu, M.A. Yıldırım, A. Ates¸ S. Acar, J. Electron. Mater. **46**(7) (2017)
57. M. Ali Yıldırım, Y. Akaltun, A. Ateş, Solid State Sci. **14**, 1282 (2012)
58. Y. Caglar, D.D. Oral, M. Caglar, S. Ilican, M.A. Thomas, K. Wu, Z. Sun, J. Cui, Thin Solid Films **520**, 6642 (2012)
59. T.V.K. Karthik, M. de la Luz Olvera, A. Maldonado, H. Gómez Pozos, Sensors **16**, 1283 (2016).
60. Z. Zhang, J.B. Yi, J. Ding, L.M. Wong, H.L. Seng, S.J. Wang, J.G. Tao, G.P. Li, G.Z. Xing, T.C. Sum, C.H. Alfred Huan, T. Wu, J. Phys. Chem. C **112**, 9579 (2008)
61. Q. Ma, D.B. Buchholz, R.P.H. Chang, Phys. Rev. B **78**, 214429 (2008)
62. R. Yoo, S. Cho, M.J. Song, W. Lee, Sens. Actuat. B **221**, 217 (2015)
63. J. Panda, I. Sasmal, T.K. Nath, AIP Adv. **6**, 035118 (2016)
64. I. Karaduman Er, A. O. Çagırtekın, Tugba Çorlu, M. A.Yıldırım, A. Ateş, S. Acar, **Bull. Mater. Sci. 42**(32) (2019)
65. A.A. Ziabari, F.E. Ghodsi, Thin Solid Films **520**, 1228 (2011)
66. N.G. Deshpande, J.C. Vyas, R. Sharma, Thin Solid Films **516**, 8587 (2008)
67. D. Miao, Q. Zhao, S. Wu, Z. Wang, X. Zhang, X. Zhao, J. Non-Crystall, Solids **356**, 2557 (2010)
68. S. Ilican, Y. Caglar, M. Caglar, M. Kundakci, A. Ates, Int. J. Hydrog. Energy **34**, 5201 (2009)
69. M. Ajili, M. Castagné, N. Kamoun Turki, Superlattices Microstruct. **53**, 213 (2013)
70. A. Sharma, M. Tomar, V. Gupta, Sens. Actuat. B **156**, 743 (2001)
71. R.K. Sonker, S.R. Sabhajeet, S. Singh, B.C. Yadav, Mater. Lett. **152**, 189 (2015)
72. M. Reddeppa, T.K. Phung Nguyen, B.-G. Park, S.-G. Kim, M.-D. Kim, Physica E **116** 113725 (2020)
73. S. Zhang, M. Yang, K. Liang, A. Turak, B. Zhang, D. Meng, C. Wang, F. Qu, W. Cheng, M. Yang, Sens. Actuat. B **290**, 59 (2019)
74. N. Barsan, U. Weımar, J. Electroceram. **7**, 143–167 (2001)
75. G. Korotcenkov, Mater. Sci. Eng. B **139**, 1–23 (2007)

76. L. Yu, F. Guo, S. Liu, B. Yang, Y. Jiang, L. Qi, X. Fan, J. Alloys Compd. **682**, 352 (2016)
77. K.-M. Sun, X.-Z. Song, X.-F. Wang, X. Li, Z. Tan, Mater. Chem. Phys. **241**, 122379 (2020)
78. I. Karaduman , A. Alaylı Güngör, H. Nadaroğlu, A. Altundaş, S. Acar, J. Mater. Sci: Mater. Electron. **28**, 16094 (2017)
79. Q. Geng, Z. He, X. Chen, W. Dai, X. Wang, Sens. Actuat. B **188**, 293 (2013)
80. N.M. Vuong, D. Kim, H. Jung, H. Kim, S.K. Hong, J. Mater. Chem. **22**, 6716 (2012)
81. S. Nasirian, Appl. Surf. Sci. **502**, 144302 (2020)
82. H. Ahn, Y. Wang, S. Hyun Jee, M. Park, Y. Soo Yoon, D.-J. Kim, Chem. Phys. Lett. **511**, 331–335 (2011)
83. I. Karaduman, O. Barin, D.E. Yıldız, S. Acar, J. Appl. Phys. **118**, 174501 (2015)
84. W. Xie, B. Liu, S. Xiao, H. Li, Y. Wang, D. Cai, D. Wang, L. Wang, Y. Liu, Q. Li, T.W. Pen, Sens. Actuat. B **215**, 125 (2015)
85. R. Gao, Z. Ying, W. Sheng, P. Zhen, Mater. Lett. **229**, 210–212 (2018)
86. H. Wang, L. Zhou, Y. Liu, F. Liu, X. Liang, F. Liu, Y. Gao, X. Yan, G. Lu, Sens. Actuat. B **305**, 127498 (2020)
87. X. Wang, M. Xia, H. Li, X. Zhu, B. Leng, T. Tao, W. Xu, J. Xu, Appl. Surf. Sci. **493**, 862 (2019)
88. R. Gao, Z. Ying, W. Sheng, P. Zheng, Mater. Lett. **229**, 210 (2018)
89. T. Saidi, D. Palmowski, S. Babicz-Kiewlicz, T. Geremariam Welearegay, N. El Barib, R. Ionescu, J. Smulko, B. Bouchikhi, Sens. Actuat. B, **273**, 1719 (2018)
90. Y. Chen, X. Li, X. Li, J. Wang, Z. Tang, Sens. Actuat. B **232**, 158 (2016)
91. Y.-T. Tsai, S.-J. Chang, L.-W. Ji, Y.-J. Hsiao, I.-T. Tang, H.-Y. Lu, Y.-L. Chu, ACS Omega **3**, 13798–13807 (2018)
92. N.D. Chinh, N.D. Quang, H. Lee, T.T. Hien, N.M. Hieu, D. Kim, C. Kim, D. Kim, Sci. Rep. **6**, 35066 (2016)
93. M. Reddeppa, B.-G. Park, G. Murali, S.H. Choi, N. Duc Chinh, D. Kim, W. Yang, M.-D. Kim, Sens. Act. B **308**, 127700 (2020)
94. X. Wang, X. Zhu, T. Tao, B. Leng, W. Xu, L. Mao, J. Alloys Compd. **829**, 154445 (2020)
95. G. Kumar, X. Li, Y. Du, Y. Geng, X. Hong, J. Alloys Compd. **798**, 467–477 (2019)
96. C.-L. Hsu, B.-Y. Jhang, C. Kao, T.-J. Hsueh, Sens. Actu. B **274**, 565–574 (2018)
97. J. Zhai, T. Wang, C. Wang, D. Liu, Appl. Surf. Sci. **441**, 317–323 (2018)
98. P. Srinivasan, D. Prakalya, B.G. Jeyaprakash, J. Alloys Compd. **819**, 152985 (2020)

An Ab-initio Study of Structural, Elastic, Electronic, Vibrational and Optical Properties of Semiconductor NaAlSi Compound for Optoelectronic Applications

Yasemin O. Ciftci

Abstract Semiconducting ternary half-Heusler (HH) compounds with narrow band gaps have turned out to be an important class of materials in recent years due to their wide range of interesting physical properties (e.g., high melting point, large thermal conductivity, large thermopower and unusual magnetic properties, etc.). Compounds formed in the half-Heusler structure have proven to be an important class of materials in recent years. Due to their potential applications in cooling and power generation, high-performance thermoelectric materials have attracted a lot of attention from the science community. They crystallize in MgAgAs structure with the $F43$m space group. This structure consists of four interpenetrating face-centered cubic (FCC) sublattices where three of them contain one of the three elements each, and the fourth sublattice is vacant. In this study, we focus on structural, electronic, elastic, lattice dynamic and optical properties of NaAlSi using *ab-initio* density-functional theory (DFT) within the generalized gradient approximation (GGA) for the exchange-correlation potential. The calculated lattice constants were in good agreement with the available data. Our band structure calculations show that NaAlSi has an indirect energy band gap of 0.39 eV at Γ and X Brillouin zone. In addition to the electronic structure, elastic and optical properties, phonon dispersion curves and phonon density of states (DOS) were calculated. The elastic constants C_{ij} are calculated using the total energy variation with strain technique. Calculated elastic results show that this compound is mechanically stable. The obtained phonon spectra for NaAlSi do not exhibit any significant imaginary branches of the phonon spectrum showing to be dynamically stable. Further analysis of the optical response of the dielectric functions, optical reflectivity, refractive index, extinction coefficient and electron energy loss delves into for the energy range of 0–20 eV. Our computed structural results are in reasonable agreement with the literature.

Keywords Elastic properties · Optical properties · Phonon dispersion curves

Y. O. Ciftci (✉)
Department of Physics, Gazi University, Teknikokullar, 06500 Ankara, Turkey
e-mail: yasemin@gazi.edu.tr

S. J. Ikhmayies et al. (eds.), *Advances in Optoelectronic Materials*,
Advances in Material Research and Technology,
https://doi.org/10.1007/978-3-030-57737-7_5

1 Introduction

Owing to the serious energy crisis and environmental pollution, the demand for sustainable and clean energy sources is becoming more and more important. Thermoelectric materials can directly convert waste heat into electricity, which has attracted intensive attention from the science and industry communities. Thermoelectric materials transform waste heat into electric energy based on thermoelectric effects with low noise, high stability, and reliability because there are no moving parts. The widespread applications of thermoelectric devices are limited now mainly due to their lower transforming efficiency [1].

Half-Heusler compounds as potential candidates for high-performance thermoelectric materials have attracted much attention from the scientific community that is committed to the development of sustainable and renewable energy [2–4]. Ternary half-Heusler compounds with a narrow band gap at the Fermi level are promising for different applications such as optoelectronics, thermoelectrics, and spintronics [5–14]. These alloys have composition XYZ, where X and Y denote a transition or rare Earth elements and Z denotes a main group element. They crystallize in MgAgAs structure.

The development and optimization of optoelectronic devices depend crucially on the availability of suitable semiconductor materials. During the last years, there has been a clear trend toward multinary compounds, providing a wider range of electrical, optical and chemical properties. Obviously, by going from silicon and binary semiconductors such as GaAs to ternary or higher multinary materials, the number of possible compounds grows strongly, providing a great variety of material properties together with a growing difficulty to choose the right one for the respective applications. The search for new semiconductor materials with structural and electronic properties for the advancement of optoelectronic applications, such as thin-film solar cells or laser diodes, constitutes one of the central challenges in materials science. II–VI and III–V compound semiconductors have long been viewed as promising materials for the device applications in electronic and optoelectronic technologies. Many of these binary semiconductors crystallize in a cubic zinc-blende structure. The limited applications of some binary semiconductors in various fields have prompted scientists to look for alternative axes of promising materials. A particularly interesting class of materials belongs to half-Heusler compounds or "Nowotny–Juza,"[15] with a chemical composition XYZ. For example, in III–V compounds, the crystal structure of Nowotny–Juza compounds can be derived by transmuting the group III atom into an isovalent pair I + II and are found to crystallize mostly in cubic zinc-blende-type-related structures [15–17]. So, if the elements X, Y and Z have a total number of eight valence electrons, they form a particularly stable ground-state structure, and the most electropositive element X donates n valence electrons to the more electronegative elements Y and Z. Therefore, the class of eight-electron half-Heusler compounds includes a large number of semiconductors, whose band gaps vary in a wide range [18]. Similar stable eight-electron states cause the band gaps in binary semiconductors of the III–V or II–VI types. Up to now, only a small number of the huge class of possible eight-electron half-Heusler compounds of I–II–V and I–III–IV types have been synthesized [15, 19–36] and have been also investigated with ab-initio calculations [18, 37–47]. The similarity of physical properties and information obtained for both of the half-Heusler compounds and of the binary semiconductors helped us to find new half-Heusler semiconductors favorable for optoelectronic applications.

In this study, we have researched on a structure, electronic (such as electronic band structure, partial and total electronic density of states and charge density), elastic (such as elastic constants (C_{ij}), bulk modulus (B), shear modulus (G), Young's modulus (E), Cauchy pressure, Poisson's ratio (σ), elastic anisotropy (A), brittle/ductile characteristic (B/G) and hardness (H_V)) vibrational and optical properties (such as the real and imaginary parts of the complex dielectric constant, refractive index, extinction coefficient, energy loss function and reflectivity) of NaAlSi compound which belong to half-Heusler family (space group $F\bar{4}3m$; No: 216) and compared our calculated results with experimental studies. We have also investigated the pressure effect of above-considered electronic and elastic properties. The method of calculation is given in Sect. 2; the results are discussed in Sect. 3. Finally, the summary and main conclusions are presented in Sect. 4.

2 Method of Calculation

All the calculations were conducted using the Vienna ab-initio simulation package (VASP) [48–50] based on the density-functional theory (DFT) and the electron–ion interaction was considered in the form of the projector augmented wave (PAW) method [50–52]. The electron–electron interaction was also being considered within the generalized gradient approximation (GGA) [53, 54]. The exchange-correlation potential was considered by the generalized gradient approximation (GGA) in the scheme of Perdew–Burke–Eruzerhof (PBE) [31] using Vanderbilt ultrasoft pseudopotential. The valence-electron configurations for the Na, Al and Si were generated as s^1p^0, s^2p^1 and s^2p^2, respectively. The 15 × 15 × 15 Monkhorst and Pack [55] grid of k-points was used for integration in the irreducible Brillouin zone. The plane-wave basis set cutoff was set as 700 eV for all calculations. This cutoff (700 eV) was found to be adequate for the structural, mechanical properties as well as for the optical ones. Convergence test demonstrated that the values of the k-point and the cutoff energy were selected in the range that guarantees the convergence of the total energy. Atomic coordinates and axial ratios have been relaxed for different volumes of the unit cell. With the above parameters, convergence minimum with respect to atomic relaxations is assumed to have been attained when the energy difference between two successive iterations is less than 10^{-10} eV per unit cell and the forces acting on the atoms are less than 0.0001 eV/Å.

3 Results and Discussion

3.1 Structural and Electronic Properties

As mentioned previously, NaAlSi crystallizes in the cubic MgAgAs-type structures with space group$F\overline{4}3m$ and four formula units, where Na atoms occupy the sites (0.5, 0.5, 0.5) and the atomic positions of Al are (0, 0, 0) and Si (0.25, 0.25, 0.25) as depicted in Fig. 1. The optimization of the geometrical structure by the approach of GGA with the PBE function is necessary for the sake of obtaining the ground state.

Within DFT calculations, the structural properties (crystal structure, equilibrium lattice constant, bulk modulus) are a very important first step toward understanding the material properties from a microscopic viewpoint. The calculated ground-state energies are plotted as a function of the lattice constants for NaAlSi. The common procedure for the determination of the structural properties near equilibrium consists in computing the total energy at different values of the lattice parameter and in fitting the results to a semiempirical equation of states. With the calculated equilibrium lattice constant, the Murnaghan equation of state [56] was fitted to the E(V) curves (see Fig. 2) in order to determine B and its pressure derivative $B' = dB/dP$. The obtained results of all investigated compound are given in Table 1 and compared with the available theoretical results. The calculated value of lattice parameter is underestimated as ~1.7% for compared to Ref. [57] for NaAlSi compound. The differences between our values and the experimental data may be due to the use of an approximate DFT. It is well known that the GGA leads to a volume slightly overestimated in relation to the experimental value. The total energy values as a function of volume for NaAlSi computed using GGA-PBE schemes are shown in Fig. 2. This is within the accuracy range of calculations based on density-functional GGA. The bulk modulus (B) at the calculated equilibrium volume is 50.27 GPa, while the first pressure derivative of the bulk modulus (B') is 4.638.

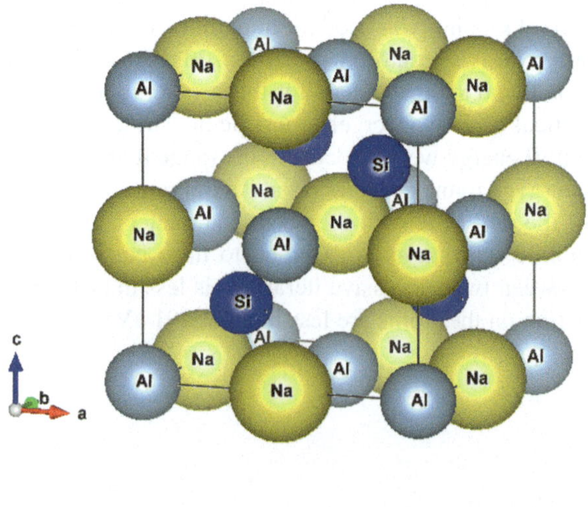

Fig. 1 Unit cell of the NaAlSi

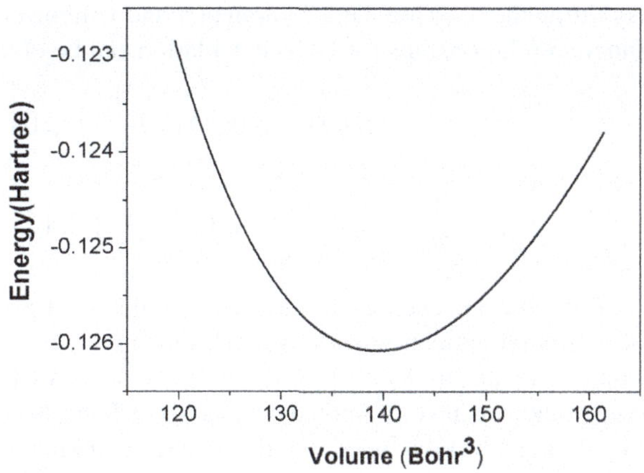

Fig. 2 Energy–volume curve for NaAlSi

Table 1 Calculated equilibrium lattice constants (a_0), bulk modulus (B) and pressure derivative of bulk modulus (B') for NaAlSi in MgAgAs ($F\bar{4}3m$) type structure at 0 GPa and 0 K with available data

Material	Reference	a_0 (Å)	V(Å^3)	B (GPa)	B'
NaAlSi	Present theory [57]	6.283 6.294	62.0081	50.27	4.638

The pressure dependence of the normalized lattice constants a/a_0 and the normalized cell volume V/V_0 of NaAlSi are illustrated in Fig. 3, where a_0 and V_0 are the zero pressure equilibrium structure parameters. It is shown that as the applied pressure increases from 0 to 50 GPa (25, 50 GPa), the ratio a/a_0 decreases quickly (8.6, 23.79%). Presently, there are no experimental data to be compared with our results.

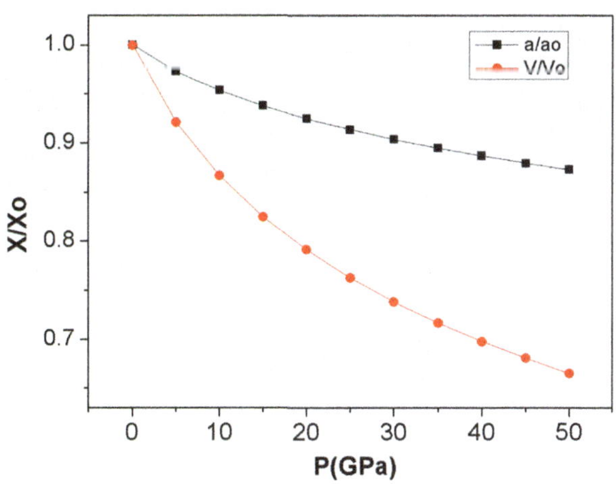

Fig. 3 Variations of lattice parameter and cell volume of NaAlSi with pressure

By fitting the pressure, lattice parameter and volume curves, the following quadratic function relationships for GGA calculation can be obtained as

$$\frac{a}{a_0} = 0.995 - 0.00421 \cdot P + 3.6403 \times 10^{-5} \cdot P^2 \tag{1}$$

$$\frac{V}{V_0} = 0.9847 - 0.01177 \cdot P + 1.113 \times 10^{-4} \cdot P^2. \tag{2}$$

The optical properties are correlated to the band structure and to the probabilities of interband optical transitions. Consequently, it is of interest to evaluate the electronic structure in detail. In order to make sense of the electronic behavior of this compound, we have computed the electronic band structure as shown in Fig. 4. From Fig. 4, it can be concluded that the valence band maximum occurred at the Γ point, while the conduction band minimum located at the *X*-point for NaAlSi compound. This indicates that NaAlSi compound is an indirect band gap semiconductor. It is also consistent with the previous theoretical study [57]; however, our obtained band gap value 0.40 eV is lower than the theoretical one (0.462 eV). We recall that the theoretical values for the band gap energy are smaller when compared with the experimental one, due to the well-known underestimation of conduction band state energies in ab-initio calculations which are performed within DFT.

The total and partial density of states are also plotted in Fig. 5. Clearly, the valence bands locate in two regions. In the low energy region of the valence band is mixed state of Al-*s* and Si-*p* atoms. The next bands below the Fermi level originates from the hybridization of Al-p states and Si–*p* states. In the conduction band, above the Fermi energy level, the major contribution comes from Na, Al and Si s and p states.

For the sake of comprehending the variations of the DOS of NaAlSi compound with increasing external pressure, the total density of states (TDOS) under different pressures is analyzed. As illustrated in Fig. 6, by the comparison of the DOS of NaAlSi at 0, 25 and 50 GPa, it can be found that the shapes of the curve with the

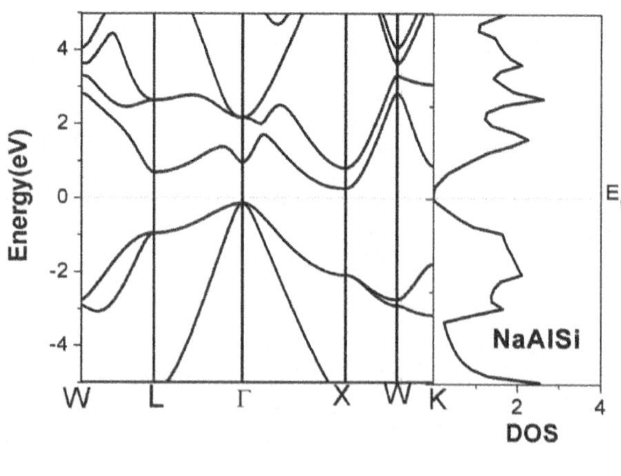

Fig. 4 Electronic band structure of NaAlSi

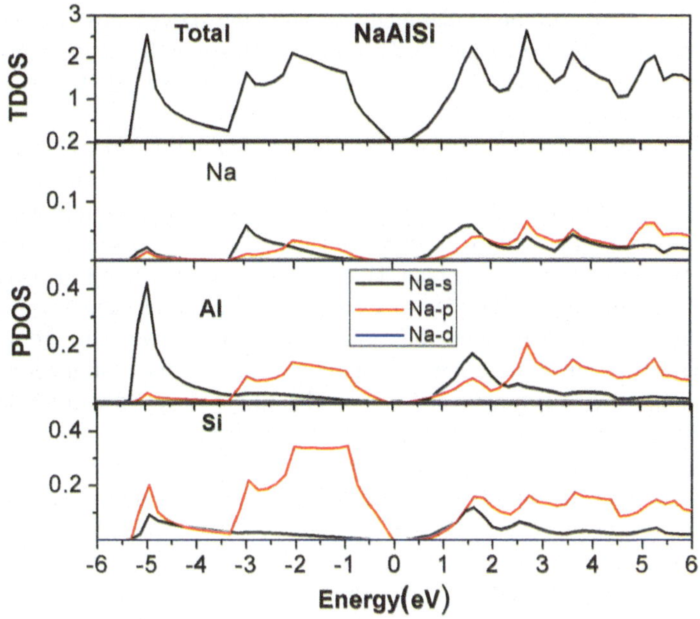

Fig. 5 Total and partial density of states of NaAlSi

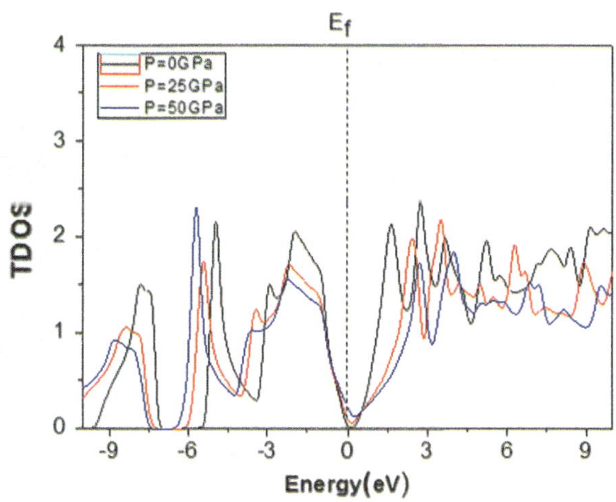

Fig. 6 Total density of states under pressure for NaAlSi

pressure increasing have no obvious change, indicating that the structure of NaAlSi compound remains stable and there is no structural phase transformation that occurs at the considered pressures. The values of TDOS at Fermi level become smaller, and the energy regions near the Fermi level become wider with the increasing pressure. In addition, we also can note that when the external pressure is applied, the TDOS above the Fermi level shifts to the higher energies while the states below the Fermi level are pushed to lower energies as shown in Fig. 6. This may be due to the shrinking between the distance of atoms under compression, which leads to the change of interaction potentials.

For the better understanding of the electronic properties of the half-Heusler compounds, the electron density distributions were also calculated in order to understand the bond nature of the investigated compound. To visualize the nature of the bonding characteristics in NaAlSi compound, the charge density distributions for the MgAgAs-type structure are examined. The contour map of the charge density in (100) plane is shown in Fig. 7. The electronegativity of the crystal constituent elements plays the main role in determining the nature of the bonds between the atomic species where a large difference in the electronegativity causes a charge transfer among the different atoms, thus resulting in an ionic bond nature. Conversely, small electronegativity difference leads to charge sharing and is responsible for the covalent bond nature in the investigated compound. Deformation of charge distribution of the Al and Si atoms indicates the existence of a directional bonding between these two atoms due the hybridization of their upper valence bands. Since there is no obvious electron density accumulation in the regions of Na–Si and due to the different electronegativity between Na and Si, the bonding between the Na and Si atoms has ionic nature. Therefore, the bonding in NaAlSi may be expressed as a combination of covalent–ionic behavior. A similar feature is also observed for the other half-Heusler compounds in Ref. [58].

Fig. 7 Charge density map in (100) direction for NaAlSi

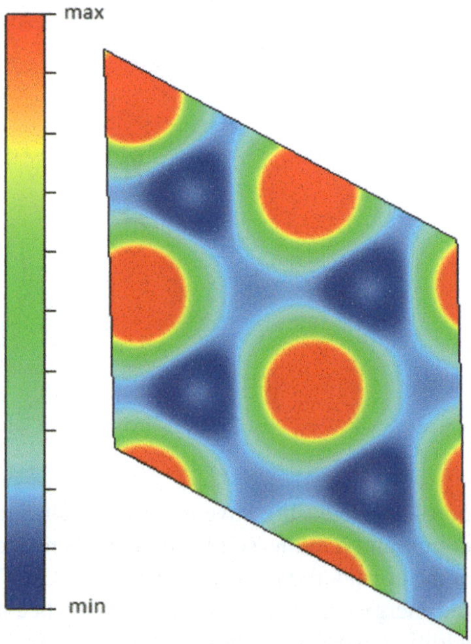

3.2 Elastic Properties

Elastic properties of a solid are important because they relate to various fundamental solid-state properties such as interatomic potentials, the equation of state and phonon spectra. Elastic properties are also linked thermodynamically to the specific heat, thermal expansion, Debye temperature, melting point and Gruneisen parameter. So, it is important to calculate elastic constants of solids.

Elastic constants are one of the important fundamental physical properties of solids whose knowledge is essential to the understanding of many of their properties. In particular, elastic constants determine the elasticity and mechanical stability of crystals. Determination of the elastic constants requires knowledge of the curvature of the energy curves as a function of strain of selected deformation of the unit cell [59]. For a small strain on a solid, Hook's law is applicable and the elastic energy ΔE is a quadratic function of strains:

$$\Delta E = V_0 \sum_{i,j=1}^{6} \frac{1}{2} C_{ij} \delta_i \delta_j \tag{3}$$

where C_{ij} is the elastic constants, V_0 is the equilibrium cell volume, and δ_i is the six components of strain tensor. Symmetry can be used to design an optimal set of deformations under two sometimes competing requirements: (i) use as few deformations as possible and (ii) use deformations that keep the highest possible symmetry on the deformed cell.

The full potential linearized augmented plane waves method allows total energy calculations to be done for arbitrary crystal structures. We can, therefore, apply small strains to the equilibrium lattice, determine the resulting change in the total energy and from this information deduce the elastic modulus. The calculation of the modulus allows us to garner information about non-equilibrium properties from our simple initial system. So the calculated elastic constants can then be used to check the experimental bulk and shear modulus, if available. Since the elastic constants of solids provide a link between the mechanical and dynamical behavior of crystals and give important information concerning the nature of the forces operating in solids, we have calculated the second-order elastic constants (C_{ij}) applying the "stress–strain" method [60] for NaAlSi compound. The obtained values of elastic constants are listed in Table 2 under zero pressure and zero temperature. For further study to explain the effect of pressure on the elastic constants, which are crucial for understanding interatomic interactions, mechanical stabilityand phase transition mechanisms, the pressure dependence of the elastic constants curves is given in Fig. 8 for this compound at zero temperature. The pressure dependence of the elastic constants is a very important characterization of the crystals under varying pressure and/or temperature. Moreover, the change of the elastic constants at the transition point may drastically influence the macroscopic properties, such as the velocity of sound or reflective properties of the crystal [61]. Unfortunately, there are no measurements or theoretical data on these compounds to compare with our calculations.

Table 2 Calculated elastic constants (C_{ij}), Cauchy pressure (C_{12}–C_{44}), bulk modulus (B), Voigt shear modulus (G_v), Reuss shear modulus (G_R), shear modulus (G), ratio of G/B, the Young modulus (E), Poisson's ratio (σ), microhardness (H), anisotropy factors (A) at zero pressure

MgAgAs	NaAlSi
C_{11} (GPa)	112.96
C_{12} (GPa)	20.19
C_{44} (GPa)	28.91
C_{11}–C_{12}	92.77
C_{12}–C_{44}	-8.72
B	51.12
G	34.97
B/G	1.46
$E = 9\,GB/(G + 3B)$	85.42
$\sigma = (3B - 2G)/(6B + 2G)$	0.221
$H = (E - 2\upsilon E)/(6 + 6\upsilon)$	9.68
$A = 2C_{44}/(C_{11}–C_{12})$	0.632

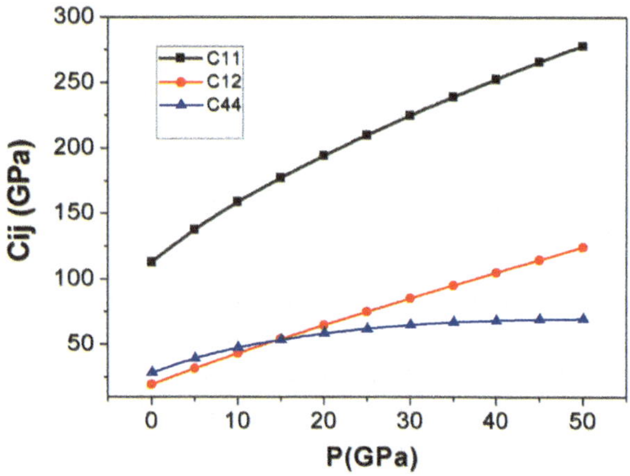

Fig. 8 Dependence of elastic constants on pressure for NaAlSi

The mechanical stability of a cubic crystal lattice was examined using a Born criteria: C_{11}–$C_{12} > 0$, $C_{11} > 0$, $C_{44} > 0$, $C_{11} + 2C_{12} > 0$, and these restrictions are satisfied for all considered compounds in the pressure range 0–50 GPa, implying mechanical stability in this phase. In Fig. 8, it is observed that C_{11} and C_{12} show monotonic increase with increasing pressure compared to C_{44} and the impact of pressure on C_{11} is also more predominant than C_{12} and C_{44}. It can be seen from Fig. 8 that after 15 GPa C_{12} is greater than C_{44} which provides positive Cauchy pressure. From the linear fitting, we can obtain the normal equation of the aggregate elastic moduli for NaAlSi, as follows:

$$C_{11} = 123.986 + 3.225 \times P \tag{4}$$

$$C_{12} = 22.608 + 2.081 \times P \tag{5}$$

$$C_{44} = 38.495 + 0.769 \times P \tag{6}$$

From the above relationships, it is expected to obtain elastic constants under even more high pressures. It is found that C_{11} and C_{12} vary largely when compared with the changes in C_{44} when the pressure increases and that C_{11} is larger than C_{12} under pressure, which shows that the hardness of NaAlSi in the direction of the *x*-axis is larger than that in the direction of the *y*- and *z*-axes.

The elastic properties such as the Zener anisotropy factor (A), Poisson's ratio (σ), Young's modulus (E) are often measured for polycrystalline materials, which are the most interesting elastic properties for applications, were also calculated in terms of the computed constants using the following relationships [62]:

$$A = \frac{2C_{44}}{C_{11} - C_{12}} \tag{7}$$

$$\sigma = \frac{1}{2}\left[\frac{\left(B - \frac{2}{3}G\right)}{\left(B + \frac{1}{3}G\right)}\right] \tag{8}$$

and

$$E = \frac{9GB}{G + 3B} \tag{9}$$

where $G = (G_V + G_R)/2$ is the isotropic shear modulus, G_V is Voigt's shear modulus corresponding to the upper bound of G values and G_R is Reuss's shear modulus corresponding to the lower bound of G values which can be written as $G_V = (C_{11} - C_{12} + 3C_{44})/5$ and $5/G_R = 4/(C_{11} - C_{12}) + 3/C_{44}$. In order to gain more information about noteworthy mechanical properties such as the averaged B, G, E A and σ, we calculated them by inputting the elastic constants into Voigt–Reuss–Hill equation as collected in Table 2. It should be emphasized that the calculated bulk modulus from the Voigt–Reuss–Hill equation is very close to estimation from Murnaghan's equation of state for this compound.

Young's modulus E, the ratio between stress and strain, is required to provide information about the measure of the stiffness of the solids. The bulk modulus is a measure of resistance to volume change by applied pressure, whereas the shear modulus is a measure of resistance to reversible deformations upon the shear stress [63]. Therefore, by using the isotropic shear modulus, the hardness of a material can be determined more accurately than by using the bulk modulus. The calculated bulk modulus (B), the Young modulus (E) and isotropic shear modulus (G) are 51.12 GPa, 34.97 GPa and 85.42 GPa for NaAlSi at 0 GPa, respectively. Pressure dependence of B, G and E up to 50 GPa is given in Fig. 9. It can be seen from Fig. 9 that the calculated values B, G and E increase with increasing pressure up to 50 GPa. In other

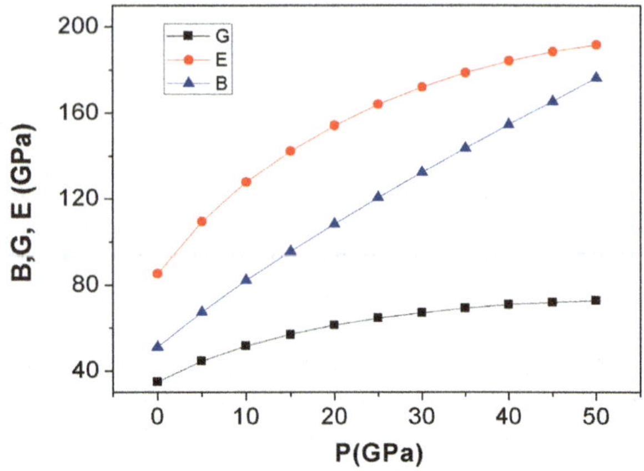

Fig. 9 Pressure dependence of bulk modulus (*B*), shear modulus (*G*) and the Young modulus (*E*) for NaAlSi

words, the increase of external pressure up to 50 GPa is the benefit to the hardness of NaAlSi.

The quotient of bulk to shear modulus (*B*/*G*) of polycrystalline phases is obtained by considering that the shear modulus *G* represents the resistance to plastic deformation, while the bulk modulus *B* represents the resistance to fracture [64]. We predicted the brittle or ductile nature of this compound on the basis of the Pugh criteria [65–67] (*B*/*G* ratio). That is, when *B*/*G* ratio is higher (less) than 1.75, then a material shows ductile (brittle) behavior. It is found that NaAlSi exhibits brittle behavior. Alternatively, we can also discuss the brittle/ductile manner of this compound from Cauchy pressure (C_{12}–C_{44}). The Cauchy pressure may be used to describe the angular character of atomic bonding in metals and compounds [65, 66]. For materials with metallic bonding and a ductile manner, the Cauchy pressure is typically positive, while those with directional covalent bonding and angular character often have a negative value [68]. For the central force to be considered, the deviation from Cauchy relation $\delta = C_{12}–C_{44}$ is a measure of the contribution from the non-central many-body forces, since the Cauchy relation should be satisfied when interatomic potentials are purely central. NaAlSi has a negative Cauchy pressure which means that this compound has the brittle character under 0 GPa. It is noticeable that we reached the same conclusion from Pugh criteria and Cauchy pressure. Remarkably, NaAlSi has positive Cauchy pressure after 15 GPa, turning brittle to ductile manner.

The Zener anisotropy factor *A* is a measure of the degree of elastic anisotropy in solids. The *A* takes the value of 1 for a completely isotropic material. The calculated Zener anisotropy factor is 0.632 for this compound, which indicates that NaAlSi is completely anisotropic material**.** Obviously, it can be seen from Fig. 10 that the value of A increases with the increase of pressure up to 25 GPa and it closes 0.93 value, which reveals that the degree of elastic anisotropy is getting lower. However, the value of A is nearly constant when the pressure is above 25 GPa.

Poisson's ratio σ is a very important property for industrial applications. Poisson's ratio (σ) represents the negative ratio of transverse and longitudinal strains, which

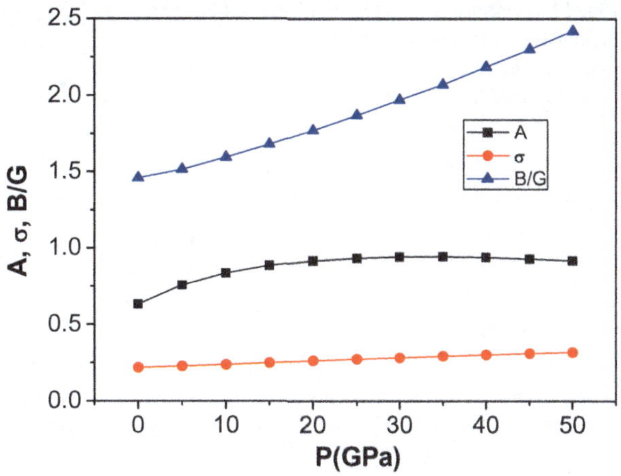

Fig. 10 Pressure dependence of Anisotropy factor (A), the Poisson's ratio (σ) and B/G ratio for NaAlSi

plays a vital role in mechanical engineering design. It provides more information about the characteristics of the bonding forces than any of the other elastic constants. The lower limit and upper limit of σ are given 0.25 and 0.5 for central forces in solids, respectively [69]. The values of σ are typically between 0.1 and 0.25 for covalent materials, and interatomic forces are non-central forces [70]. For ionic crystals, the lower and upper limits of σ are 0.25 and 0.5, respectively, and interatomic forces are central forces [71]. Calculated σ value is equal to 0.221 at 0 GPa for NaAlSi. It shows that the interatomic forces in the NaAlSi are central forces and NaAlSi has covalent-ionic bonding properties, which is in agreement with Fig. 7. It can be seen from Fig. 10 that Poisson's ratio increases with increasing external pressure. The values of σ from 0 to 50 GPa are in the range of 0.221–0.317, indicating that the interatomic forces are central forces. On the other hand, after 15 GPa, high Poisson's ratio σ (>0.25) [72] corresponds to the large deformation of volumetric change and high anisotropy.

The investigation of the stiffness can be completed by providing the microhardness parameter (H), given by the following relationship [73]:

$$H = \frac{(1 - 2\sigma)E}{6(1 + \sigma)} \tag{10}$$

It can be seen from Table 2 that the calculated H value of NaAlSi at 0 GPa is 9.68 GPa, indicating that this material is almost hard materials. The obtained microhardness parameter for NaAlSi increases with pressure up to 15 GPa, and then, it is decreasing from 20 to 50 GPa (see Fig. 11). Unfortunately, as far as we know, there is still no data reported in the literature about Poisson's ratio and microhardness parameter of this material for comparison. All the above values need to be experimentally confirmed in the future works.

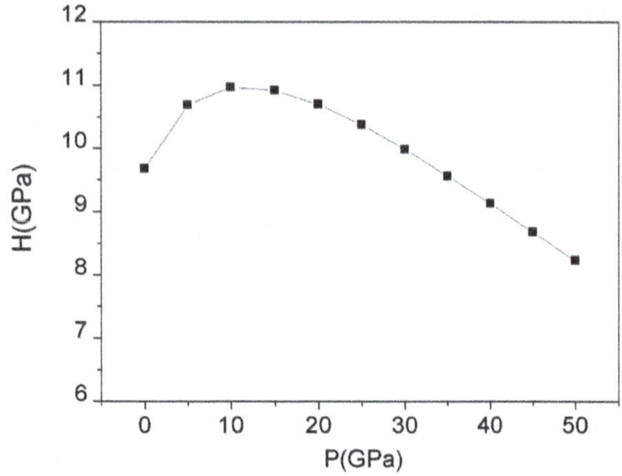

Fig. 11 Pressure dependence of microhardness parameter (H) for NaAlSi

3.3 Phonon Dispersion Curves

Interatomic interactions can be studied by investigating the lattice dynamics. In order to get an idea about the behavior of phonons and role played by them in the stability of the structures and phase transformation in NaAlSi, we have calculated the phonon dispersion curves and phonon density of states for MgAgAs phase of NaAlSi using the PHONOPY code [74]. This code calculates force constant matrices and phonon frequencies using the density-functional perturbation theory as described in Ref. [75]. We have used a $2 \times 2 \times 2$ supercells containing 24 atoms for phonon dispersion calculations. The phonon dispersion curves along with the several high symmetry directions were illustrated in Fig. 12. As this compound has three atoms in the primitive cell, the phonon spectra have nine phonon branches, including three acoustic branches and six optical branches. One of the remarkable observations is that none of the branches involves a soft mode in the whole Brillouin zone, indicating

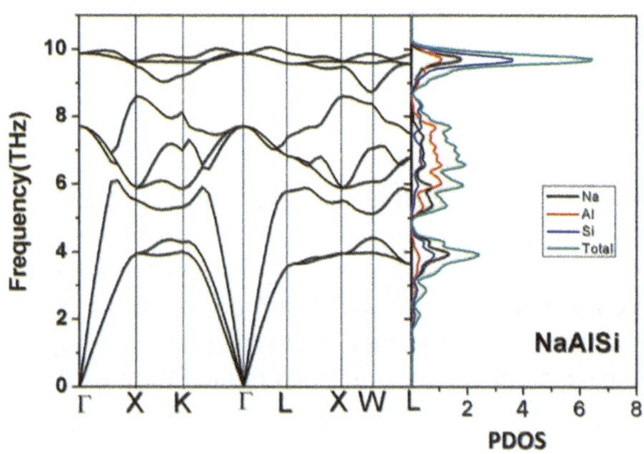

Fig. 12 Phonon dispersion curves at 0 GPa pressure for NaAlSi in MgAgAs structure

that this compound is vibrational stable. The standard group theory analysis yields the following decomposition of the optical vibration representation in irreducible representations of group T_d at Γ point: $\Gamma^{optic} = 2T_2$. Two main points of optical branches of NaAlSi at Γ point are found to be 7.69 and 9.88 THz.

As shown on the right side of Fig. 12, the phonon spectrum of the NaAlSi compound consists of three groups of bands. The lower modes with frequencies up to 4.5 THz are predominantly determined by the vibrations of the Na atoms. The modes in the range 4.5–9.0 THz are predominantly determined by the vibrations of the Al atoms, while the vibrations of the Si atoms are dominant in the range of 9–10.5 THz. The general features are consistent with phonon dispersion relations reported for other HH compounds such as ZrNiSn and ZrCoSb [76, 77].

3.4 Optical Properties

Optical properties are of fundamental importance, since these involve not only the occupied and unoccupied parts of the electronic structure, but also carry information on the character of the bands. It is well known that the interaction of a photon with the electrons in the system can be described in terms of time-dependent perturbations of the ground-state electronic states. Transitions between occupied and unoccupied states, including plasmons and single particle excitations, are caused by the electric field of the photon. The spectra resulting from these excitations can be described as a joint density of states between the valence and conduction bands. The optical phenomena can be quantified by a number of parameters that determine the properties of the medium at the macroscopic level. The study of the optical properties is extremely important for the identification of the internal structure of the materials. The half-Heusler compound NaAlSi with the cubic crystal structure only requires the electric tensor components to determine its optical properties via the frequency dielectric function [78]. The optical properties of a material must be investigated to determine its utility in solar energy technologies and optoelectronic applications.

The frequency-dependent dielectric function $\varepsilon(\omega)$ helps in explaining the optical properties of a variety of materials including photovoltaics and optoelectronics. The optical properties of a solid material are defined by the complex dielectric function $\varepsilon(\omega)$, which characterizes the linear response of the material to an electromagnetic radiation and therefore governs the propagation behavior of radiation in a medium. The frequency-dependent real $\varepsilon_1(\omega)$ and imaginary $\varepsilon_2(\omega)$ parts of the complex dielectric function $\varepsilon(\omega) = \varepsilon_1(\omega) + i\varepsilon_2(\omega)$ are mainly connected to the electronic structure. The fine structure of the energy distribution of the electron states in the valence and conduction bands can be obtained from the knowledge of the optical functions. The imaginary part of the dielectric function $\varepsilon_2(\omega)$ represents the optical absorption in the crystal, which can be calculated from the momentum matrix elements between the occupied and unoccupied states [79, 80] as follows:

$$\varepsilon_2(\omega) = \frac{Ve^2}{2\pi \hbar m^2\omega^2} \int d^3k \sum_{n,\acute{n}} |kn|p|k\acute{n}^2| f(kn)(1 - f(k\acute{n}))\delta(E_{kn} - E_{k\acute{n}} - \hbar\omega) \tag{11}$$

where e is the electronic charge, V is the unit cell volume, p is the momentum operator, $|kn>$ is a crystal wave function, $f\,(kn)$ is the Fermi distribution function, and $\hbar\omega$ is the energy of the incident photon. The real part $\varepsilon_1(\omega)$ can be evaluated from the Kramers–Kroning relationship as:

$$\varepsilon_1(\omega) = 1 + \frac{2}{\pi} M \int_0^{\infty} \frac{\varepsilon_2(\acute{\omega})\acute{\omega}}{\acute{\omega}^2 - \omega^2} \mathrm{d}\omega \tag{12}$$

where M is the principal value of the integral. The optical properties at the equilibrium lattice constants of these compounds were analyzed in energy range up to 20 eV in MgAgAs structure. The obtained real and imaginary parts of the dielectric function are displayed for the linear polarization along the [100] direction. The present results at 0 GPa are depicted in Fig. 13. An important quantity of $\varepsilon_1(\omega)$ is the zero frequency limit $\varepsilon_1(0)$, which is the electronic part of the static dielectric constant. It depends strongly on the band gap. The precise determination of $\varepsilon_1(0)$ is complicated by the fact that the band gap obtained by GGA calculations is underestimated because GGA Kohn–Sham states do not take into account the quasi-particle self-energy correctly. Dielectric constant $\varepsilon_1(0)$ is found to be 15.89 for NaAlSi at zero pressure and zero temperature. In Fig. 13a, the real part of the dielectric function is negative for 2.97–10.98 eV. Negative dielectric function means that the light would not be penetrating the matter and the incident electromagnetic waves are totally reflected; hence, the material exhibits metallic character. Thus, these fluctuations in the real part of the dielectric function correspond to the local maxima of absorption or reflectivity. The main peak of the $\varepsilon_2(\omega)$ occurs in 2.73 eV at 0 GPa pressure. Interpretation of the imaginary part of the dielectric function is in terms of optical transitions between

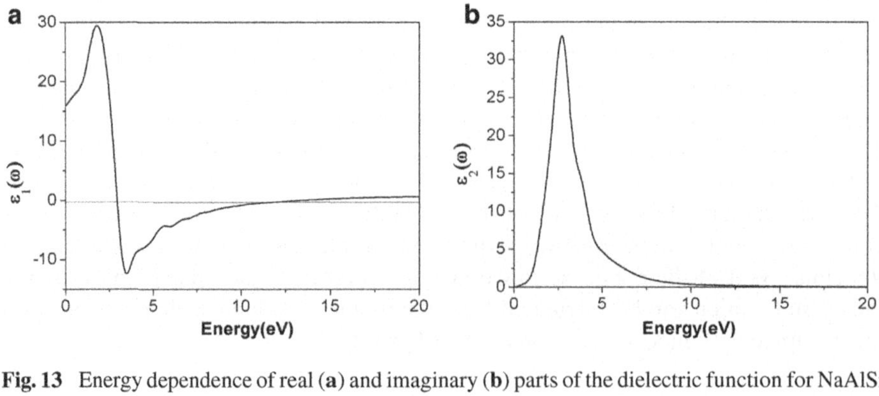

Fig. 13 Energy dependence of real (**a**) and imaginary (**b**) parts of the dielectric function for NaAlSi

states based on band structure calculations. In this compound, the characteristic peak in this spectrum is due to the transition of electrons from Na, Al and Si-p states of the valence band to the unoccupied states in the conduction band.

The refractive index of a material is a very useful optical parameter to determine the propagation of light through the optical medium. The complex refractive index describes the refraction as well as the absorption of compounds and plays a vital role in devices such as wave guides, photonic crystals, solar cells and detectors. The real part of the complex refractive index $n(\omega)$ describes the ordinary refractive index while the imaginary part which also known as the extinction coefficient $k(\omega)$ interprets the photon energy loss as propagates through the medium.

We first employ these complex dielectric function data sets to construct the optical parameterizations, namely the refractive index $n(\omega)$, extinction coefficient $k(\omega)$, reflectivity $R(\omega)$ and the energy loss function $L(\omega)$ using the Kramers–Kronig equation [81, 82]. Our calculated results for $n(\omega)$, $k(\omega)$, $R(\omega)$ and $L(\omega)$ are summarized in Fig. 14a–d under zero pressure, respectively. As shown in Fig. 14a, refractive index increases in energy in the transparency region, reaching a maximum value of 6.183 at about 2.025 eV and then decreases to a minimum level. The value of refractive index at zero pressure $n(0)$ is found as 3.98. The maximum extinction coefficient $k(\omega)$ is estimated as 3.91 at about 3.44 eV.

Lastly, $L(\omega)$ is also an important parameter for describing the energy loss of a fast electron traversing in a material. From the real and imaginary parts of the complex dielectric response function, the electron energy loss function $L(\omega)$ can be easily obtained. The peaks in $L(\omega)$ spectra, displayed in Fig. 14d, represent the characteristic associated with the plasma resonance, and the corresponding frequency is the so-called plasma frequency, above which the material exhibits the dielectric behavior $[\varepsilon_1(\omega) > 0]$ while below which the material behaves the metallic property $[\varepsilon_1(\omega) < 0]$ [83]. In the 10–17.62 eV, there is a peak in the energy loss spectrum. At higher energies, the imaginary part of the dielectric function becomes smaller while the amplitude of the energy loss function becomes large. This function has a main peak at the so-called Plasma frequency at 13.52 eV for NaAlSi. We could not check our results due to the absence of experimental and other theoretical optical data.

4 Summary and Conclusions

The basic physical properties of NaAlSi compound have been carried out through the density-functional theory in MgAgAs structure. These calculations show good agreement with previous studies. In the electronic band structure, NaAlSi should be denoted as a semiconductor and has indirect band gap. It is concluded from calculated elastic constants that the NaAlSi is mechanically stable. The interatomic forces in the NaAlSi are central forces, and NaAlSi has covalent-ionic bonding properties. It is also observed that NaAlSi is an anisotropic material. The positive phonon frequencies indicate that they provide a dynamical stable condition in considering phase. The obtained elastic constants increase from 0 to 50 GPa under pressure. These

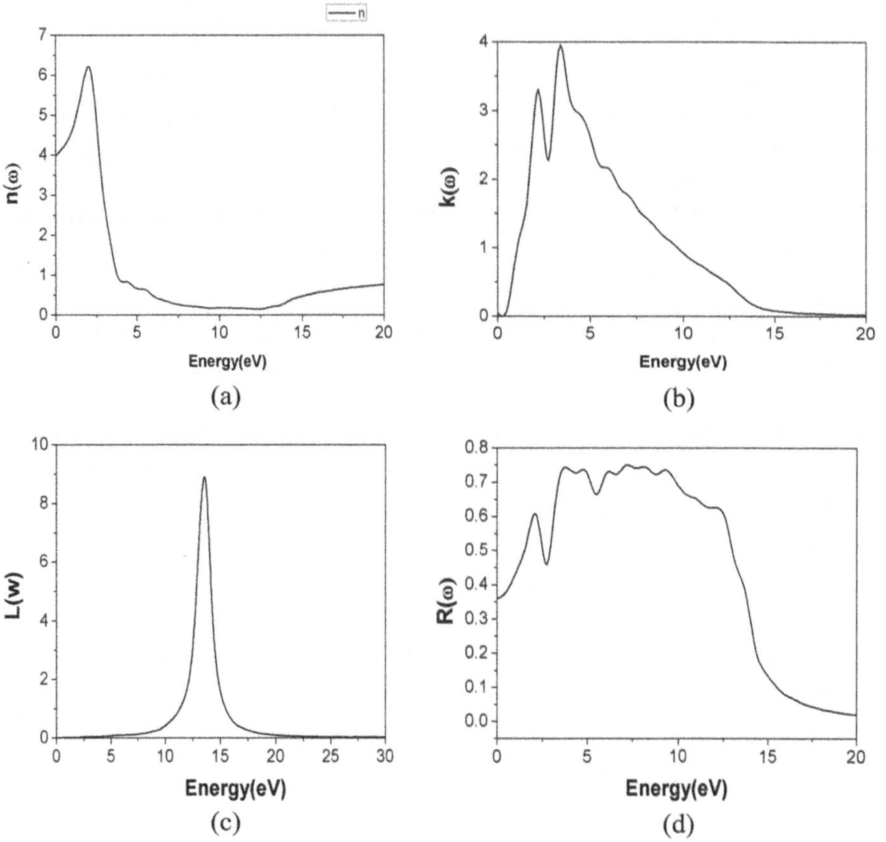

Fig. 14 Computed **a** refractive index $n(\omega)$, **b** extinction coefficient $k(\omega)$, **c** reflectivity $R(\omega)$, *d* electron energy loss spectrum $L(\omega)$ for NaAlSi zero pressure

calculations provide a beneficial information about this compound which could be useful for further investigation for optoelectronic applications.

References

1. M. Cutler, J.F. Leavy, R.L. Fitzpatrick, Phys. Rev. **133**, A1143–A1152 (1964)
2. F. Casper, T. Graf, S. Chadov, B. Balke, C. Felser, Semicond. Sic. Technol. **27**, 063001 (2012)
3. M. Zou, J.-F. Li, P. Guo, T. Kita, J. Phys. D: Appl. Phys. **43**, 415403 (2010)
4. T. Garf, C. Felser, S.P.P. Parkin, Prog. Solid State Chem. **39**, 1 (2011)
5. C. Uher, J. Yang, S. Hu, D.T. Morelli, G.P. Meisner, Phys. Rev. B **59**, 8615 (1999)
6. H. Hohl, A.P. Ramirez, C. Goldmann, G. Ernst, B. Wolfing, E. Brcher, J. Phys. Condens. Matter. **11**, 1697 (1999)
7. P. Larson, S.D. Mahanti, S. Sportouch, M.G. Kanatzidis, Phys. Rev. B **59**, 15660 (1999)

8. R.A. de Groot, F.M. Mueller, P.G. Van Engen, K.H.J. Buschow, Phys. Rev. Lett. **50**, 2024 (1983)
9. J. Pierre, R. Skolozdra, J. Tobola, S. Kaprzyk, C. Hordequin, M.A. Kouacou, I. Karla, R. Currat, E. Lelievre-Berna, J. Alloy. Compd. **262**, 101 (1997)
10. H.C. Kandpal, C. Felser, R. Seshadri, J. Phys. D: Appl. Phys. **39**, 776 (2006)
11. P. Larson, S.D. Mahanti, M.G. Kanatzidis, Phy. Rev. B **62**, 12754 (2000)
12. J. Tobola, J. Pierre, S. Kaprzyk, R.V. Skolozdra, M.A. Kouacou, J. Magn. Magn. Mater. **159**, 192 (1996)
13. J. Tobola, J. Pierre, S. Kaprzyk, R.V. Skolozdra, M.A. Kouacouz, J. Phys. Condens. Matter **10**, 1013 (1998)
14. J. Tobola, J. Pierre, J. Alloys Comp. **296**, 243–252 (2000)
15. H. Nowotny, K. Bachmayer, Monatsch. Chem. **81**, 488 (1950)
16. H. Nowotny, B. Glatzl, Monatsh. Chem. **82**, 720 (1950)
17. R. Juza, F. Hund, Z. Anorg, Chemistry **257**(1) (1948)
18. R. Juza, W. Dethlefesen, H. Seidel, K. Benda, Anorg. Allg. Chem. **356**, 253 (1968)
19. C. Kandpal, C. Felser, R. Seshadri, J. Phys. D **39**, 776 (2006)
20. R. Juza, K. Langer, K.V. Benda, Angew. Chem., Int. Ed. Engl. **7**, 360 (1968)
21. K. Kuriyama, T. Kato, K. Kawada, Phys. Rev. B **49**, 11452 (1994)
22. K. Kuriyama, R. Taguchi, K. Kushida, K. Ushiyama, J. Cryst. Growth **802** 198 (1999)
23. K. Kuriyama, Y. Yamashita, T. Ishikawa, K. Kushida, Phys. Rev. B 75 (2007)
24. H. Nowotny, F. Holub, Monatsch. Chem. **91**, 877 (1960)
25. R. Bacewicz, T.F. Ciszek, Appl. Phys. Lett. **52**, 1150 (1988)
26. L. Spina, Y.-Z. Jia, B. Ducourant, M. Tillard, C. Belin, Z. Kristallogr. **218**, 740 (2003)
27. W. Blase, G. Cordier, R. Kniep, Z. Anorg, Allg. Chem. **619**, 1161 (1993)
28. W. Bockelmann, H.-U. Schuster, Z. Anorg, Allg. Chem. **410**, 241 (1974)
29. U. Eberz, W. Seelentag, H.U. Schuster, Z. Naturforsch., B **35**, 1341 (1980)
30. R. Marazza, D. Rossi, R. Ferro, J. Less-Common Met. **138**, 189 (1988)
31. K. Kuriyama, K. Nagasawa, K. Kushida, J. Cryst. Growth **237**, 2019 (2002)
32. R. Bacewicz, T.F. Ciszek, Mater. Res. Bull. **23**, 1247 (1988)
33. K. Kuriyama, K. Kushida, R. Taguchi, Solid State Comm. **108**, 429 (1998)
34. K. Kuriyama, R. Taguchi, K. Kushida, K. Ushiyama, J. Cryst. Growth **802**, 198 (1999)
35. K. Kuriyama, T. Kato, T. Tanaka, Phys. Rev. B **49**, 4511 (1994)
36. K. Kuriyama, T. Katoh, Phys. Rev. B **37**, 7140 (1988)
37. P. Villars, L.D. Calvert, in *Pearson's Handbook of Crystallographic Data for Intermetallic Phases*, 2nd edn. (ASM International, Materials Park, OH, 1996)
38. H. Mehnane, B. Bekkouche, S. Kacimi, A. Hallouche, M. Djermouni, A. Zaoui, Superlattices Microstruct. **51**, 772–784 (2012)
39. F. Kalarasse, B. Bennecer, J. Phys. Chem. Solids **67**, 1850 (2006)
40. G. Jaiganesh, T.M.A. Britto, R.D. Eithiraj, G. Kalpana, J. Phys.: Condens. Matter **20**, 181 (2013)
41. L.H. Yu, K.L. Yao, Z.L. Liu, Y.S. Zhang, Phys. Lett. A **367**, 389 (2007)
42. S.-H. Wei, A. Zunger, Phys. Rev. Lett. **56**, 528 (1986)
43. S. Curtarolo, G.L.W. Hart, M.B. Nardelli, N. Mingo, S. Sanvito, O. Levy, Nat. Mater. **12**, 181 (2013)
44. S. Ogut, K.M. Rabe, Phys. Rev. B. **51**, 10443 (1995)
45. D. Kieven, R. Lenk, S. Naghavi, C. Felser, T. Gruhn, Phys. Rev. B. **81**, 125210 (2010)
46. T. Gruhn, Phys. Rev. B. **82**, 125210 (2010)
47. F. Casper, R. Seshadri, C. Felser, Phys. Status Solidi A **206**, 1090 (2009)
48. A. Roy, J.W. Bennett, K.M. Rabe, D. Vanderbilt, Phys. Rev. Lett. **109**, 037602 (2012)
49. G. Kresse, J. Hafner, Phys. Rev. B **47**, 558(R) (1993)
50. G. Kresse, J. Furthmüller, Comp. Mat. Sci. **6**, 15 (1996)
51. G. Kresse, D. Joubert, Phys. Rev. B **59**, 1758 (1999)
52. G. Kresse, J. Furthmüller, Phys. Rev. B **54**, 11169 (1996)
53. P.E. Blöchl, Phys. Rev. B **50**, 17953 (1994)

54. J.P. Perdew, A. Zunger, Phys. Rev. B **23**, 5048 (1981)
55. J.P. Perdew, J.A. Chevary, S.H. Vosko, K.A. Jackson, M.R. Pederson, D.J. Singh, C. Fiolhais, Phys. Rev. B **46**, 6671 (1992)
56. J.P. Perdew, K. Burke, M. Ernzerhof, Phys. Rev. Lett. **77**, 3865 (1996)
57. H.J. Monkhorst, J.D. Pack, Phys. Rev. B **13**, 5188 (1976)
58. S. Kacimi, H. Mehnane, A. Zaoui, J. Alloy Comp. **587**, 451 (2014)
59. D. Bende, F.R. Wagner, Y.I. Grin, Inorg. Chem. **54**(8), 3970 (2015)
60. M.J. Mehl, Phys. Rev. B **47**, 2493 (1993)
61. Y.L. Page, P. Saxe, Phys. Rev. B **65**, 104104 (2002)
62. B.B. Karki, L. Stixrude, J. Crain, Geophys. Res. Lett. **24**, 3269 (1997)
63. B. Mayer, H. Anton, E. Bott, M. Methfessel, J. Sticht, P.C. Schmidt, Intermetallics **11**, 23 (2003)
64. A.F. Young, C. Sanloup, E. Gregoryanz, S. Scandolo, R.E. Hemley, H.K. Mao, Phys. Rev. Lett. **96**, 155501 (2006)
65. P. Ravindran, L. Fast, P.A. Korzhavvyi, B. Johnson, J. Wills, O. Eriksson, J. Appl. Phys. **84**, 4891 (1998)
66. D.G. Pettifor, Mater. Sci. Technol. **8**, 345 (1992)
67. S.F. Pugh, Philos. Mag. **45**, 823 (1954)
68. I. Papadimitriou, C. Utton, A. Scott, P. Tsakiropoulos, Metal. Mater. Trans. A **46**, 566 (2015)
69. H. Chen, X. Lei, J. Long, W. Huang, Mater. Sci. Semicond. Process. **27**, 207 (2014)
70. H. Fu, D. Li, F. Peng, T. Gao, X. Cheng, Comput. Mater. Sci. **44**, 774 (2008)
71. J. Haines, J.M. Leger, G. Bocquillon, Ann. Rev. Mater. Res. **31**, 1 (2001)
72. H. Fu, D. Li, F. Peng, T. Gao, X. Cheng, Comp. Mat. Sci. **774**, 23 (2003)
73. S. Aydin, M. Simsek, J. Alloys Compd. **509**, 5219 (2011)
74. Y. El Sayed, E.-A. Amin, El-KheshKhany N. Solid State Commun. **139**, 108 (2006)
75. A. Togo, Phonopy manual Release 1.6.2 (2012)
76. X. Gonze, C. Lee, Phys. Rev. B **55**, 10355 (1997)
77. H. Ozisık, K. Colakoglu, H.B. Ozisik, Fizika **16**, 154 (2010)
78. J. Shiomi, K. Esfarjani, G. Chen, Phys. Rev. B **84**, 104302 (2011)
79. C. Ambrosch-Draxl, J.O. Sofo, Comput. Phys. Commun. **175**, 1 (2006)
80. M. Gajdoš, K. Hummer, G. Kresse, J. Furthmüller, F. Bechstedt, Phys. Rev. B **73**, 045112 (2006)
81. B. Amin, I. Ahmad, M. Maqbool, S. Goumri-Said, R. Ahmad, J. Appl. Phys. **109**, 023109 (2011)
82. S. Adachi, in *Properties of Semiconductor Alloys: Group-IV, III-V and II-VI Semiconductors* (Wiley, 2009)
83. P. Ravindran, A. Delin, B. Johansson, O. Eriksson, J.M. Wills, Phys. Rev. B **59**, 1776 (1999)

Optoelectronic Properties of GaAs, GaP, and ZnSe Cathodes in a Plasma–Semiconductor Cell

H. Hilal Kurt

Abstract This chapter is focused on the optical and electrical characterization of semiconductor materials in a plasma–semiconductor cell because of their applications in electronic devices and infrared image converter cells. The main focus was to get the discharge current and discharge light emission of the plasma in the structure in the case of various interelectrode distance *d*, gas pressure *p* and cathode diameter *D*. The gas discharges systems with semiconductor cathode are the subjects of interest in recent years due to their applications such as semiconductor processing, surface treatment, and sterilization. Even though semiconductor electronic materials have been known for years, more information is needed to completely understand the mechanisms controlling the plasma process. A great effort has been made to get knowledge about these materials during the years in the fields of optics and electronics. In that manner, the present chapter concerns with the optoelectronic characterization of some infrared-sensitive semiconductors to explore the physical parameters having great importance in the investigated structure. The optoelectronic properties of the III–V (GaAs and GaP)and II–VI (ZnSe) group semiconductor electronic materials have been briefly studied in a plasma–semiconductor structure. The interest in those materials has increased mainly due to their emergence of important applications over a wide range of areas such as power amplifier, light-emitting diodes, detectors, photovoltaic cells and microwave. GaAs, GaP and ZnSe are significant materials that are used as infrared detectors in device applications. They exhibit good performance and fast response against infrared radiation.

Keywords Cold plasma · Infrared image converter · Breakdown · Gas discharges · Semiconductor cathodes

H. H. Kurt (✉)
Department of Physics, Faculty of Science, Gazi University, 06500 Teknikokullar, Ankara, Turkey
e-mail: hkurt@gazi.edu.tr

S. J. Ikhmayies et al. (eds.), *Advances in Optoelectronic Materials*,
Advances in Material Research and Technology,
https://doi.org/10.1007/978-3-030-57737-7_6

1 Introduction

Semiconductors are key components in advanced electronic technology, and in the high-speed and frequency semiconductor applications, that are associated with the response to the optical signal and electric field in a strong and fast mode [1]. A high-speed device is a system that produces, modulates, or records the incoming light at microwave frequencies in optoelectronics. Optoelectronics implies the relationship between electronics and optics and the synergies in many areas like science and engineering, for example, materials, optics, electronics, and microtechnology [2, 3]. It also contains different areas such as light generation, modulation, transmission, detections in communications, and also signal processing. The application of these functions are based on the optical, electrical, electro-optical, and also acusto-optical properties of the materials like semiconductors, dielectrics, ferroelectrics and polymers [4]. Optoelectronic semiconductor devices have many applications from information storage such as memory effect, communication, and radar to imaging systems especially as infrared (IR) detectors, and sensors even at the laser printers [5]. Furthermore, optical fiber communication systems and high-power infrared laser diodes are the main factors to improve in the optoelectronic device technology.

Silicon (Si) and III–V group semiconductors covering gallium arsenide (GaAs), Indium phosphide (InP) and their compounds are crucial materials for high-speed applications and their importance results from their direct bandgap [1]. II–VI group materials are usually used for specific long wave applications. On the other hand, most of the important high-speed optoelectronic devices are constructed using the gallium arsenide (GaAs), gallium phosphide (GaP), Indium phosphide (InP) and Zinc selenide (ZnSe) [5, 6].

Since semiconductors have a key role in photonic applications such as photonic switching—because of their large third-order nonlinearities—they have been the subject of extensive studies, from the past to present for both experimental and theoretical studies [7–11].

Because of all these features, semiconductors are excellent and attractive materials. Semiconductors are of interest for a variety of applications such as photodetectors, LEDs, PV cells, and transistors. To improve the performance of the devices based on the semiconductors, low-cost and low power consumption must be provided in the field of optoelectronic materials. The main optoelectronic devices implemented with semi-insulating materials are described in the context of photodetectors in the plasma–semiconductor structure. Optoelectronic properties of semiconductor materials include band structure, transport, and optical, electrical, and magnetic properties.

III–V group semiconductor compound technology dominates the field of optoelectronic device applications when compared to the II–VI group semiconductor materials. The performance of the devices based on the III–V group materials is pretty good compared with materials based on AII–BVI semiconductors. For example, zinc sulfide (ZnS) and zinc selenide (ZnSe) is not fully compatible with gallium arsenide(GaAs). However, ZnSe–ZnS based photodetectors can work as

Table 1 Division of infrared radiation [17]

Region (abbreviation)	Wavelength range (μm)
Near-infrared (NIR)	0.78–1
Short wavelength IR (SWIR)	1–3
Medium wavelength IR (MWIR)	3–6
Long-wavelength IR (LWIR)	6–15
Very long-wavelength IR (VLWIR)	15–1000

light-emitting diodes (LEDs), high-power terahertz (THz) pulse generators, blue-green laser diodes (LDs), photodetectors (PD), modulators operating on the mid-infrared regions, solar cells, and transistors [12]. In addition to their infrared (IR) and visible range optical applications, these wide bandgaps (WBG) materials have also received big attention for the manufacture of heterostructures and waveguides [13–15]. Both ZnSe and ZnS are suitable for high-temperature and high-power applications due to their bandgap energy compared to Si and GaAs materials [16]. The progress in semiconductor physics has been so fast due to their applications in the optoelectronic technologies.

Apart from the Si-based solar cells and specific long-wavelength applications based on II–VI and III–V group semiconductor materials, these materials represent the scope of our studies to realize the optimal response to optical stimulation in the infrared region.

This chapter is devoted to IR-sensitive detectors including III–V group materials such as GaAs and GaP and II–VI group materials such as ZnSe, to pay particular attention to plasma–semiconductor interactions in the optoelectronic device technology and underlying electrical and optical characteristics in a plasma–semiconductor structure. The plasma–semiconductor structure has an advantage for providing low power consumption, high-speed response, and highly efficient discharge light emission from the micro discharge plasma cell, where the power consumption of the structure is lower than that of the other plasma cells [4]. The semiconductor materials under investigation are attractive because they can work at room temperature and they have electronic and optical stability up to atmospheric pressure. Further, ZnSe exhibits good performance for IR detection. The IR sensitivity performance of ZnSe photodetector can be compared with that of GaAs. The progress in the IR detector technology has made many invisible objects to be visible. The objective of this chapter is to review infrared (IR) detectors which are sensitive to infrared wavelengths in a microplasma system [17]. Table 1 lists the types of infrared radiation according to wavelength range.

2 Theory

Plasma technology has been growing rapidly in both industrial and commercial applications. Plasma technology has a powerful application for nanotechnology tools and has attracted a great deal of interest in recent years [18]. Also, there has been growing interest to develop the IR converter cells. Infrared converter cells use the plasma as a visualization media in the conversion process of the infrared images. Many different materials were used in the field of IR conversion. Interest has focused mainly on the wavelengths of the two atmospheric windows 3–5 and 8–14 μm, even though the recent researches are increasing in longer wavelengths stimulated by space applications. The advance in IR detector technology is connected to the development of semiconductor infrared detectors. IR systems have found great potential in the enhancement of automobile and aircraft safety and automatic supermarket door openers [18].

GaAs and GaP detectors are sensitive to near and middle IR wavelength regions whereas ZnSe has IR sensitivity up to far IR region. Various authors made studies on IR detectors such as Sharma [17]. GaAs, GaP and ZnSe are photon detectors which respond to IR radiation by generating free electric charges according to the photo effect mechanism. Photon detectors make the conversion of the IR light beam to free current carriers along the bandgap of the semiconductor material. The sensitivity of the photon detectors directly depends on the spectral absorption efficiency when the photon hits and excites the electrons in the conductive band. The spectral response is also powerfully related to the energy of the incoming IR light beam [17].

In this chapter, some IR-sensitive materials will be briefly discussed. Electronic materials can be classified in physics as metals, insulators, and semiconductors. The conductivity of semiconductors is between that of metals and insulators, and its value increases with the increasing temperature. Si, Ge, GaAs, InP, GaP, InSb, ZnSe, ZnS and CdS are most commonly used semiconductors for optoelectronic materials in UV–Visible and IR regions. The resistivity of the semiconductor materials is between 10^{-2} and 10^{9} Ω cm and if the bandgap $E_G > 3$ eV at $T = 0$, that means the conduction band is empty and resistivity of the material is so high and this situation corresponds to the insulator. There exist some exceptions to few semiconductors, such as ZnS (whose energy gap is 3.6 eV), gallium nitride (GaN) with a 3.5 eV energy gap, etc. [4]. The bandgap of the semiconductors changes with the lattice constant. The decrease in the lattice constant causes the increase in the energy gap which is inversely proportional to the interatomic distance. Another factor that affects the energy gap is the dielectric constant so that the energy gap is inversely proportional to the dielectric constant [19].

Si is the best-known element semiconductor with 1.12 eV energy gap which exist in sand and glass and that is used for the Light-emitting diodes (LEDs), photovoltaic cells (solar cells), switching systems, field-effect transistors (FETs), power electronics, and the photodetectors. Germanium (Ge) is also very well-known semiconductor whose band structure is similar to the Si with 0.7 eV bandgap. Ge is also very important, non-toxic semiconductor material which is used as an IR detector in the imaging systems for military application, furthermore in the transistors, and as a substrate for lenses [4].

GaAs is a compound semiconductorwhich belongs to III–V group elements in the periodic table. It is a direct bandgap semiconductor with a zinc blende crystal structure, and it is used for infrared detectors in the near-infrared region, in the field of

LEDs, optical communications, control systems, solar cells, and for the high-speed applications such as fast electronic switching in the optoelectronic technologies. The interactions between plasma and semiconductorhave an influence on the system characteristics [4].

ZnSe is an II–VI group infrared material, and preferred due to its low absorptivity at infrared wavelengths, and its transmission is in the wavelength range 0.5–15 μm. ZnSe has great importance due to its ability in high-resolution thermal imaging systems, and it can operate in wavelengthsup to 20 μm. ZnS is an II–VI inorganic mineral compound semiconductor that has been used for IR windows in the range of (8–14 μm). Even though the large bandgap of II–VI compound semiconductors find application for displays and lasers, semiconductors with the smaller bandgap are of great importance for the fabrication of infrared detectors [4]. GaP is a polycrystalline compound semiconductor with an orange color and has an indirect bandgap of 2.26 eV and is used for the production of the LEDs in the visible range.

CdS is an important II–VI semiconductor material with a direct wide bandgap of 2.43 eV [20]. CdS has good optical properties and candidate to be used as a window layer in solar cells, nano piezo-electrical devices, field-effect transistors, gas sensing application, light-emitting diodes, photodetectors, photoelectrochemical cells, photoresistors, etc. [21]. CdS has been used for the photosensitive and photovoltaic devices, or as photoresistors [22]. Its usage areas cover laser light-emitting diodes, and non-linear optical optoelectronic devices.

Semiconductors with narrow bandgap operate in the near and mid-infrared (MIR) region, whereas those having wide-band gaps can find application in the far-infrared region (FIR) [23]. Table 2 lists the bandgap type and energy for various semiconductors at room temperature [24].

From an important perspective, the carrier transport mechanism [25] in the semiconductors are written by the following well - known formula:

$$\sigma = en\mu \tag{1}$$

where σ is the conductivity of material, e is the fundamental charge, n is the conductive charge density, and μ is the mobility of the conductive charge carrier. Mobility is

Table 2 Bandgap type and energy for various semiconductors at room temperature [24]

Material	Bandgap type	Bandgap energy (ev)	Bandgap wavelength
Gallium arsenide (GaAs)	Direct	1.441	857 nm
Gallium phosphide (GaP)	Indirect	2.24	551 nm
Zinc selenide (ZnSe)	Direct	2.70	480 nm
Cubic zinc sulfide (ZnS)	Direct	3.54	349 nm
Cadmium sulfide (CdS)	Direct	2.42	510 nm
Gallium nitride (GaN)	Direct	3.4	363 nm
Silicon (Si)	Indirect	1.12	1.1 μm

defined as the ability of the electrons to move across the materials when the electric field is applied. Before an understanding of the transport mechanism in materials, one should get information about σ in terms of n and μ and their interplay that results in semiconductors with different conducting properties, which have made the semiconductor-based electronics industry possible.

When the semiconductors are doped with impurities which artificially add free electrons to the material, conductivity changes dramatically (they can cause transition of electrons from the valance band-creating 'holes' with an effective positive charge which are also capable of electrical conduction-to conduction band [26]).

There have been few electrons with sufficient energy that can pass the forbidden bandgap and occupy the higher-lying energy levels in the conduction band at low temperatures. Therefore, the electrical conductivity is conducted primarily by the free carrier charge concentration which depends exponentially on temperature and the bandgap energy [26]:

$$\sigma \propto n \propto \exp(-E_g/2kBT) \tag{2}$$

where n is charge carrier concentration, σ is conductivity, E_g is the bandgap of the material ad k is Boltzmann constant.

In semiconductors, an electronic bandgap is formed as a result of electrons get scattered by the atoms in the crystallographic lattice points separated by a few nanometers. The bandgap width has a decisive role in the transport mechanism when an external electric field is applied to the material or optically stimulated. When the light hits the material having sufficiently high energy compared to the forbidden bandgap, energetic electrons makes the transition from the valance band (a band is full of the electrons at zero kelvin) to conduction band (a band is thought as completely free at zero kelvin) as a result of photo effect mechanism in semiconductors. In this way, the carrier concentration and the conductivity of the material can be controlled depending on the energy of incoming light [25].

Many materials have been presented to provide the development in the IR detector technology [27] in the past and as Norton [28] stated that all physical situations in the range of 0.1–1 eV may be designed for IR detectors. Considerable effort has been made to improve IR detector technology in the course of World War II and all authorities accept this date as an origin of the modern IR detector technology. During this period, it was especially focused IR detectors that can be operated between 3–5 and 8–14 μm wavelength of the atmospheric window range although there has been a growing interest in detectors, which have recently been exploited in the far IR regions that find applications in the space research [29].

Development in the IR detector technology is relevant to the IR semiconductor including photon detectors. In these type of detectors, photons transfer its energy to the material, or rather to the electrons as a result of light-material interaction. The acquired electrical output signal is derived from the changed electronic energy distribution in the materials. Photon detectors are sensitive only to the certain wavelengths per unit incident radiation power in the spectrum. We encounter some restrictions at this point as cooling requirements which cause the heavy, expensive, and inconvenient to use. Besides, signal-to-noise performance is another drawback in the IR detectors.

Depending on the nature of the interaction, photon detectors are classified as intrinsic detectors, extrinsic detectors, photoemissive detectors, and quantum well detectors [29]. Another IR detector consists of thermal detectors which are sensitive to the change for the temperature of the materials and absorbed photon flux result in the change on the material temperature and hence all these changes in the material properties are used to produce an electrical output. Thermal detection of the detectors is independent of the wavelength of the incident photon. Contrary to photon detectors, thermal detectors operate at room temperature. When compared to photon detectors, thermal detectors remain to have unobtrusive sensitivity and a low response to the incoming light beam.

The earlier studies on plasma–semiconductor systems have been done by Yu. A. Astrov with many experimental studies. In their recent study, they have investigated plasma–chemical anodic oxidation of GaAs under dc Townsend discharge conditions in the case of two discharge gaps [30]. Townsend discharge in argon and nitrogen were also studied by Astrov et al. [31]. Astrov et al. used a short plane-parallel discharge gap and a high-resistivity photosensitive semiconductor electrode to measure the noise of optical emission from gas excited by the discharge using a photomultiplier [32].

Microdischarge unit is the main part of an IR image converter and it was invented in the 1970s [33]. In such converter systems, a micro gap filled by gas, a high-resistivity semiconductor cathode (sensitive to IR region), and a glass plate coated with a transparent film is mainly used. It has crucial importance to choose the convenient semiconductor materials specified by good sensitivity and high-speed of operation [34]. The operation principle of such a device is based on controlling the gas discharge via a high resistivity semiconductor electrode [4]. The spatial distribution of the discharge light emission and current in the discharge gap is proportional to the semiconductor conductivity, which is controlled through incident IR radiation projected onto the photodetector; it is converted into the visible glow of the gas [34]. Gas discharges are mainly classified as a Townsend, glow, and arc discharges [35].

To understand the plasma related features and the effect of the gas pressure on plasma current, a non-thermal plasma condition has been sustained in the IR converter system shown in Fig. 1a [36]. Inside the plasma, electrons which are emitted from the semiconductor cathode make successive collisions with gas atoms and that process determines the value of the plasma current depending on the number of secondary electrons and their energies [37]. Once these secondary electrons contribute to the ionization process in plasma, current can increase suddenly and if the increase in current cannot be controlled, the surface of the material can be damaged by the filament (nonuniform current density). Secondary electrons are produced with the positive charges accumulated in front of the cathode [37, 38].

The shape of the I–V characteristic in the system depends on the voltage increment of the power supply and the illumination intensity [36]. Current-voltage characteristics (CVCs) and discharge light emission (DLE) in the cell depend on the discharge

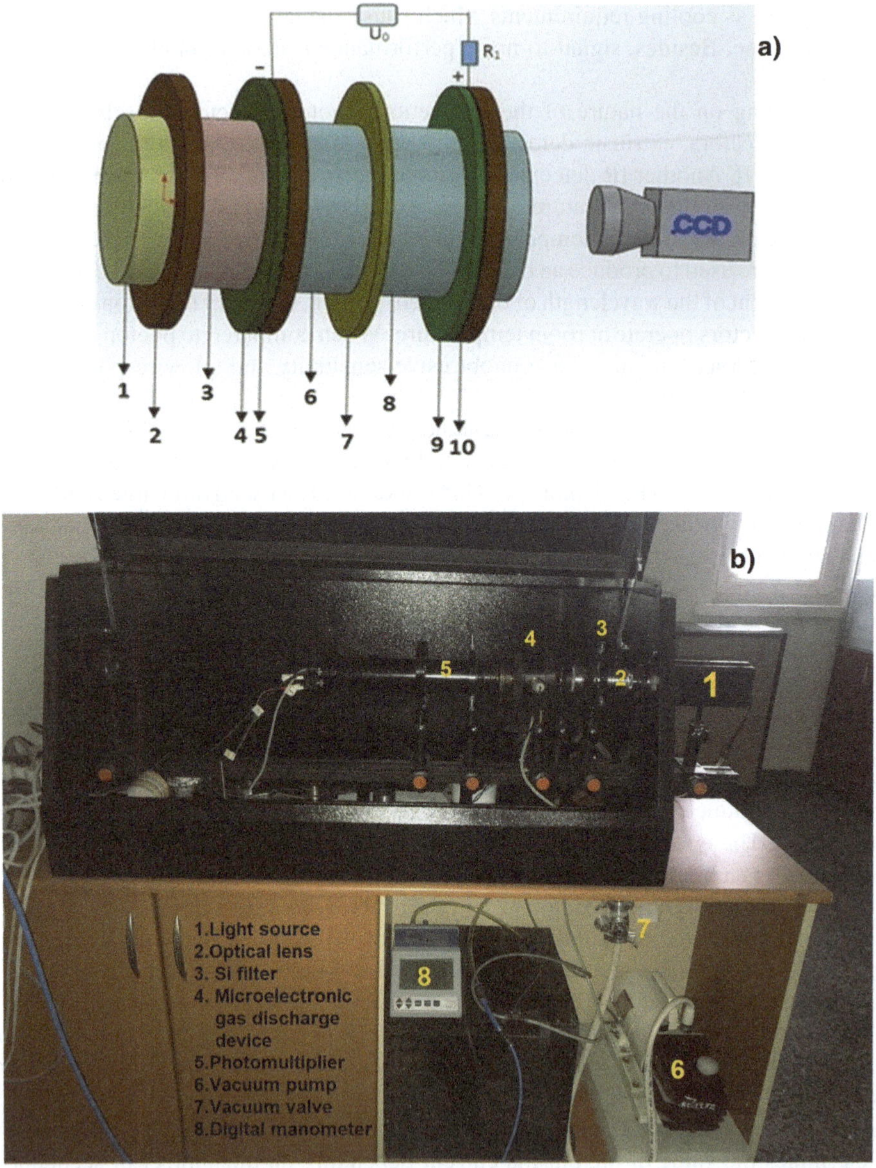

Fig. 1 **a** Semiconductor-plasma cell which is the main part of IR image converters. 1. light source, 2. Si filter, 3. IR light beam, 4. semi transparent Au contact, 5. GaAs, GaP and ZnSe materials, 6. gas discharge gap, 7. Mica foil, 8. UV-visible light beam, 9. Transparent conductive SnO_2 contact, 10. glass disc [36]. **b** Photo of the gas plasma system with GaAs, GaP, and ZnSe electrodes [41]

parameters, such as the gas pressure, the feeding voltage, interelectrode distance, cathode diameter, etc.

Many different studies made in different gases show that plasma depends on the operating conditions. The pattern formation in plasma–semiconductor structure has been investigated by Purwins Group at Munster University. Their research interest focused on understanding the complex behavior of the plasma process in an IR image converter cell. They obtained spatially and temporally changing inhomogeneous patterns in the discharge gap. They followed the different approaches to understand the microscopic process on a very fast time scale. They used the Monte-Carlo methods to understand the observed spatiotemporal patterns [39, 40].

3 Experimental Procedure

The present work deals with an experimental study of Townsend and glows discharge in air and argon media to explore the properties of IR-sensitive semiconductors at the stable gas discharge conditions. Effect of interelectrode distance d and gas pressure p on the optical and electrical propertiesof GaAs, GaP and ZnSe in the case of different IR illumination intensities were studied in the plasma–semiconductor system. CVCs and DLEs were determined using multimeter and electrometer and also photomultiplier. Figure 1a shows a schematic diagram of the parallel plate dc gas discharge device. The device consists of two electrodes made of a semiconductor and SnO_2evaporated glass disc. The cathode has a large diameter, large resistivity, and thickness of 1 mm. The anode has 1 mm thickness and very low and negligible resistance when compared with the cathode. These electrodes are separated from each other by insulating mica foil which determines the active discharge area in the plasma device. 99.9% purified argon is used as the working gas and its pressure (p) is varied from 10 to 760 Torr, and the gas pressure was measured using a digital manometer[36].

The discharge current varies according to the type of the cathode material and plasma parameters including pressure p, gap distance d, and the gas type. The value of current is restricted with high-resistivity semiconductor electrode to avoid filamentation which causes the damage to the cathode materials. The current does not exceed 10^{-4} A in the stable discharge conditions. Dc power supply is connected to the gas discharge cell and is operated up to 2500 V.

A parallel plate micro discharge cell is used in the measurements of current and radiation values and the setup has been explained in detail in our previous studies [26]. The anode is made of SnO_2 evaporated on the glass disc to provide conduction of the charge carriers, and the cathode is made of high-resistivity semiconductor plates. The gaps between the electrodes have micro sizes and the system was evacuated to pressures on of the order of 10^{-1} Torr and filled with air and argon gases. The electrical and optical measurements have been used to understand the properties of semiconductors according to their IR responses. The gap between electrodes changes in the range of 50 and 600 μm. The electrical field in the cell can be regarded as homogeneous since the maximum current was limited to 100 μA.

When a potential difference is applied between two electrodes in the micro discharge plasma cell, gas pressure causes the phenomenon is known as a plasma discharge beyond the critical voltage value which is called breakdown voltages when a sufficiently strong electric field is applied. After breakdown occurs, the gas becomes conducting. Electrons gain energy and result in the successive ionization in the plasma cell with a semiconductor electrode and they transfer their energy to the gas atoms [42]. Electrons collide with neutral gas atoms causing ionization in the discharge cell. In addition, electrons can be released from a neutral atom with the interaction of a photon of cosmic rays or other natural radiation.

Figure 1a, b illustrates the basic physical structure of the plasma–semiconductor cell. There is a well-known relationship, called Paschen's Law in which U_B is a function of the *(p.d)* and this important equation implies the required condition to ignite gas discharge between the electrodes. CVCs and DLEs depend on the plasma parameters such as the pressure, and the applied voltage and interelectrode gap *d*. The change in these parameters is effective on the type and character of the plasm [29–44]

When the discharge starts in the cell, a pair of electrons enter into the discharge from the cathode, they accelerate in the electric fieldtowards the anode, and they can recombine with positive ions. Furthermore, those electrons can ionize the neutral atoms by successive collisions, and both of them (the original electron and newly released electron) leave the gas discharge plasma with each other, and move to the anode. Both ions move toward the cathode, and electrons directed to the anode will contribute to the plasma current in the cell.

When the currents formed by ions and electrons are sufficiently high, ionized gas may emit light in the visible range. On the other hand, secondary electron emission takes place from the cathode and these electrons are mainly caused by positive charges accumulating in front of the cathode over a wide range of operating conditions. When the ions strike the cathode, secondary electrons are produced and the generated secondary electrons are accelerated towards the anode, and contribute to the ionization by collisions with gas atoms. As a result of the multiplication of the ionization process, breakdown, that is, self-sustaining discharge is obtained in the plasma cell and external ionization is not necessary for the discharge current to flow [43].

Electrical breakdown of gases is the transition process from an insulator to a conducting state and the minimum voltage at which this transition occurs is called the breakdown voltage V_B. The physics of the electrical breakdown has a great significance because of its wide applications in electronics and technology [43].

Impact ionization and secondary electron generation are two important factors for the occurrence of gas discharge.

The breakdown is considered to occur when Townsend's criterion is fulfilled [43].

$$\gamma_e(e^{\alpha d} - 1) = 1 \tag{3}$$

Townsend's criteria for avalanche breakdown as follows:

$$i = i_o \frac{e^{\alpha d}}{[1 - \gamma_e(e^{\alpha d} - 1)]} \tag{4}$$

where α is the Townsends first ionization coefficient which represents the impact ionization γ is the Townsends second ionization coefficient related to the secondary electron generations from the cathode [30, 31], and d is the interelectrode distance between the electrodes.

Townsend's first ionization coefficient $\boldsymbol{\alpha}$ represents the probability per unit length of ionization occurring during an electron–gas atom collision. The critical breakdown voltage (V_B) is a function of the product of pressure pP and interelectrode distance (d).

$$V_B = \frac{APd}{\ln(Pd) + B} \tag{5}$$

The above equation is known as Paschen law, and the optimum pressure and the value of the voltage to initiate gas discharge are determined by the Paschen law, where A and B are constants based on the gas composition [43]. When the conditions for the breakdown mechanism are fulfilled, discharge can be sustained, and Paschen curves are obtained by different gases' ionizations in the following format [43, 44] shown in Fig. 2.

Paschen's curves describe the breakdown voltage as a function of the electrode spacing or gap distance d, operating pressure p, and gas composition as shown in Fig. 2. The mathematical equation of the Paschen lawwas derived from the Townsend criteria for the charge multiplication process and secondary electron emission by ion bombardment accumulated in front of the cathode (space charges) even though the other bombardment process plays a role in the production process of the electrons [46, 47].

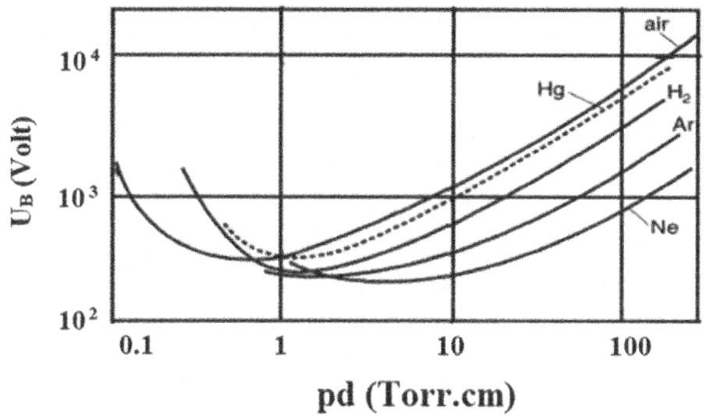

Fig. 2 Paschen curves for different gases [43, 45]

Different researchers have discussed various plasma structures such as optical-electronic and elastic of these semiconductors. Glow discharge between the electrodes can be used widely in different technological plasma processes [48].

Breakdown voltage depends on the pressure, interelectrode distance and gas type. Even though all parameters are kept constant, (*p.d*) value cannot be held constant for characterizing breakdown voltage and as can be seen from the Paschen curves, V_B depends on the product of *pd* exponentially [49].

On the other hand, the breakdown expresses the transition of a non-sustaining discharge into a self-sustaining discharge when the appropriate conditions are fulfilled in the plasma in which ionization process takes place depending on the interelectrode distance *d* and gas pressure *p* which was discussed by Panchen's Law. Many processes are responsible for ionization phenomena such as pressure, temperature, plasma species, electrode configuration, nature of electrode surfaces, and the availability of the initial conducting particle in the plasma [44, 50].

Despite that, there exist very few electron-ion collisions in plasma at low-pressure ranges and the number of the secondary electrons is too low to sustain ionization in the discharge cell.

It becomes frequent collisions at higher pressure values because electrons do not obtain sufficient energy to ionize gas atoms, the discharge is quenched and non-sustainable [51]. Thus, in both situations, the ion production rates are low and high voltages are required to provide discharge.

On the right side of the Paschen curve, the U_B voltage shows a proportional increase with *pd*. This is due to the probability of an electron to produce ionization, which is very high for the relatively large *pd* even in moderate reduced electric field *E/p*. On the left, the U_B voltage increases sharply when *pd* decreases. Higher electric fields are needed, because low *pd* values are less likely to ionize due to the low ionizing collisions of electrons on the left-hand side of the curve. Therefore, the Paschen curve shows a minimum point where the ionization probabilities of the electrons are maximum at this point, and optimal conditions to sustain gas discharge is provided around the minimum [44].

According to widely accepted common opinion [52–56], self-sustained gas discharge plasma occurring in micro discharge gap with parallel plane electrodes can be divided into Townsend and Glow discharges. If the current is very low and in the order of (a few μA or less) and space charge effect is very small and it does not deteriorate the homogeneity of the electric field, the discharge is called "the Townsend Discharge". Such discharge is observed after breakdown voltage and discharge is independent of the external plasma parameters in this discharge mode. Micro discharges are characterized by a characteristic dimension less than 1 mm and they have the advantage that the plasma can operate in atmospheric pressure. If the interelectrode distance between the electrodes is (<10 μm), field emission takes an active role, so that is not important at larger scales. In the Townsend discharge mode, the current is in the order of 10^{-12}–10^{-5} A. Townsend mechanism depends on the type of the cathode material in the plasma cell.

Moreover, the production of the secondary electronsis needed for the Townsend breakdown when the electric field is low and not enough to sustain the discharge.

When the plasma current increases in the cell, space charge effect on the electric field uniformity becomes important and it distorts the homogeneity of the electric field between electrodes, and gas discharge begins to contract. As a result of this, the CVCs tend to be negative in this mode leading to the current instabilities in the system. The discharge is now in glow mode. In the glow discharge mode, the current is in the order of 10^{-4}–10^{-1} A, and discharge light emission is brighter than that of Townsend mode [55–57].

The properties of the glow discharge are changeable with accordance to its discharge parameters such as gas pressure, plasma *gas composition*, *gas pressure*and applied voltage to the electrodes (*input electrical power*). The plasma gas composition is particularly important in the excitation and ionization mechanisms including the production of active species [58].

4 Results and Discussion

The distance between the electrodes has a significant effect on the electrical and optical properties of the system. Figure 3a–c shows CVCs measured between d = 50–445 μm as a function of the pressure for different cathode diameters D (12, 15 and 18 mm). CVCs are more stable when d = 50–240 μm. Moreover, the distance d between the electrodes leads to differences in the discharge current due to the plasma behavior in the cell. The difference in CVCs results from the presence of the electric field domains (regions) moving between cathode and anode in the plasma cell. While CVCs show linear behavior at 50 and 240 μm, whereas it behaves nonlinearly at d = 445 μm (Fig. 3).

As shown in Fig. 3a–c, different conductivities are obtained under different interelectrode distances at various gas pressures. Plasma current is measured at the wide range pressure of 28–100 Torr.

On the other hand, stable constant conductivities were obtained for low interelectrode distances (d = 50 μm and 240 μm). In the experiments, the interelectrode distance between two electrodes is fixed at a certain value, and gas pressure is changed in the range 28–80 Torr. It has been found that cathode diameter plays an important role in plasma current. A maximum current is recorded at 2×10^{-4} A for 18 mm at d = 50 μm, where the plasma is in the glow mode. It can be seen that the plasma current decreases with the increasing interelectrode distance d. Current changes from 2×10^{-4} A to 3×10^{-5} A when d is changing from 50 to 445 μm. Based on the results, plasma oscillations have been observed for 445 μm. The non-linear behavior of the CVCs is resulted of the space charge effect in the plasma, because of the imbalance in the excitation and ionization process. In all states, plasma currents proportionally increase with the cathode diameter D.

According to our experimental findings, when the appropriate pressure value is set, the optical properties of GaP can be improved and the IR sensitivity of the material can be optically increased. Figure 4a–c shows the relationship between the current and the interelectrode distance d between the electrodes for three different

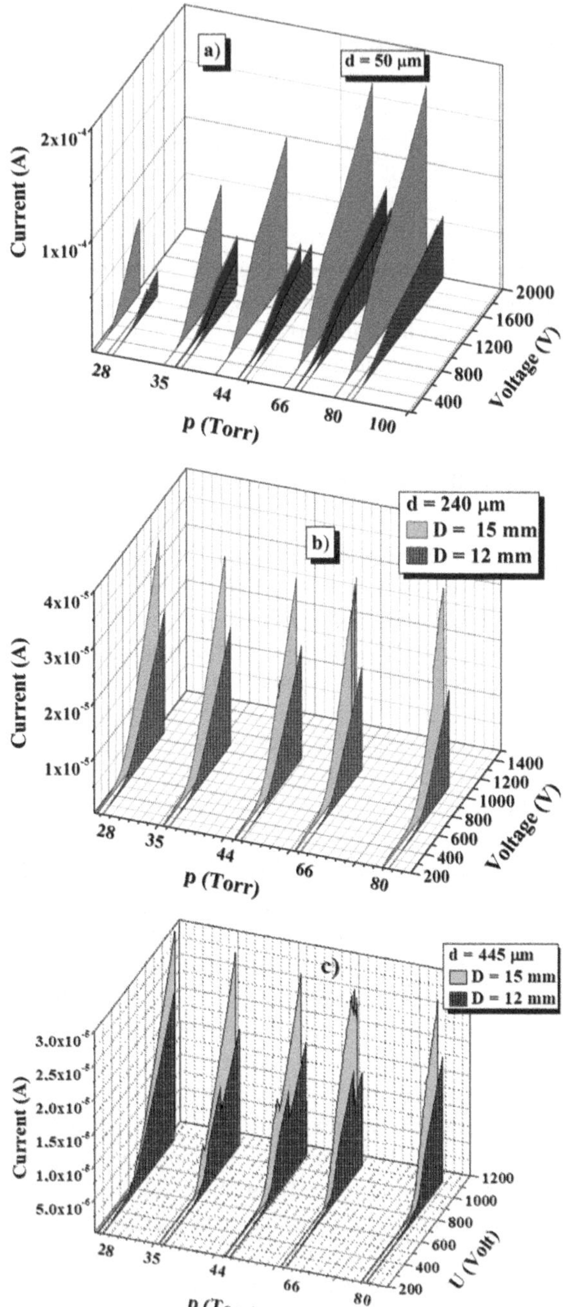

Fig. 3 **a–c** CVCs as a function of the gas pressure: **a** for $D = 12$ mm and $D = 15$ mm at 240 μm; dark grey corresponds to 18 mm, blue corresponds to 15 mm and light grey corresponds to 12 mm; **b** for $D = 12$ mm and $D = 15$ mm at 445 μm; **c** for $D = 12$–18 mm at 50 μm

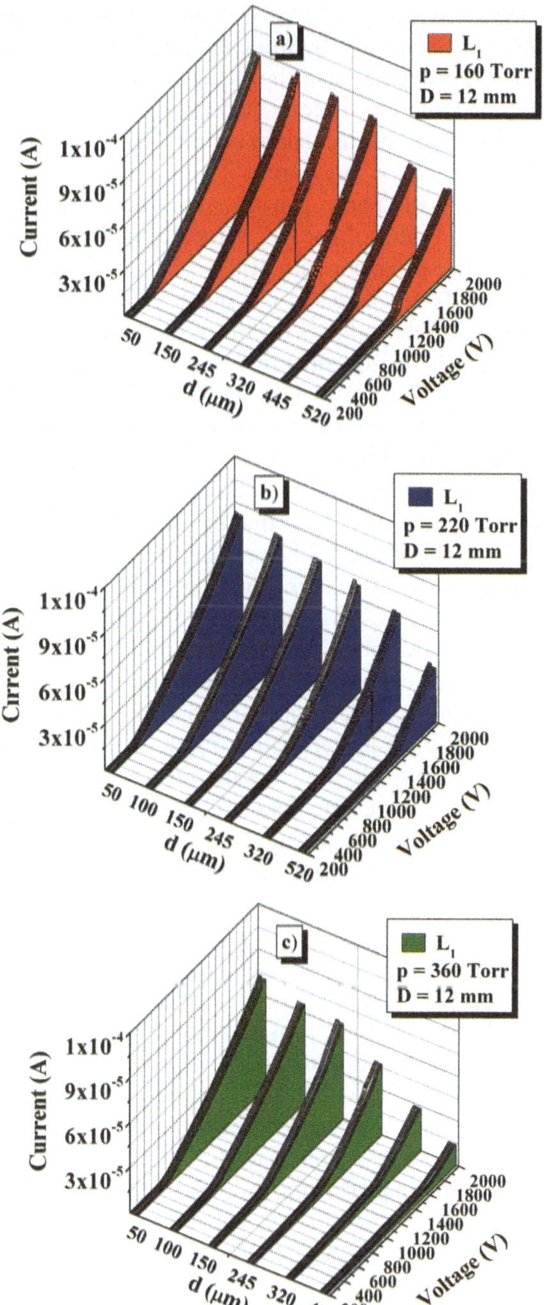

Fig. 4 **a–c** CVCs as a function of interelectrode distance d for three different pressure values under weak IR illumination intensity L_1. **a** for 160 Torr, **b** for 220 Torr, and **c** for 360 Torr

pressure regimes. The measurements were carried out for 160 Torr, 220 Torr, and 360 Torr, respectively, while the interelectrode distance varied from 50 to 520 μm. The experimental data shows that different magnitudes of currents are obtained for different interelectrode distances. The comparison of the currents at different pressure values shows that the maximum current values will be obtained at a low interelectrode distance and low pressures.

As the discharge gap decreases from 520 to 50 μm, CVC curves reach the maximum value of 1×10^{-4} A for all pressures. Besides, CVCs show a linear decrease with increasing pressure at the same d. The plasma current is around 1×10^{-4} A for 160 Torr, whereas it is around 7×10^{-5} A for 220 Torr at $d = 50$ μm. Several factors influence the behavior of the plasma current. The characteristics of the gas discharges are governed by potential difference, gap distance, the gas type, and gas pressure in the cell. Plasma can be obtained in a wide range of pressure. However, the quality of the semiconductor electrode is important to ensure the optimal discharge conditions in the cell.

In Fig. 4a–c, CVC measurements are performed to show the effect of pressure on hysteresis. It is seen that the application voltage changes the hysteresis width (ΔV) as a function of the pressure in the forward and reverse bias. Hysteresis width becomes narrow with the pressure increasing. In addition, discharge light emission graphs within CVCs confirm the situation. The distance between the pressure and the electrodes plays an important role in determining the true value of hysteresis. As the application voltage is increased, the kinetic energy of the electrons increases which leads the electrons to the higher energy bands.

Further, the hysteresis curves show that the discharge shows oscillations in the range of 160–220 Torr. As the pressure increases, the homogeneous discharge is obtained. The electric field redistribution due to space charges accumulated in front of the cathode causes the bistable behavior of the plasma current and discharge emission. Both the semiconductor electrode and the gas discharge gap are responsible for the bifurcation process when the stationary bias voltage is applied to the plasma cell.

The infrared response of GaAs and GaP electrodes are compared in Fig. 5a, b. Negative differential resistance effect (NDR) in the CVC is associated with the bistability of the electrons in the plasma when the electrical field domains move from the cathode towards anode.

Large and reproducible NDR has been observed for GaP in the bias ranging from 1400 to 1700 V. The measured peak current for GaP electrode is 3.7×10^{-5} A under strong illumination intensity L_3 with a valley current value of 2.2×10^{-5} A when the maximum peak is around 4.8×10^{-5} A with a valley current value of 3.5×10^{-5} A without IR illumination. The defect states including impurities during the growth process of the material are responsible for the existence of the NDR together with the plasma effect. Similar bistability is observed for DLE curves in the same voltage range. Experiments have shown that electric field emission plays a crucial role in the electrical bistability in a plasma–semiconductor structure because of the high electric field in the microscale gaps [47]. The NDR region disappears after 1700 V as shown in Fig. 6 for GaP. It is observed from the figures that GaAs has straight

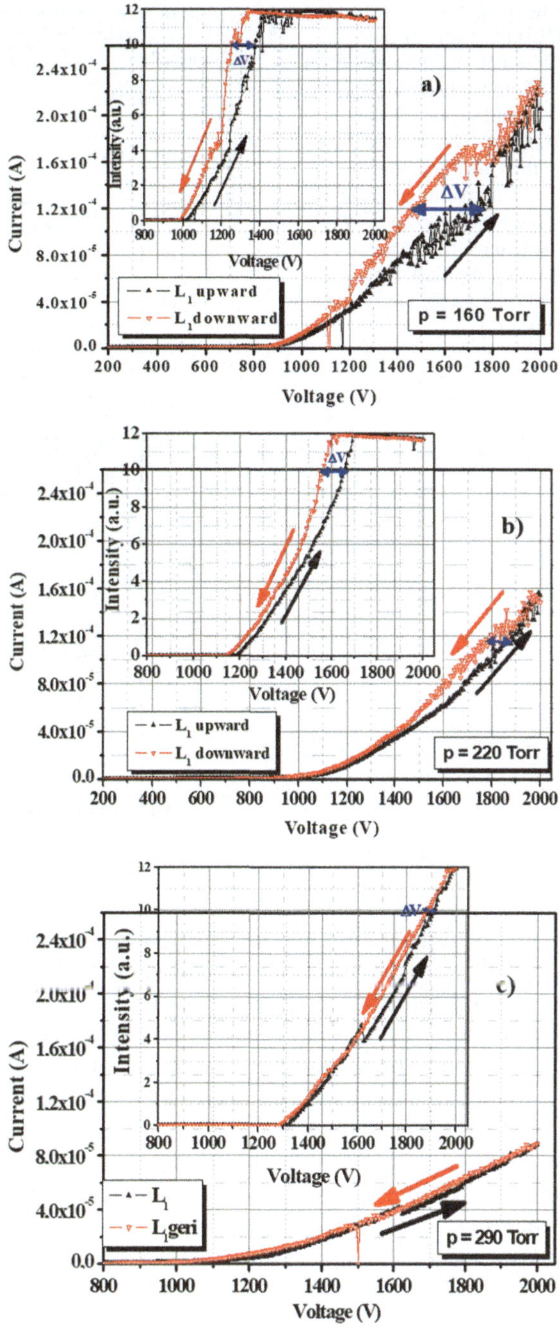

Fig. 5 a–c The pressure dependence of the hysteresis graphs for the forward and backward state when the GaP photodetector is subjected to weak illumination intensity L_1: **a** for $p = 160$ Torr; **b** for $p = 220$ Torr; **c** for $p = 290$ Torr. The system parameters are given as follows: $D = 22$ mm, $d = 330$ μm

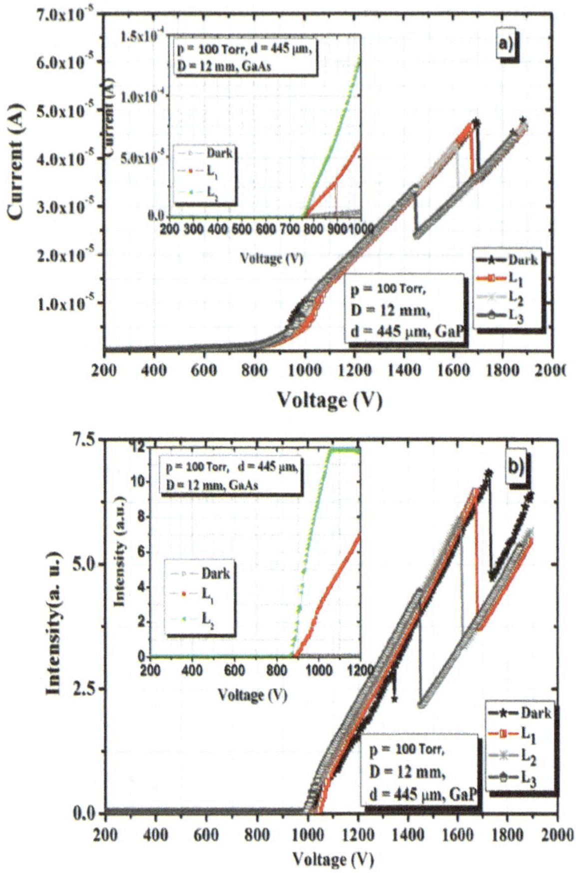

Fig. 6 **a**, **b** The CVC and DLE plots for various illumination intensities from dark, to weak (L_1), moderate (L_2) and strong (L_3) at 100 Torr. The insets belong to the GaAs specimen in (**a**, **b**). Other parameters are the same for both graphs. *D* is the diameter of materials. **a** CVCs for GaAs and GaP and **b** DLEs for GaAs and GaP [35]

CVC, but higher current and radiation emissions at the same conditions as in the case of various IR illuminations. Based on these measurements, it can be concluded that GaP is sensitive to IR radiation inside the NDR region. Stable I–V curves are measured for GaAs. GaAs causes higher plasma current and radiation compared to GaP. The maximum current is around 1.5×10^{-4} A, and maximum radiation intensity is around 12 (a.u.) for GaAs electrode. It is around 7.5 (a.u.) for GaP electrode.

In Fig. 7, the hysteresis graphs are recorded to investigate the effect of inter-electrode distance *d* on the hysteresis behavior. CVC measurements show that the hysteresis width varies due to the trapping centers in the GaP material, especially when this effect is reduced by a certain amount of thickness. The distance between the narrow electrodes supports the hysteresis formation; whereas the distance between the large electrodes prevents hysteresis formation.

Also, the curves are shifting from linear to non-linear with increasing distance *d*. The electric field zones cause unsteady current behavior with increasing distance *d* between the electrodes. This result indicates that *d* distance plays an important

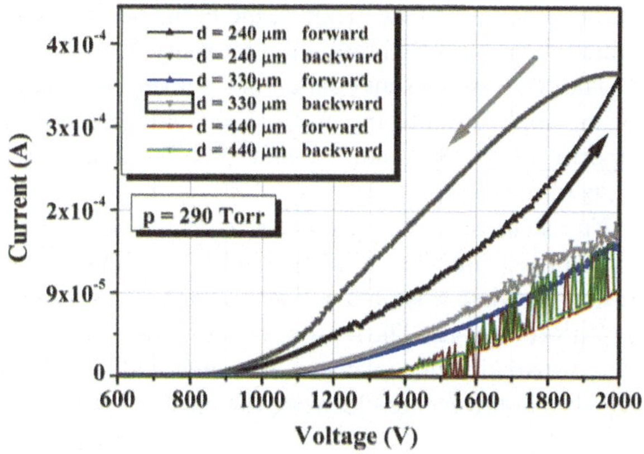

Fig. 7 Hysteresis graphs at $p = 290$ Torr for various d distances

role in the optimization of the system. This oscillating conductivity at high distances limits the use of GaP material as a detector.

The plasma medium is very chaotic and complex so that the plasma current is influenced by many parameters such as filamentary conducting in the cell. Furthermore, charge trap centers in the semiconductor electrode cause bistable situations like hysteresis under different IR illumination as seen in Fig. 8. Secondly, charge trap centers are associated with the hysteresis curves for the case of upward and downward voltage change. In other words, the current has two different behaviors for forwarding and backward bias voltage. The hysteresis behavior has been obtained for ZnSe at only $d = 330$ μm and $p = 500$ Torr and $D = 18$ mm. In other plasma parameters, hysteresis did not appear in the plasma. As can be seen in Fig. 8, breakdown occurs earlier for the backward voltage. Because the gas ionization has already occurred when the voltage increases from 200 to 1400 V for the first stage. This situation affects the lower breakdown voltage for the gas ionization. Based on the experimental results, it can be concluded that IR illumination is the main factor for

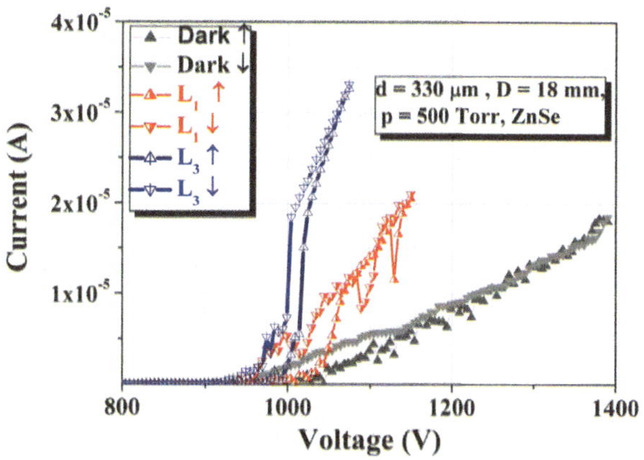

Fig. 8 Hysteresis curves for ZnSe electrode

the breakdown voltage. That is, the breakdown has been observed at earlier voltage value for the strong illumination intensity L_3.

The NDR in the semiconductors is a very useful physical event in device physics for the application of the high- speed switches and high-frequency oscillators. The experimental results indicate that the NDR behavior is optically stimulated at room temperature [59].

The sharp increase in current and DLE is followed by a sharp decrease to the contrary to Ohm's law as shown in Fig. 9. NDR type CVCs are found for higher gap distances. NDR was obtained at $d = 525$ μm for GaP electrode and at $d = 330$ μm for ZnSe electrode. The reason for the NDR-branches is a modification of the electric field distribution due to avalanche-generated free charge carriers in the plasma. NDR vanishes at lower interelectrode distances [59].

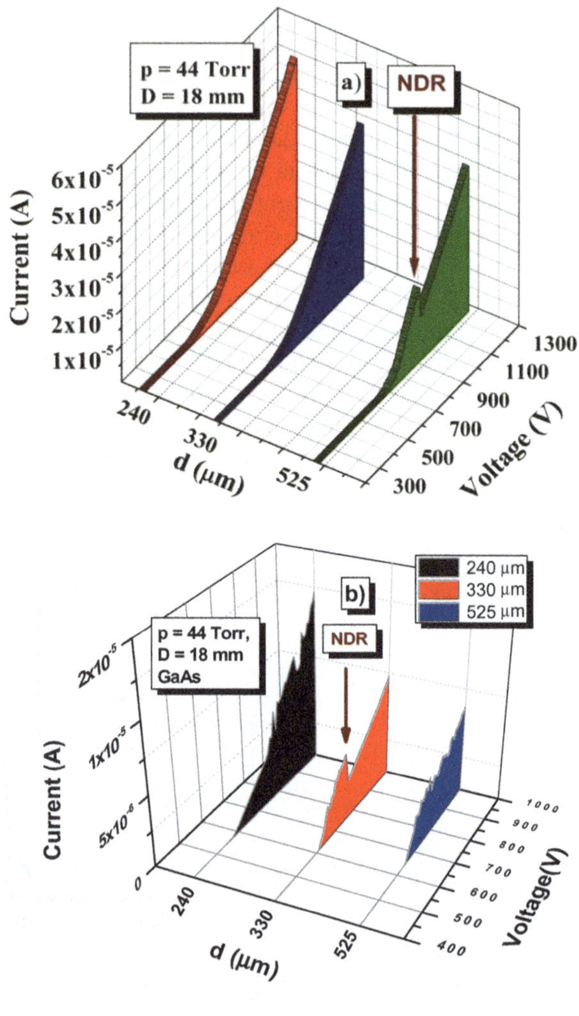

Fig. 9 **a**, **b** CVCs with NDR for different discharge gaps ($d = 240$, 330 and 525 μm) at 44 Torr without IR illumination. **a** for GaP, **b** for GaAs

These NDR effects have attracted a lot of interest related to the basic physical mechanisms involved and to their potential applications in the development of high-speed logic devices [59, 60]. NDR can cause current filaments and those filaments give damage to the semiconductor electrode when they move between cathode and anode. Besides, the ion–electron emission can affect the NDR related oscillations. In this case, electric field changes in space and time. Therefore, optimal plasma conditions require a stable discharge. Stable plasma discharge can be done by adjusting the plasma parameters such as pressure and interelectrode distance. Figure 10 shows plasma currents for ZnSe and GaP electrodes under various IR stimulations. Plasma currents increase resulting in the lower conductivity in the semiconductor electrodes with increasing IR illumination intensity as shown in Fig. 10. For strong illumination intensity, gas becomes conductive earlier as shown in Fig. 10 (i.e. 519 V for ZnSe, 645 V for GaP). The analyses of the CVCs for different electrodes indicate that ZnSe provides larger plasma currents under the same circumstances even at lower voltage values. It can be related to its direct bandgap. CVCs for different electrodes show that plasma current is governed by both IR stimulation and electrode type. The

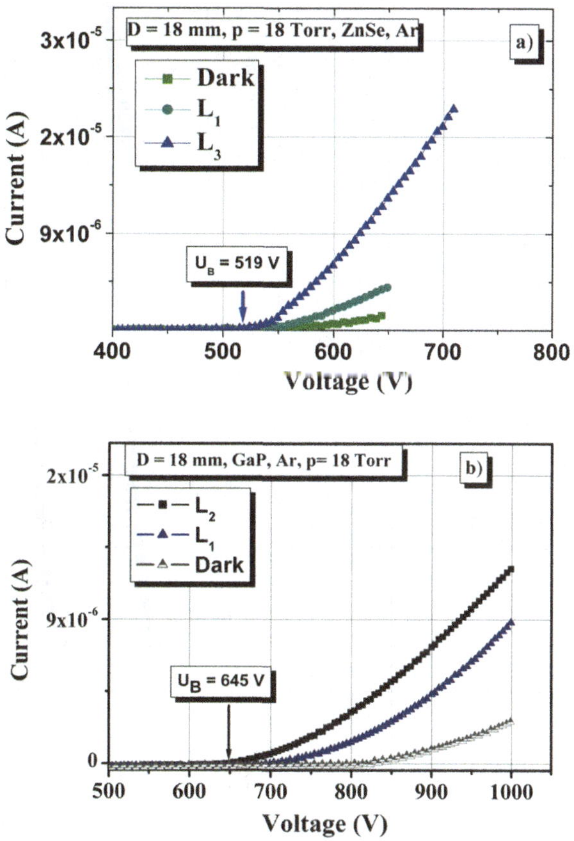

Fig. 10 **a**, **b** CVCs under different IR illumination intensities in which L_1 corresponds to weak and L_2 corresponds to moderate IR illumination intensity and L_3 corresponds to strong IR intensity. **a** for ZnSe, **b** for GaP

current is in the order of 2.3×10^{-5} A for ZnSe electrode and 1×10^{-5} A for GaP at illumination intensities L_3 and L_2 respectively.

The infrared absorption of semiconductors has been a topic of research for almost as long as the topic of semiconductors itself. Effect of IR illumination on the optical and electrical properties of GaP and ZnSe electrodes have shown that the optical response of the materials can be efficiently controlled by the gas discharge plasma. Present investigations have shown that CVCs are strongly dependent on the IR stimulation in the plasma–semiconductor system.

Figure 11 shows CVCs for air and Ar media concerning the diameter of the GaP photodetector at 35 Torr. The current values in the air media are greater than that of the argon media at the same conditions. Argon has advantages in comparison with air because it is a noble gas and more stable. The reactive species like oxygen in air

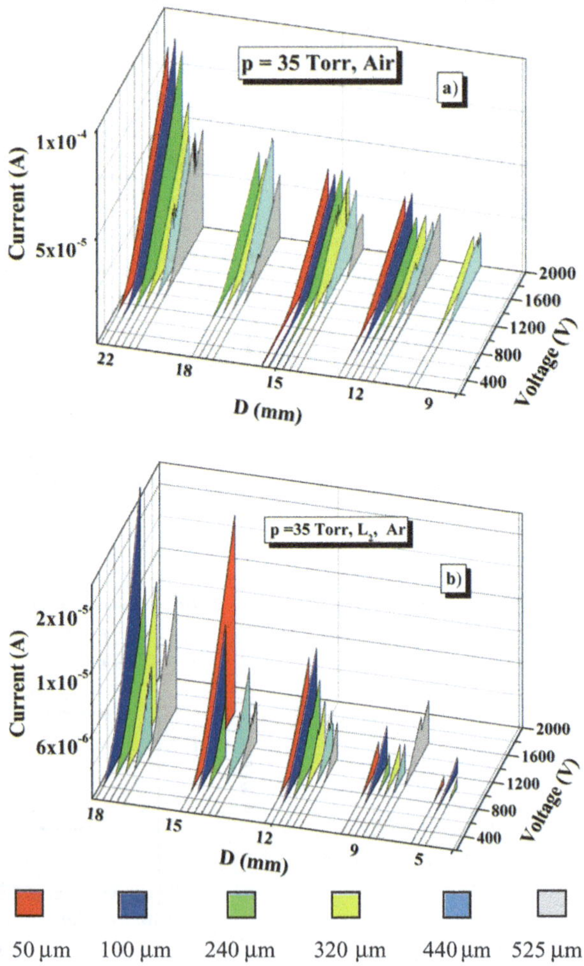

Fig. 11 Comparison of the CVCs for different *d* and cathode diameter *D* in air and Ar media at 35 Torr

plasma can cause a negative effect like etching of the electrode surface with plasma. For this reason, argon is preferred to avoid the oxidizing of the material. It is very crucial to get the stable plasma inside the cell for controlling the plasma current to suppress the instabilities. The maximum plasma current is 5×10^{-5} A for air plasma while it is around 1×10^{-5} A for Ar in the case of $D = 18$ mm and $d = 440$ μm.

Reactive oxygen species can destruct the cathode and anode by causing high energetic filaments in the discharge plasma system. For this reason, Ar is used as a working gas for many plasma related applications. The experimental results will provide significant contributions to a scientist working in this area especially for the understanding of the optimum plasma conditions to the device characterization for different gas media. Since the physical and chemical processes that occur in the plasma are quite complicated, a good understanding of the species present in the plasma and their energy levels is necessary [7]. For this reason, the experimental characterization of the Ar and air discharge operated at low and high pressure is very important in the field of gas discharge

To test the sensitivity of GaAs and GaP in the region of IR, CVCs were measured under three different IR illumination intensities, where the results clearly show that GaAs is more sensitive to IR illumination than GaP (Fig. 12). The results were repeated and confirmed for different *D* cathodediameters of semiconductors. Since GaAs is a direct bandgap semiconductor and its mobility is higher than GaP, the current values are higher as shown in Fig. 12. In fact, both semiconductor materials have critical importance to device performance in the IR detector applications.

The effect of IR stimulation on the CVCs was mainly investigated for two different electrodes. In Fig. 12, IR sensitivity of GaAs and GaP have been compared based on the CVCs measurements for different cathode diameter *D* under various IR stimulation. In terms of IR sensitivity, GaAs gives better response for different IR intensities. As a result of the plasma currents, higher current values also are observed for GaAs. The current for the plasma cell with GaP electrode is 10 times smaller than that of the cell with GaAs electrode.

The development of semiconductor performance requires the ability to capture charge carriers from lower energy levels to the higher levels that are exposed to the IR illumination and it is also related to the better electrical properties. The problem was how to control and convert the optical energy to the electrical energy efficiently without any loses at critical parameters before the devices had been used for an optoelectronic system. The influence of the gas type on the discharge current is shown in Fig. 13. In the cases studied above, the effect of the electrode diameter *D* on the CVCs are also reported for Ar and air-filled microplasma cell. The more electrode diameter increases, the higher plasma current can be obtained for both cases. This is related to the larger active plasma areas. The micro discharge systems also allow obtaining portable and flexible plasma cells for the technology and industry at room temperature.

The effect of the pressure on the system characteristics are seen in an argon and air media (Fig. 13). The current values change depending on the pressure. At the lower pressures in the argon, even up to 2 Torr, current can be obtained; whereas unstable current behaviors have been observed for air media. One of the biggest

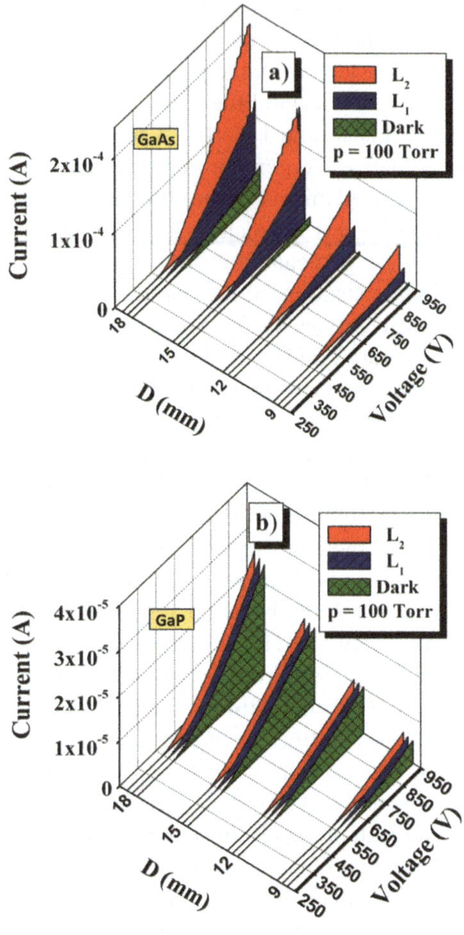

Fig. 12 **a**, **b** CVCs for the cathode diameter D (9–18 mm) in the case of different IR stimulations. **a** for GaAs and **b** for GaP cathodes. $d = 100\ \mu m$

problems when working in the air is the glow-to-arc transition which develops after critical electric fields at high applied voltages. Further, the electric field along with the discharge gap changes by the value of interelectrode distance. The current was measured around 2×10^{-5} A for Ar and around 5×10^{-5} A for air at $D = 18$ mm and $p = 35$ Torr. The plasma current and discharge light emissions are drastically changed by its discharge parameters such as gas type, gas pressure and applied voltage and semiconductor material used in the system as a cathode. Specifically, the gas type plays a key role in the excitation and ionization mechanisms in the ionization cell with the semiconductor electrode.

New plasma trends are focused on stable discharges. Nevertheless, the optimization of the plasma requires understanding the complex physical and chemical process. Microdischarge systems with semiconductor electrodes promise the uniform plasma by suppressing the instabilities in the cell. ZnSe and GaP are chosen to compare their

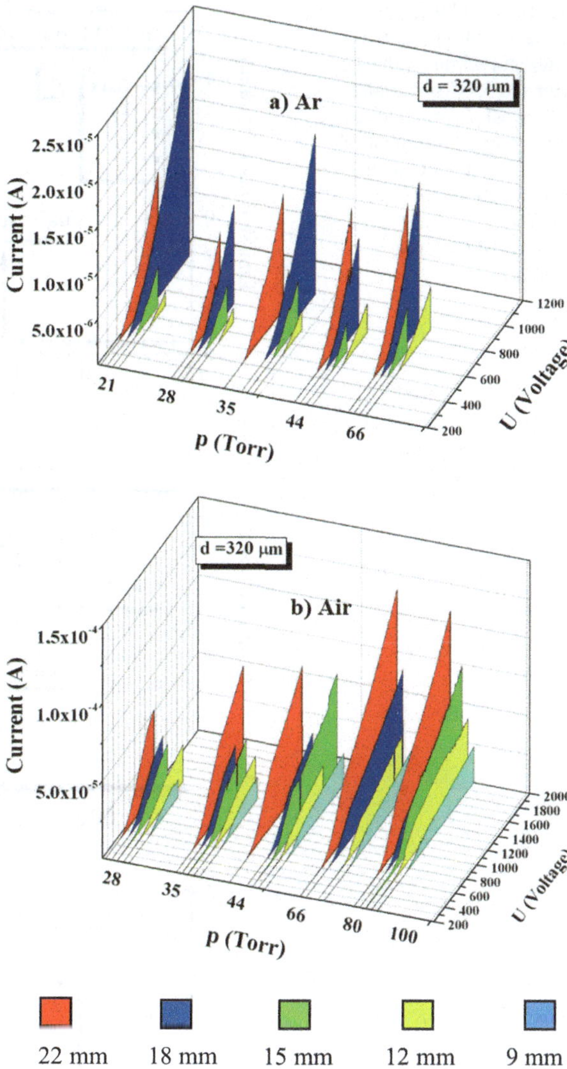

Fig. 13 **a**, **b** Current versus pressure curves for GaP in the case of Ar **a**, **b** air media

plasma radiations with the aim of optimization of the plasma cell. ZnSe is a direct bandgap material, whereas GaP is an indirect bandgap material.

The graphs below show the intensities of the gas discharge radiation for two different pressures when the GaP and ZnSe semiconductor electrodes are exposed to different IR illumination intensities (Fig. 14).

When comparing the radiation values for the ZnSe and GaP cathodes, it is seen that the plasma emission values are higher in the gas discharge cell with ZnSe cathode. Also, although it has a much larger bandgap than GaP, the ZnSe electrode exhibits greater sensitivity against IR illumination. According to DLE graphs, ZnSe produces

Fig. 14 **a–d** DLEs for ZnSe and GaP cathodes at two different pressure values (100 and 550 Torr) under dark, weak L_1, moderate L_2, and strong L_3 illumination intensities. **a** for GaP at 100 Torr; **b** for ZnSe at 100 Torr; **c** for GaP at 550 Torr; **d** for ZnSe at 550 Torr

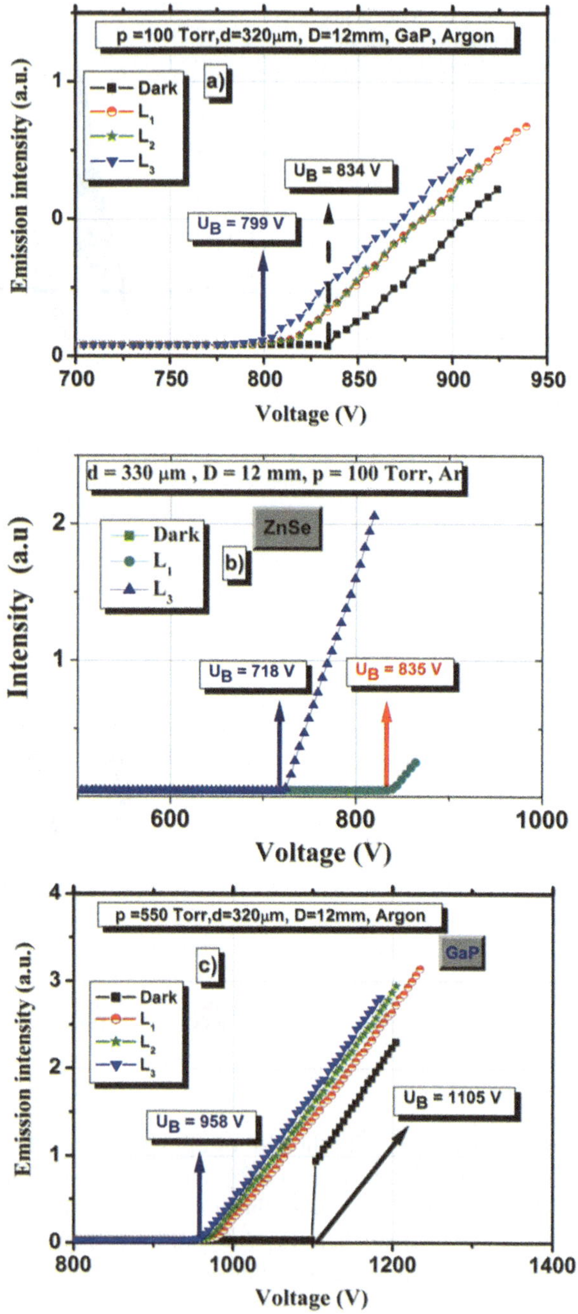

Fig. 14 (continued)

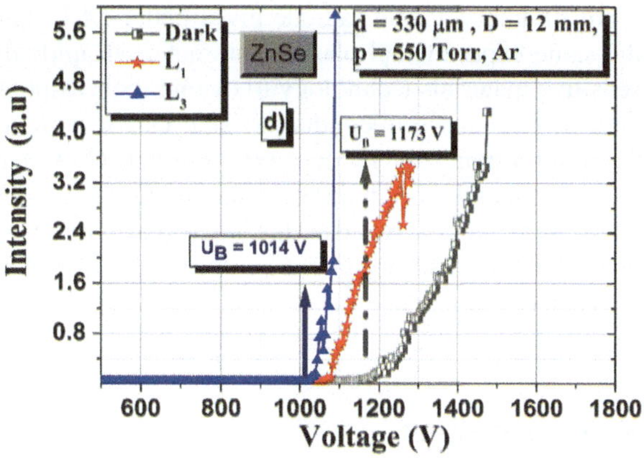

more luminous plasma because of the high electron energy in the evolution of the plasma discharge. The secondary electrons due to charge accumulation are related to the ionizing collisions between cathode and anode. Besides, 100 Torr gives more stable discharge characteristics compared to 550 Torr. Further, there is a shift in the CVCs for GaP electrode at 100 Torr under dark discharge.

When DLE intensities are measured in the plasma cell withGaAs cathode under the same experimental conditions above, it is clear that the radiation value is higher than the value obtained in cells with ZnSe and GaP cathode for 100 Torr. In this case, the highest DLE is obtained for the plasma cell with GaAs electrode (Fig. 15).

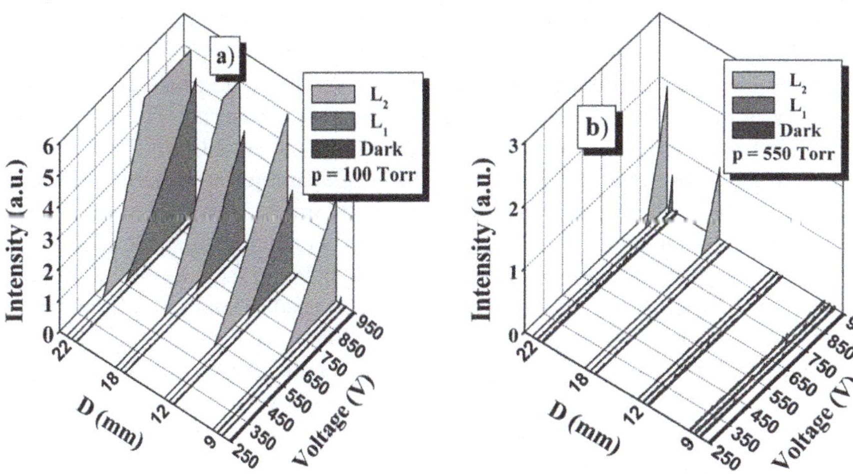

Fig. 15 **a**, **b** DLEs for the various cathode diameter *D* of GaAs electrode under different IR intensities a for 100 Torr; **b** for 550 Torr

In summary, GaAs cathode gives better DLE than GaP and ZnSe cathodes for the same experimental plasma parameters. Rapid advances are being made in IR-sensitive materials technology to extend their response in a long-wavelength scale. Narrow bandgap semiconductors like GaAs show a fast response by providing a good opportunity for IR detection. Further, GaAs have high electron mobility and absorption, and these features make the GaAs an ideal IR detector in a wide range of the wavelength. Besides, the microplasma cell with those semiconductors can provide better discharge light emissions for material processing technology in the microelectronic industry.

5 Conclusions

Several discharge experiments have been done for plasma related studies in the conventional plasma cells. The major difference of this study from the conventional plasma investigations that the plasma is produced between the semiconductor electrode and thin-film covered glass electrode for the micro-dimensional discharge gap. In this chapter, our motivation focused on the electrical gas discharges with various semiconductor electrode. Ar and air discharge was extensively examined. Despite its growing potential importance in physics and engineering, there does not exist enough information in the field of plasma–semiconductor systems and their optoelectronic application under different gas discharge parameters. Experiments were conducted by gas discharge cells with ZnSe, GaAs and GaP semiconductor electrodes. Highly important semiconductor materials and their CVC and DLE characteristics have been presented and the role of plasma on those characteristics has been clarified. The results show that the system characteristics strongly dependendt on both the semiconducting electrode and the plasma parameters. At the same time, IR sensitivity is strongly influenced by pressure variation and cathode material. Those materials are called photon detectors since they convert IR optical signal into the current in the system and thus act as a kind of transducer. Generation, transport, and multiplication are three important phenomena in the IR image converter systems. The qualities of the photodetectors are measured with their uniform spectral response to the optical signal, and their high-speed and low noise in the conversion process. The semiconductors with high-resistivity have been used to suppress plasma instabilities and to get more stable plasma. GaAs and ZnSe give response quickly to the optical signal because they are direct bandgap semiconductor materials and are therefore preferred in the optoelectronics. Technological improvements made IR-sensitive materials important. Valuable information can be obtained by using an IR converter system to investigate the surface properties of materials. This work can lead to many experimental studies in the field of plasma–semiconductor structure to know about the complex plasma and radiation behaviors.

Acknowledgements This research was funded by two grants, namely BAP No. 05/2012-47 and 05/2012-72 from the Gazi University-Scientific Research Project Unit.

References

1. J.M. Longras Figueiredo, Dissertation, Faculdade Ciencias da Üniversidade do Porto, Apr 2000
2. M.L. Riaziat, *Introduction to High Speed Electronics and Optoelectronics* (Wiley, New York, 1996), p. 79
3. A. Katz, *Indium Phosphide and Related Materials: Processing, Technology and Devices* (Artech House, London, 1992), p. 65
4. Y.Y. Peter, C. Manuel, *Fundamentals of Semiconductors Physics and Materials Properties*, 4th edn. (Springer, 2010), p. 1
5. M. Dagenais, R.F. Leheny, J. Crow, *Integrated Optoelectronics* (Academic Press, London, 1995), p. 178
6. S. Adachi, *Physical Properties of III–V Semiconductor Compounds, InP, InAs, GaAs, GaP, InGaAs, InGaAsP* (Wiley, New York, 1992), p. 145
7. D.A. Fishman, C.M. Cirloganu, S. Webster, L.A. Padilha, M. Monroe, D.J. Hagan, E.W. Van Stryland, Nat. Photonics **5**, 561 (2011)
8. P. Genevet, J.P. Tetienne, E. Gatzogiannis, R. Blanchard, M.O. Scully, F. Capasso, Nano Lett. **10**, 4880 (2010)
9. J.A. Schuller, S.B. Edward, C. Wenshan, C.J. Young, S.W. Justin, B.L. Mark, Nat. Mater. **9**, 193 (2010)
10. F.S. Capasso, A.Y. Cho, IEEE J. Quantum Electron. **30**, 1313 (1994)
11. S.A. Haque, J.T. Nelson, Science **327**, 1466 (2010)
12. R. Khenata, A. Bouhemadou, M. Sahnoun, A.H. Reshak, H. Baltache, M. Rabah, Comput. Mater. Sci. **38**, 29 (2006)
13. M.A. Hasse, J. Qui, J.M.D. Puydt, H. Cheng, Appl. Phys. Lett. **59**, 1272 (1991)
14. H. Kinto, M. Yagi, K. Tanigashira, T. Yamada, H. Uchiki, S. Iida, J. Cryst. Growth. **117**, 348 (1992)
15. M.C. Tamargo, M.J.S.P. Brasil, R.E. Nahory, R.J. Martin, A.L. Weaver, H.L. Gilchrist, Semicond. Sci. Technol. **6**, 6 (1991)
16. E. Monroy, F. Omnes, F. Calle, Semicond. Sci. Technol. **18**, R33 (2003)
17. U.C. Sharma, *Dissertation* (Electronic Systems Group, EE Dept, IIT Bombay, 2004)
18. A. Rogalski, *Infrared Detectors*, 2nd edn. (CRP Press, Boca Raton, 2010), p. 309
19. ReseachGate Database: https://www.researchgate.net/post/How_does_band_gap_of_a_semiconductor_change_with_the_lattice_constant_Please_explain_the_physical_significance. Accessed 14 Feb 2014
20. T. Peng, H. Yang, K. Dai, X. Pu, K. Hirao, Chem. Phys. Lett. **379**, 432 (2003)
21. B.R. Kumar, S.R. Meher, Appl. Phys. **8**, 47 (2016)
22. A.F. Holleman, E. Wiberg, N. Wiberg, W. Holleman, Die Zinkgruppe, in *Lehrbuch der Anorganischen Chemie*, 101th edn. (de Gruyter, Berlin, New York, 1995), p. 1375
23. F. Johannes, in *Dissertation zur Erlangung des Doktorgrades der Naturwissenschaften* (Doctor rerum naturalium, 2007)
24. R. Paschotta, in *RP Photonic Encyclopedia* (2001). https://www.rp-photonics.com/band_gap.html. Accessed 2011
25. P. Meredith, C.J. Bettinger, M.I. Vladu, A.B. Mostert, P.E. Schwenn, Rep. Prog. Phys. **76**, 034501 (2013)
26. J.J.L. Morton, in *Electrical and Optical Properties of Materials* (June 2019), pp. 1–29. https://www.ucl.ac.uk/qsd/people/teaching/EOPM-Part5.pdf
27. A. Rogalski, Infrared detectors: an overview. Infrared Phys. Technol. **43**, 187 (2002)
28. P.R. Norton, Infrared detectors in the next millennium. Proc. SPIE **3698**, 652 (1999)
29. A. Rogalski, *Infrared Detectors* (Gordon and Breach Science Publishers, Amsterdam, 2000), p. 232
30. Y.A. Astrov, A.N. Lodygin, L.M. Portsel, A.A. Sitnikova, J. Appl. Phys. **124**, 103303 (2018)
31. A.N. Lodygin, L.M. Portsel, E.V. Beregulin, Y.A. Astrov, J. Appl. Phys. **126**, 173302 (2019)
32. Yu.A. Astrov, A.N. Lodygin, L.M. Portsel, E.V. Beregulin, Phys. Rev. E **95**, 043206 (2017)

33. S.S. Kasymov, L.G. Paritskii, Device for tracking images, in *Russian Authors' Certificate* 1798020/18–10 (1973)
34. Y.A. Astrov, A.N. Lodygin, L.M. Portsel, J. Phys. D Appl. Phys. **49**, 095202 (2016)
35. H.H. Kurt, E. Tanrıverdi, J. Electron. Mater. **46**, 4024 (2017)
36. H.H. Kurt, E. Tanrıverdi, B.G. Salamov, JOM **71**, 644 (2019)
37. H. Yasuda, L. Ledernez, F. Olcaytug, G. Urban, Pure Appl. Chem. **80**, 1883 (2008)
38. E. Wagenaars, Dissertation, Technische Universiteit Eindhoven (2006)
39. Yu.A. Astrov, H.G. Purwins, Phys. Lett. A. **283**, 349 (2001)
40. A.W. Liehr, A.S. Moskalenko, Yu.A. Astrov, M. Bode, H.G. Purwins, Eur. J. Phys. B. **37**, 199 (2004)
41. B.G. Salamov, N.N. Lebedeva, H.Y. Kurt, V.I. Orbukh, E.Y. Bobrova, J. Phys. D Appl. Phys. **39**, 2732 (2006)
42. Z. Nikitovic, A. Strinic, V. Samara, G. Malovic, Z. Petrovic, Acta Chim. Slov. **52**, 463 (2005)
43. Y.P. Raizer, *Gas Discharge Physics* (Springer, 1997), p. 76
44. M.C. Penache, Dissertation, Physik der Johann Wolfgang Goethe Universität in Frankfurt am Main (2002)
45. J.M. Meek, J.D. Graggs, *Electrical Breakdown of Gases* (Clareendon, Oxford, 1953), p. 78
46. L. Mangolini, K. Orlov, U. Kortshagen, J. Heberlein, U. Kogelschatz, Appl. Phys. Lett. **80**, 1722 (2002)
47. D.B. Go, D.A. Pohlman, J. Appl. Phys. **107**, 103303 (2010)
48. V.A. Lisovskiy, P.P. Platonov, S.V. Dudin, Plasma Phys. **25**, 156 (2019)
49. I.H. Hutchinson, *Principles of Plasma Diagnostics*, 2nd edn. (Cambridge University Press, Cambridge, 1987), p. 2
50. H. Kurt, E. Koc, B.G. Salamov, IEEE Trans. Plasma Sci. **39**, 1086 (2010)
51. M. Ohring, *Materials Science of Thin Films*, 2nd edn. (Academic Press, 2001), p. 794
52. B.N. Klyarfeld, L.G. Guseva, A.S. Pokrovskaya-Soboleva, Sov. Phys. Tech. Phys. **11**, 520 (1966)
53. D.D. Sijacic, U. Ebert, Phys. Rev. E **66**, 066410 (2002)
54. H.Y. Kurt, J. Electron. Mater. **47**, 4486 (2018)
55. V.I. Kolobov, A. Fiala, Phys. Rev. E **50**, 3018 (1994)
56. R. Arslanbekov, V. Kolobov, J. Phys. D Appl. Phys. **36**, 2986 (2003)
57. M.T. Jamil, J. Ahmad, S.H. Bukhari, M.E. Mazhar, U. Nissar, A.J. Rao, H. Ahmad, G. Murtaza, Dig. J. Nanomater. Biostruct. **12**, 595 (2017)
58. A. Qayyum, M. Ikram, M. Zakaullah, A. Waheed, G. Murtaza, R. Ahmad, A. Majeed, N.A.D. Khattak, K. Mansoorand, K.A. Chaudhary, Int. J. Mod. Phys. B. **17**, 2749 (2003)
59. L.W. Yu, K.J. Chen, J. Song, J.M. Wang, J. Xu, W. Li, X.F. Huang, Thin Solid Films **515**, 5466 (2007)
60. H. Inokawa, A. Fujiwara, Y. Takahashi, Appl. Phys. Lett. **79**, 3618 (2001)

Charge Transport Mechanisms in the Silver-Modified Zeolite Porous Microstructure

B. G. Salamov

Abstract The interaction between microdischarge and microporous zeolite electronic material (ρ~10^6–10^{11} Ω cm) modified by silver (Ag^0) nanoparticles cold plasma generated at the atmospheric pressure is investigated experimentally as a function of pressure p (8–760 Torr), electrode gap d (50–250 μm), and diameters D (9–22 mm) of the cathode areas in the gas discharge electronic device (GDED) with nanoporous zeolite cathode (ZC). The role of charge carriers in mixed conductivity processes, electrical and dielectric features of zeolite, is analyzed in air microplasmas. The results obtained from experiments indicate that Ag^0 nanoparticles play a significant role in reducing the breakdown voltage (U_B) in GDED with ZC. It was found that with increasing silver concentration, resistance of zeolite plate monotonically decreases and at the same time the capacitance is increased. The observed frequency dependence (1–200 MHz) of the capacitance and resistance of ZC on the silver concentrations may be explained on the basis of an electrode–dielectric interface gap model.

Keywords Microporous zeolite cathode · Atmospheric pressure plasma light source · Electric field · Electro-optical properties · Charge transport mechanisms · Ag-modified zeolite

1 Introduction

In recent years, there has been a great interest in an insulating nanoporous zeolite which is an improvable advanced material instead of the semiconductor. This is due to their attractive physical properties, multifunctionality, and wide applications in various fields of the modern industrial technology, optoelectronics, and atmospheric microplasmas. Nanoporous zeolites have three-dimensional framework structures.

B. G. Salamov (✉)
Physics Department, Faculty of Sciences, Gazi University, Beşevler, 06500 Ankara, Turkey
e-mail: bala@gazi.edu.tr

Institute of Physics, Azerbaijan Academy of Science, AZ 0143 Baku, Azerbaijan

S. J. Ikhmayies et al. (eds.), *Advances in Optoelectronic Materials*,
Advances in Material Research and Technology,
https://doi.org/10.1007/978-3-030-57737-7_7

The zeolite structure is built from TO_4 (T = Si, Al) tetrahedra which are linked together to form a three-dimensional framework of interconnecting pore and channel structure. The conductivity of zeolites is usually associated with ionic conductivity. Charge is carried by means of the motion of mobile charge-balancing ions in the anion framework. Hydrated zeolites are also known as proton conductors and hydroxyl ion conductors. It is usually presumed that zeolites do not act as electronic conductors. Furthermore, dielectric materials such as zeolites have the ability to store energy upon the application of external electric field.

In recent years, different studies are conducted for natural zeolite modified by ions and nanoparticles of various metals, not only for medical purposes, but also as an element of the electrical circuit. Specific electronic properties of nanopores in zeolite structure make zeolites good candidate materials in the electronic industry so that silicon-based technologies could be replaced or supplemented by nanoporous zeolitic materials. The dielectric response of zeolites is one of the key parameters for applications. Such knowledge about zeolites has become important because of their potential application in electrochemical devices.

At the same time, this study has been undertaken to better understand the effect of geometrical (discharge gaps d) and operating (different residual pressures and applied voltages U_0) parameters on the current behavior of the system and operation principle of charge transport in nanoporous materials. This study follows the previous work [1] which described the gas breakdown processes in DC microdischarges with a ZC. In Ref. [1], authors established that the electric field is a more effective parameter on the discharge characteristics in the cell with ZC than active area diameters D. Therefore, the main aim of this study is to investigate the enhanced effect of electric field on the fundamental characteristics of GDED with constant D of ZC. Transport of charge carriers, conductivity measurements, and stable discharge glow regions depending on the electric field and gas pressure were investigated.

2 Electro-optical Properties of Porous Zeolite Cathode in the Plasma Light Sources

Cold plasma processing techniques can be preferred for the active operation at atmospheric pressure applications due to the development in many areas of the modern industrial technology. These plasma processing techniques are not required for costly vacuum technology since the high rates of thin-film deposition are easier at atmospheric pressure. Recently, many studies have been performed to produce and sustain a homogeneous-stable atmospheric pressure plasma [2, 3]. There are two approaches based on the so-called Paschen similarity law ($p \times d =$ const), which reduce the electrode dimensions up to micrometer range to ignite atmospheric pressure discharges, and moderate voltages, working in the Paschen minima for different gas environments. To generate atmospheric pressure glow discharges, a microhollow cathode discharge (MHCD) is used by many authors. In Ref. [4], a gas discharge electronic device with a high-resistivity *GaAs* semiconducting photocathode was suggested as alternative large emitting area UV plasma source. Earlier, this idea was suggested for the electrode configurations including the MHCD, operating with continuous pulsed DC power supplies [5]. Simultaneous operation of multiple discharges in some circuit has been achieved, but there is a natural problem in large-scale scheme because of the requirement of series resistors in order to balance all microdischarges in MHCDs. Semiconductor gas discharge electronic device with multichannel spacer can notably prevent this problem.

In a plane-parallel discharge chamber, the uniform plasma emission and distributed discharge current can be observed for photocathode with resistivity of more than 10^6 Ω cm [4]. The plasma emission uniformity is controlled by the resistivity distribution over the semiconducting cathode. Such a discharge is supported by the electrode processes and by the multiplication of the number of charge carriers in the gas volume due to the avalanche mechanism. We assume that a description of Townsend and glow discharge modes in a planar gas discharge system (see Fig. 1) on the right-hand branch of the Paschen curve (Paschen curve demonstrates the breakdown voltage U_B necessary to start a discharge) must take into account a change in the effective secondary emission coefficient γ_{eff} of the cathode, which is related to the electron backscattering effect [6]. The cathode boundary condition is usually taken in the form of the Townsend relation between the electron j_e and ion j_i current densities,

$$j_e(0, t) = \gamma_{\text{eff}} j_i(0, t) \tag{1}$$

where γ_{eff} is the second Townsend coefficient reflecting the contributions of various processes to the secondary electron emission from the cathode. Since the coefficient γ_{eff} depends rather strongly on the parameter E/p (E—is electric field and p—is pressure), this dependence has to be taken into account in the description of non-stationary discharges. This approach is most consistently developed by Phelps and Petrovic in [7], where the γ_{eff}values (called the apparent secondary electron emission coefficient) for argon are determined and tabulated based on a thorough analysis of

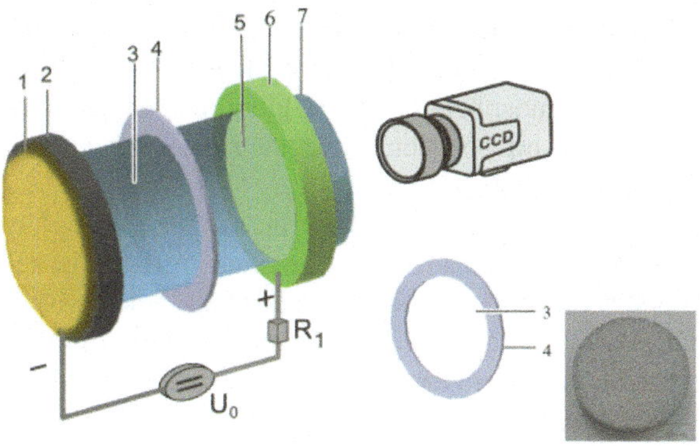

Fig. 1 Scheme of the gas discharge cell: 1-metallic Cu contact; 2-semiconductor or zeolite cathode; 3-diameter of the discharge gap; 4-thickness of the insulating mica; 5-semi-transparent conductive SnO_2 contact; 6-flat glass disc; 7-UV–visible light beam. **b** Porous structure of zeolite cathode [21]

experimental results. Thus, it has been established in [6] that on the right-hand branch of the Paschen curve, and the change in the effective secondary emission coefficient $\gamma_{\text{eff}}$of the cathode must be taken into account, which is related to the electron backscattering effect. The γ_{eff}value depends predominantly on the initial energy of emitted electrons and on the applied electric field amplitude.

The gas discharge electronic devices with a semiconducting photocathode (the same as it is shown in Fig. 1) have been used for various necessities in the IR image systems [8]. Current density J in the converter is controlled locally to focus the optical IR pattern on a photosensitive semiconductor, which gives an image converted to visible radiation by means of discharge. The I-V characteristics (CVC) allow us to determine the cell parameters: (1) breakdown voltage U_B; (2) variation of resistivity ρ due to homogeneities of the semiconducting cathode at different illumination intensities L (i.e., change of conductivity $\sigma = \partial J/\partial U$ or $\rho = \partial U/\partial J$). In order to quantify this parameter, the influence of the incident light beam on the resistance of the semiconductor is measured. This has been done as follows. We assumed that a homogeneous stationary Townsend discharge [9] is established in the gap at appropriate V_B. This mode of discharge is observed for low currents between the point of ignition and the point where negative differential conductivity is observed in the gas characteristic. The *CVC* of the system is very close to a linear curve if $U > U_B$, reflecting the ohmic behavior of the semiconductor cathode [10]. The voltage drop at the discharge gap for this discharge mode is independent of the current. Therefore, the slope of the *CVCs* provides the resistance of the *GaAs* cathode. Then, the specific conductivity can be computed from this resistance and the geometric dimensions. The resistivity of the semiconductor decreases monotonically when the intensity of the irradiation is raised. The filamentation was primarily due to the formation of a space charge of positive ions in the discharge gap, which changed the discharge from the Townsend to the glow type [11]. But since the gas discharge electronic device with semiconducting photocathode cannot effectively operate at atmospheric pressure, therefore we used for the first time an insulating nanoporous zeolite which is an improbable advanced material instead of the semiconductor. Efficient operation of the system will be based on the use of zeolite cathode, which is a good absorber of gas molecules in its nanopores. Thus, the discharge initiates from the surface and channels of the zeolite cathode, unlike conventional planar gas discharge electronic device with *GaAs* semiconducting photocathode [12]. The electrode dimensions are too small to generate sufficiently high electric field strengths to ignite atmospheric pressure glow discharges. Therefore, it is important to know the relationship between the geometrical parameters and the discharge characteristics for the cell optimization.

Modification of electrode surfaces with zeolites has aroused considerable attention [13]. The reported advantageous characteristics of zeolites, which affect the electron transfer reactions, are the size and shape selectivity, because of the framework. The ion-exchange capacity depends on the nature of the cation species, size, and charge. Therefore, using of the zeolite cathode in the gas discharge electronic devices acquires value, because of the setup, preparation, and use of different type of the zeolite cathodes (more detailed information is given in next section).

The high emissive characteristics of nanotubes and pores provide the basis for a new class of electron emitters with very low supply voltage and power consumption [14, 15]. Analysis of the emissive properties of different nanomaterials such as zeolite reveals that they are promising as field emitters [16]. The unique emissive characteristics of carbon nanotubes make them effective electrode coatings in gas discharge electronic devices. Moreover, flat-panel light sources such as cold cathode fluorescent lamp (CCFL) or LED were mainly used for the backlighting of liquid-crystal displays(LCDs). Unfortunately, the CCFLs are limited by their architecture, risky high-voltage operation, and are not cost effective [17]. We propose a design of flat-panel light source without additional optical components which has a simple configuration to achieve the low-power consumption and uniform processing element (PE). This backlight panel contains zeolite which assists igniting the plasma efficiently.

The novel results are related to the analysis of electrical breakdown processes and spatial stabilization of the current and control of it with a nanoporous zeolite cathode (ZC) which operates at atmospheric pressure more efficiently compared to a semiconducting photocathode. Natural zeolite with non-stoichiometric compounds was taken as a porous object. Zeolites are aqueous alumosilicates in which an infinite alumosilicate frame is produced by $[SiO_4]^{4-}$ and $[AlO_4]^{5-}$ tetrahedra having common vertices. These tetrahedra have communicating cavities occupied by large ions and water molecules [18, 19]. For our study, we used $(Ca,K_2,Na_2,Mg)Al_8Si_{40}O_{96}.24H_2O$ the clinoptilolite which contains on the average 90–95% of the clinoptilolite zeolitic mineral[20]. Clinoptilolite belongs to a class of zeolite minerals having the clear-cut structural topology of heulandite (HEU) and the ratio of Si/Al > 4.0. The resistivity of the ZC was 1.5×10^{10} Ω cm. The measurements of breakdown voltages are performed for different residual pressures up to atmospheric pressure. It is found that when such a ZC plate is used in a planar gas discharge electronic device, the ignition voltage of the discharge, which is uniformly distributed over the electrode surface, drops considerably. Among the tested ZCs, a cathode which has a large active area diameter (*D*) shows an important contribution to the development of electrical breakdown. For this reason, we believed that the gas discharge electronic device with ZC is highly promising as plasma light sources with a large emitting area and high spatial homogeneity of UV radiation.

At this point, we assume that a homogeneous stationary Townsend discharge, where space charge effects inside the gap are small, does not cause a distortion in the electric field between the electrodes, and a homogeneous distribution of j perpendicular to the current flow is established in the gas discharge electronic device [22]. The intensity of the homogenous plasma emission depends on the values of discharge current; however, the wavelength of the plasma emission varies with respect to the type of gas used in the discharge gap. Changing the ZC resistance locally gives rise to a change of the discharge current and the plasma emission [23]. In the gas discharge electronic device with ZC, it is found that the potential drop during the course of the discharge gap is independent on the current, which is a specific operation peculiarity of the Townsend discharge mode. Therefore, the inverse of the slopes of the I-V characteristics provides the resistance of the ZC (see Fig. 2).

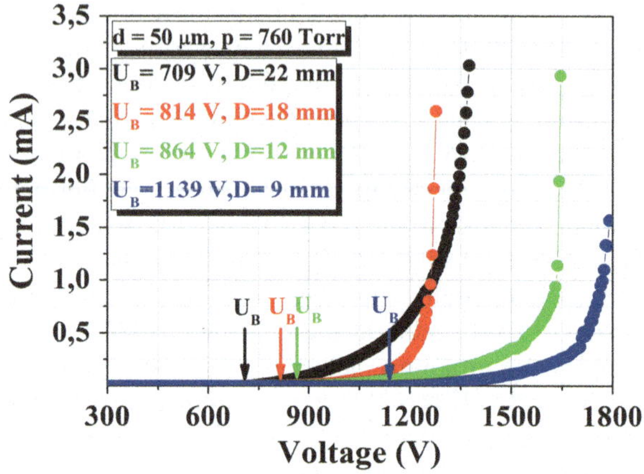

Fig. 2 I-V characteristics of the discharge system at atmospheric pressure for different values of *D* of the zeolite cathode [21]

Then, the ZC specific conductivity can be calculated from the resistance and the geometric dimensions. We remark that the current density j in the gas discharge electronic device does not exceed the limiting current density for the existence of the Townsend discharge at given experimental conditions [11]. The feeding voltage U_0 is the sum of the potential drops at the gas gap and at the zeolite component. Moreover, the potential drop remained constant through the discharge region as the current values vary, which is one of the characteristic features of the Townsend discharge mode.

The mechanism of the microdischarges inside the porous zeolite and ceramic plates are probably related to the back-corona discharge, a phenomenon often found in electrostatic precipitators [24, 25]. It occurs when charged dielectric particles are collected on the electrode and form a porous layer of high resistivity, through which the electric current must pass. The highly resistive layer does not allow the charge to decay at the desired rate, resulting in a buildup of excess charge on the layer. When the voltage drop across the layer exceeds a critical value, an ultimate breakdown through the layer occurs. The breakdown occurs in the form of fine channels (microdischarges). The exact description of entry of the electrons from the barrier discharge plasma into the capillary is very complicated. It requires tracking of the motion of charged particles, development of the geometry, the electric field intensity, and the space charge on the capillary walls [26]. With the increase of the applied voltage, the current density continues to increase. Strictly speaking, for the improvement of the gas discharge electronic device with ZC, the optimal operation value of electrode gap is $d = 50\ \mu\text{m}$ up to the atmospheric pressure which means that the better stabilization of glow discharges can be obtained. To make possible wide areas of plasma applications under those conditions, it is *reasonable* to use gas discharge electronic device because of producing gas discharges up to atmospheric pressure at moderate voltages. It can be observed that the I-V characteristics have a smooth current increase, but they are totally different from the I-V characteristics of the gas discharge electronic device with *GaAs* semiconducting photocathode (for comparison, see Fig. 2 in Ref. 3). As the D varies from 9 to 22 mm, the U_B changes and the form of the I-V characteristics is reproducible, except for small differences in the current values as shown in Fig. 2. Reference [27] shows the *comprehensive characterization* of the discharge properties with respect to the effect of the pore and channel size, discharge power, and gas mixture. From the results, it can be seen that the microplasmas inside the ceramic plate occurred only for the particular discharge power and pore size of the ceramics. A surface barrier discharge on the ceramics surface may only be observed at small voltages. Inside the ceramics, as the U_0 increases, the surface discharges pass to capillary microdischarges in which onset voltage increases with decreasing ceramics pore and channel size.

The atmospheric pressure is much more compatible and optimal for planar gas discharge electronic device with ZC, as shown in the values of current and plasma emission in Figs. 3 and 4, respectively. In addition, discharge currents and plasma

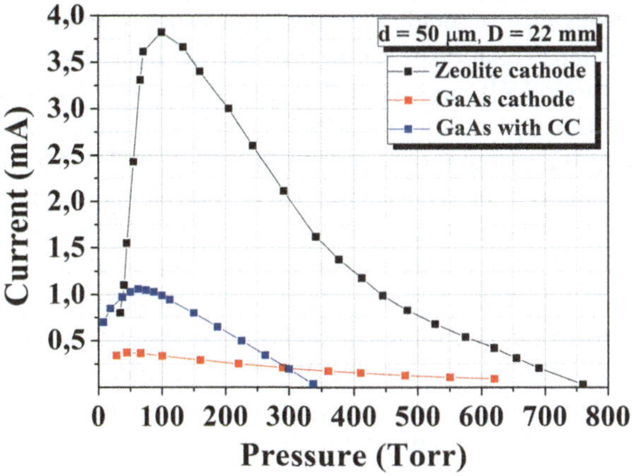

Fig. 3 Current changes as a function of pressure for discharge system with zeolite cathode, GaAs semiconducting photocathode with cold cathode (CC) [12], and GaAs semiconducting photocathode at feeding voltage $U_0 = 1$ kV [21]

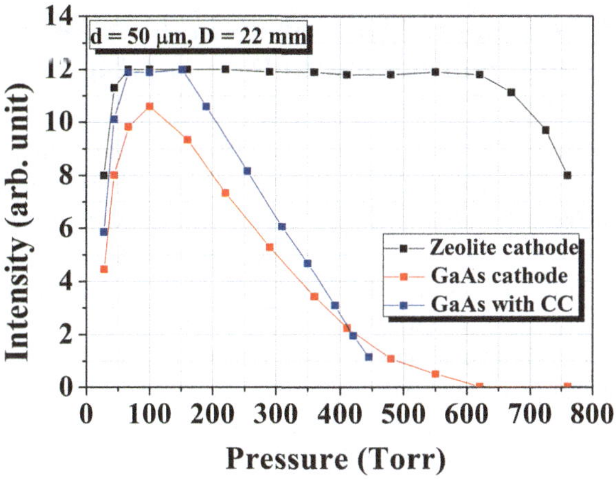

Fig. 4 Plasma emission intensity of the gas discharge gap with zeolite cathode against pressure at $U_0 =$ 1 kV [21]

emission intensities in gas discharge electronic device with ZC are much larger and intensive compared to the chamber with *GaAs* semiconducting photocathode. In the gas discharge electronic device (GDED) with *GaAs* semiconducting photocathode, plasma emission satisfies the optimal and effective conditions at lower pressures (15–350 Torr). But, this trend becomes vice versa near the atmospheric pressure. In short, higher plasma emission intensities are obtained from cell with ZC. This situation proves that one should prefer to use zeolite cathode at atmospheric pressures rather than *GaAs* semiconducting photocathode since the plasma emission values are better. At the same time, recently, it has been found that the ZC is an appropriate crystalline matrix to stabilize semiconductor particles of PbS at nanometric size (about 10 nm) [28].

Thus, porous cathodes are used for plasma light source applications with very low supply voltage and power consumption. The microstructures and performance of the zeolite cathodes are found to depend sensitively on the channel morphology and nature of the nanopores. Specific geometry of the zeolite channel structure and the strong electric field in the nanopores supply an efficient electron multiplication and related excitation of gas atoms. The use of gas discharge gap with nanoporous zeolite cathode (ZC) gives rise to the increase in plasma emission intensity. This device might find an application for generating and sustaining a stable, uniform, and homogeneous non-thermal atmospheric pressure plasma [21]. The unique electronic properties of nanopore zeolites make them also good candidate materials in the electronic industry so that silicon-based technologies could be replaced or supplemented by nanopores zeolite-like materials. We also believe that modification of the zeolite cathode as well as the geometry size in a non-thermal plasma reactor gives possibilities to build rather sensitive large emitting area plasma light source with internal emission amplification. Therefore, we confirm that the porous zeolite is a suitable electrode material for gas discharge electronic device and can serve as a source of UV radiation if pressure and electric field are sufficiently high.

3 Transport Mechanisms in the Atmospheric Pressure Plasma Device with Porous Zeolite

Zeolites are aqueous alumosilicates in which an infinite alumosilicate frame is produced by $[SiO_4]^{4-}$ and $[AlO_4]^{5-}$ tetrahedra having common vertices. These tetrahedra have communicating cavities occupied by large ions and water molecules [29] Each Al ion in the zeolite framework gives a net negative charge, which is balanced by an extra-framework ion, usually from group IA or IIA [30]. Clinoptilolite (CL), the most abundant natural zeolite mineral, has been widely used in research studies because of its abundance and considerable low cost [31, 32]. Despite the advantageous properties of CL which is a cathode material in our gas discharge cell, there is insufficient report about the electrical transport properties of CLs. This is due to a number of objective difficulties, namely formation in nature in the form of small

(0.01–0.1 mm to 1 μm) crystals, filling its cavities and channels by water molecules, the difference in the cation composition, the presence of phase boundaries, and the presence of impurities [33]. Therefore, in this study, novel results related to the transport mechanisms for GDED with ZC using the atmospheric pressure (AP) glow discharge plasma are interpreted.

Zeolite water plays a significant role for the stability of CL framework. The water can be removed by heating or vacuumizing the ZC. These processes do not affect the rigid framework: Its structure remains nearly unaffected. The hydrogen bond is formed between the O atoms from the framework, which is a coordination of positively charged cations, and H_2O molecules, located in the coordination of the cations. This hydrogen bond as believed [34] leads to the increase in mobility of cations in the pore space. Based upon this assumption, the electric current in zeolites is caused by movement of positive metal cations in the pore space, and it depends on 'the pores' content of water molecules.

Water molecules are dissociated into H and O with the plasma treatment [35–37]. The H generation rate increased with an increase in U_0, pulse repetition rate and/or number of electrode assemblies, which is related to the corresponding increase in the electrical energy deposited in the plasma. The pulsed DC discharge plasma will generate a positive surface charge on the dielectric surface [38] and consequently increase the normal component of the electric field that accelerates the electrons toward the surface. Thermionic/photoemission of electrons from the dielectric surface [39] accounts for the increase in electrical current. However, the redistribution of the electric field is still less than in the case where the cathodes are connected by a conductive layer on the backside of the dielectric layer. In this case, even without surface charge, the electrode arrangement causes a strong increase in the component normal to the surface of the dielectric that can be inferred from modeling of the electric field in such an electrode configuration [40]. The effect is the strongest at the wire where the electric field intensity is highest. This fosters the breakdown ignition. Also, the streamer plasma generates, as in any surface discharge, a positive space charge on the dielectric [38]. These charges vary dynamically and change as the discharge streamers propagate [38]. The application of electric fields in the presence of the conductive layer polarizes the dielectric and creates surface charges that lead to an additional increase in the electric field component normal to the surface[37]. The normal field component accelerates free electrons toward the surface and keeps the plasma attached to the surface[41]. It results in increased rates of secondary electron emission (*SEE*) through photoemission from the surface[39]. Likely, *SEE* due to energetic ions and metastable particles, especially close to streamer head, is even more effective, as shown in the case of plasma display panels [42]. This increased electron generation rate explains the increase in electrical power deposited in the water plasma in the presence of the conductive layer which is in agreement with similar effect for laser pumping [41] or nitrogen oxide mitigation [40] in air plasma [37].

In our case, the ions deposited on the surface of the high-ohmic ZC are not neutralized immediately, but influence the discharge through their electric field. The electric field E produced by the ions diminishes the external field in the gas. The resulting field must then be equal to the Townsend field to maintain the self-sustained discharge[43]. Another point is the discussion of the discharge using different dielectric materials: Many reasons for the different types of behavior are discussed, but the essential point is that a different dielectric will cause a completely different electric system [44]. New dielectric materials have been introduced for printed circuit board applications, such as Thermount and polyimide with the aim of matching the requirements for high speed and high density of electronic devices. Electrical and mechanical characteristics of these materials are guaranteed in standard operating conditions, but quantitative characterization of the surface resistivity for the different material samples under various testing conditions such as relative humidity, temperature, solder flux contamination, and dielectric barrier discharge and corona discharge is not yet fully investigated [45, 46]. Therefore, in this chapter, we analyze the breakdown behavior in a GDED under optimal breakdown conditions to obtain stable microplasma discharges at AP. In fact, the breakdown voltage (U_B) is an essential experimental parameter that is very useful for the understanding of microplasma discharges. It provides essential information separating the gas state from the plasma discharge state. Such measurements also help to define the *SEE* coefficient.

3.1 The Pressure-Driven DC Conduction Mechanisms

As it is well known that zeolites of aluminosilicate have negatively charged frameworks with ion-exchangeable sites, hence, they tend to favor adsorption of positively charged ions. There are two ways to move charged particles in a plasma space: Drift by an electric field and diffusion by the concentration gradients are described in a drift–diffusion equation with several assumptions [47].

$$\acute{\Gamma} p = \mathrm{sgn}(q_\mathrm{p})\mu_\mathrm{p} E n_\mathrm{p} - D_\mathrm{p}\nabla n \tag{2}$$

where $\acute{\Gamma}$ is the flux of particle *p*, *E* is the electric field, q_p is the particle charge, μ_p is the mobility, and D_p is the diffusion coefficient. Assuming all the samples have a similar electric field between the electrodes, the diffusion of charged particles into the nanosized pores of the zeolite crystal can cause a large change in the displacement current [48].

In the GDED with ZC, we intend to detect pressure-driven characteristic features of the gas discharge. In Fig. 5, I-V characteristics of a GDED for various residual gas pressures (*p*) for 9 and 18 mm diameters of ZC are shown. The results under various residual pressure conditions obtained from Fig. 5 can be summarized as follows. In the GDED with ZC, steady-state currents were observed from 8 Torr up to AP. The appearance of discharge is accompanied by an active current in the discharge circuit and by GDLE from the discharge area. Magnitudes of the steady-state current and intensity of GDLE established after some transient phase of the kinetics are determined by the amplitude of the U_0 and ρ of the ZC. The ignition of the discharge

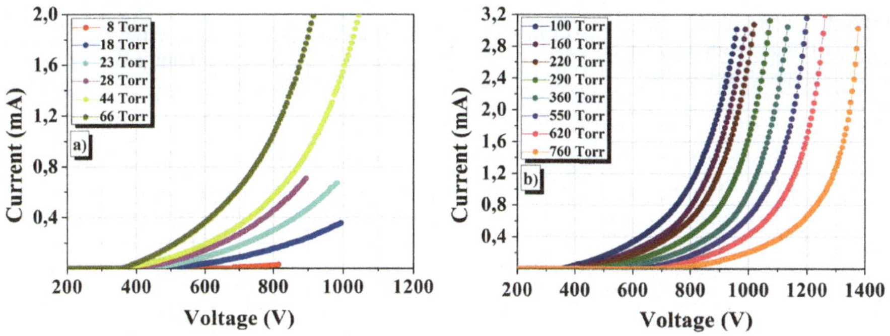

Fig. 5 I-V characteristics of a GDED for different pressures; **a** for $D = 9$ mm, **b** for $D = 18$ mm. The discharge gap $d = 50$ µm [49]

is marked with a sharp increase in the brightness of GDLE. The excess of current and brightness at the first stage of the discharge process is quite natural over their steady-state values. For the gas breakdown to be realized with a short time lag after U_0 is applied, the amplitude of the voltage on the gap must essentially exceed the critical value (i.e., the minimal value which is enough to support a self-sustained discharge). In other words, the capacity of the discharge gap shall be charged up to a value which is higher than a steady-state value [46]. As a result, the first stages of the discharge occur when the voltage on the gap (and, correspondingly, the active current) is higher in comparison with a corresponding steady-state value. When the resistance of the ZC under the action of the residual pressures decreases until 100 Torr (see curve $U_0 = 0.8$ kV in Fig. 8), an additional increase in the current and GDLE are observed. This process provides the operation of the device as a plasma light source.

The U_B values and resistivity of the ZCs exhibit ideal intersection of the two regimes, one at low voltages in which the ZCs act basically as an insulator material and one after U_B where there is a dramatic increase in the conductivity. For example, at AP, U_B is 709 V (i.e., this is the voltage value before which the steady-state current and GDLE over the whole electrode area were not observed). It can be seen from Table 1, when the gas p increases the breakdown voltages are shifted to the higher values. As the U_0 increases above the U_B, the interaction between the cations and zeolite framework is easier to break. This fact is explained by a mechanism of intrazeolite charge transport: The applied voltage after U_B is strong enough to break the interaction between ions and the zeolite framework that increases mobility and facilitates the intracrystalline charge migration. So, the needed voltage to release the ions depends only on the nature of the charge-balancing cations and the zeolite structure [50].

At the same time, a significant difference is observed in the slopes of curves with the pressure change, and current values had rapidly increased from the 44 Torr (see Fig. 5a). Analyses of physical process in the ZC and its dependence on the water content in a wide pressure range were not checked. We do not suspect

Table 1 Change of U_B and $\rho(U_0 = 0.8\ \text{kV})$ with respect to residual pressures at $d = 50\ \mu\text{m}$

p (Torr)	U_B (V)	$\rho \times 10^8$ Ohm.cm
8	679	6.54
18	459	0.97
28	394	0.33
44	354	0.27
66	354	0.14
100	344	0.14
160	364	0.19
220	400	0.25
360	494	0.57
550	614	1.64
620	639	2.06
760	709	3.04

water losses during the experiments, relying on the conductivity behavior. In water-containing zeolites, conduction is positively correlated with water content and also to an additional contribution from hydroxyl and proton conduction [51]. In Ref. [52], authors suggested results regarding mechanisms underlying conductivity increase: (i) increase in cation mobility due to OH conductance by the 'vehicle' mechanism, [51] (ii) increase in hydroxyl-proton conductivity, and (iii) enhanced mobility of hydrated cations due to pressure-induced change in the degree of hydration. In Ref. [53] considering the molecular dynamics of hydrated Na A-zeolite with the flexible lattice, it was shown that, at 1 atm, a Na^+ ion is bonded to two or three water molecules on the average. In general, applied pressure increases the coordination number of ions, which suggests a rise in the cation hydration with pressure. This leads to the formation of larger and possibly less mobile cation complexes (such as $[M(H_2O)_4]^+$), which can be compensated for by stronger screening between the cation M^+ and lattice. Conversely, a decrease in the cation hydration would result in smaller cation complexes (e.g., $[M(H_2O)]^+$) and enhanced cation-lattice interaction. Exchangeable cations seem to be more movable in hydrated zeolite than in dehydrated ones, that is, hydration of cation may decrease the activation energy of cations due to a reduction of the cation-lattice attraction. Conductivity of dehydrated zeolites is ionic, very low, and associated with diffusion mobility of cation charge carriers inside the channels. In comparison with dehydrated zeolites, conductivity of hydrated zeolites is characterized by considerably higher mobility of cations due to numerous additional diffusion paths supplied by water-cation sites and to the decrease of movable cation framework attraction by dumping H_2O molecules. Dissociated ions (H^3O^+ and OH^-) of the intra-channel water also contribute to conductivity of zeolite [54].

The formation of OH radicals depends on different mechanisms and can be expressed as follows: [55] Direct dissociative electron excitation of water,

$$e^- + H_2O \rightarrow OH(A) + H + e^- \quad (3)$$

and dissociative recombination of water ions,

$$H^2O^+(H^3O^+) + e^- \rightarrow OH(A) + H \quad (4)$$

If the ground-state OH density is sufficient, direct excitation of ground-state OH(X) to OH(A) can occur:

$$OH(X) + e^- \rightarrow OH(A) + e^- \quad (5)$$

The hydroxyl radical and atomic oxygen as plasma activated neutrals are supposed to play effective roles in microplasma generation. Hence, the hydroxyl radical and atomic oxygen can compromise the barrier function of the ZC and cause the transportation of ions and polar compounds into the cell [56, 57] giving rise to ionization process and DC conductivity. At the same time, atomic oxygen can directly react with water molecules to generate intermediate species such as hydroxyl or hydrogen peroxide [58].

It is known [59] that in a closed cell filled with zeolite, the pressure of water vapor (P_w) at equilibrium is a function of the temperature ($T_z(K)$), and the adsorption capacity (X) of the zeolite is calculated as follows:

$$X(\%) = \frac{\text{Adsorbed mass of water(Kg)}}{\text{mass of adsorbent(Kg)}} \quad (6)$$

The general trend observed showed that the adsorption capacity of the zeolite A increased with increasing water vapor pressure and decreasing zeolite temperature. Consequently, the adsorption capacity measured during adsorption/desorption process is in close agreement to the result of Ref. [60] which reports the maximal adsorption capacity of zeolite A/water pair at 30 °C ambient temperature to be 0.3 kg absorbate per kilogram of adsorbent. Also, the adsorption capacity is a weak function of the water vapor pressure at high zeolite temperatures, but its dependency on the vapor pressure is seen to be a little stronger than for a natural clinoptilolite.

Moreover, the adsorption of water by zeolites has some peculiarities. For all types of the zeolites, very sharp rise of the isotherms is observed at low concentrations of water vapor. Adsorption capacity of zeolites at ambient temperatures (T_a) and pressure of 1–2 Torr is very close to the maximum saturation P_w. Moreover, even at very low pressures, NaA zeolite exhibits significant adsorption capacity of water [61]. The second distinguishing feature of the water vapor adsorption by zeolites is to maintain the adsorption capacity even at significant change of temperatures. At $T_a = 100$ °C and $P_s = 10$ Torr, the adsorption capacity X of the zeolite is 15–16 g/100 g. Moreover, approaching 200 °C, the X is still significant and not less than—4 g/100 g [62]. With increasing T_a, the difference in the adsorption capacity X of zeolites is further increased, which is a distinctive feature of adsorption on zeolites. Oscillations of gas temperature do not introduce significant changes in the X of the synthetic zeolite. However, it should be noted the following significant disadvantage of the zeolite properties. Zeolites greedily absorb humidity, but they are hard to give it during dehydration process. At the same time, according to a global model based on He/H_2O plasma chemistry, water vapor plasmas are dominated by OH, O, and O_2 metastable species and water cluster ions depending on the concentration of water vapor in the gas [63]. An increase in water vapor concentration in the gas mixture leads to an increase in the reduced electric field strength E/N in the plasma, which results in a higher intensity of OH generations [64, 65]. Water is also an important impurity in numerous AP discharges of practical interest. For example, discharges in atmospheric air typically contain water vapor in concentrations of thousands of parts per million (air at 25° and 25% relative humidity contains ~7750 ppm of water) [63, 66].

In Ref. [1] the breakdown curves for different cathode diameters have been reported. It was obtained that as the ZC diameter increases, more channels and nanopores participate to gas discharge. Therefore, the maximum current density and minimum U_B are obtained in 22 mm diameter. As technology progresses, smaller devices are produced. This situation leads to a smaller discharge gap and thus smaller U_B. A gas conducts electricity in the presence of an electric field larger than the U_B. Devices based on gas discharge phenomena are ubiquitous in everyday applications (spark plugs, fluorescent lamps, plasma displays) and industry (arc welding, gas sensing, decomposition of toxic gases) [67–70]. Devices continue to decrease, to avoid unwanted ignition (flamentation) which could lead to device deterioration and mistakes during production, use, and operation, and understanding the mechanisms responsible for the gas breakdown has become more important. For these reasons, interactions between the zeolite cathode and the discharge plasma should be examined according to the change of cathode properties and the system parameters. The AP is much more compatible and optimal for planar GDED with ZC. Our earlier study [71] indicated that semiconductor cathode (i.e., GaAs cathode) cannot operate efficiently at AP in air media. It has been found that using zeolite as a cathode material operates at AP in air. Thus, the possibility arises of designing a portable device without the need for vacuum systems, cost reductions, and energy savings are achieved. In addition, by reducing the U_B which is required to generate the plasma light source operated at low-power consumption, miniaturization of the device, and secure mode of operation, increased efficiency, and increased life of the device are provided.

3.2 The Enhanced Effect of Electric Field on DC Conduction

Figure 6a shows a plot of the current as a function of electric field measured for different gas discharge gap d. The most homogeneous increase in current by the increase of electric field is observed for the smallest d. At AP, threshold electric field (E_{th}) for $d = 50$ μm is found $E_{th} = 151.8$ kV/cm. As the d increases, this value decreases. Moreover, current control and discharge problems occur after 150 μm.

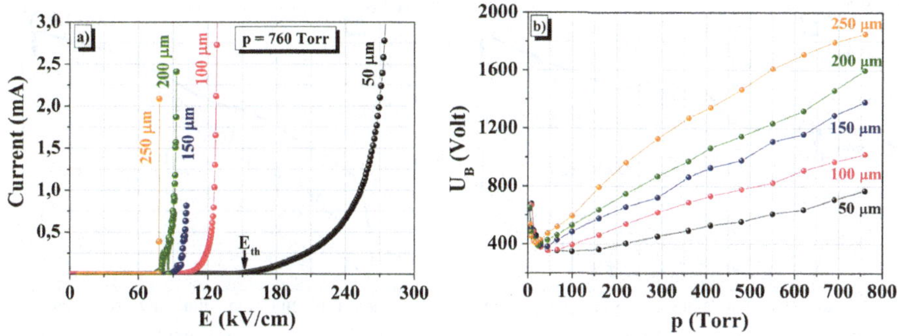

Fig. 6 **a** Current versus electric field for different *d* values at atmospheric pressure, **b** measured breakdown curves with different *d* values. The diameter of the ZC area was $D = 22$ mm [49]

As *d* increased, current quickly increased and a significant difference is occurred in the slopes of curves. The reason for this homogeneous increase in current for small *d* is due to two components, i.e., gas discharge gap and ZC's nanopores. The high electric fields obtained in small *discharge gaps* may enhance the *SEE* coefficient and that such enhancement would lead to a lowering of the U_B and a departure from the Paschen curve[47, 48]. Theoretical research on *SEE* from a *porous surface* shows a marked increase in *SEE* due to the accumulation of charged particles near the pores [72, 73].

Unlike *SEE* from solid materials, in which slow *SEE*s are emitted only from a thin surface layer in porous dielectrics, the thickness of the layer from which slow *SEE*s are emitted reaches tens and hundreds of microns, and an electric field of 10^4–10^5 V/cm, which expels the slow electrons from the material, forms within these layers [74–76]. The role of the electric field is to balance the losses of the slow electrons due to scattering by phonons within the pore walls by accelerating these electrons inside the pore, thereby lowering their recombination probability[77]. When the field in a layer of porous dielectric increases above a certain critical value, the secondary emission current increases sharply and the coefficient of *SEE* may reach hundreds or thousands [76, 78]. In fairly strong electric fields, the slow electron energy collected inside the pore is sufficient for impact ionization of the dielectric material and an avalanche-like increase in the density of the slow electrons in the porous layer parallel to the electric field. A characteristic feature of this type of emission is its slow response: After the primary electron beam has been switched off, the emission decreases very slowly and in some cases does not cease for many hours, i.e., it is self-sustaining [76, 78]. In Ref. [73], the authors showed the enhancement and the broadening in frequency of the yield of *SEE* in a porous medium, due to the growth of strong electric fields in the vacuum pores as a result of the small sizes of these pores.

Figure 6b shows pressure-driven U_B curves in the GDED with a large diameter *D* of the ZC areas for different *d* values. U_B characterizes as a function of *p* when the interelectrode distances are in the range of 50–250 μm. In addition, as shown in

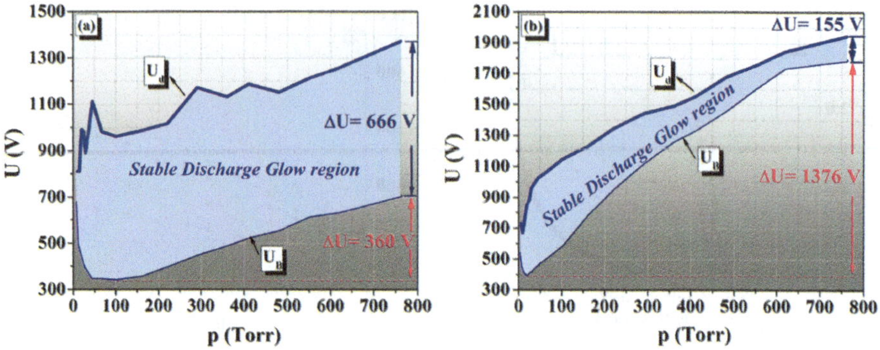

Fig. 7 Stable discharge glow regions of GDED with ZC for d = 50 μm(a) and for d = 250 μm **b** The diameter of the ZC area was D = 22 mm [49]

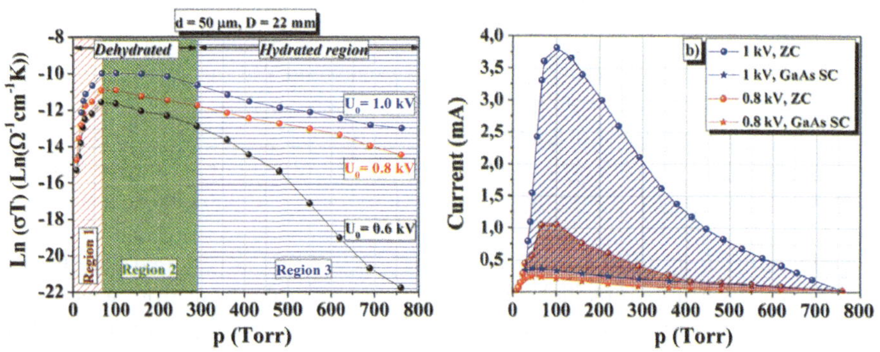

Fig. 8 **a** Natural logarithm of conductivity multiplied by temperature versus gas pressure in CL. Dehydrated regions **R1 (8–66 Torr)** and **R2 (66–290 Torr)**; Hydrated region **R3 (290–760 Torr)**, **b** Current as a function of pressure for discharge system (d = 50 μm) with ZC, GaAs semiconducting cathode at feeding voltage U_0 = 1 kV and 0.8 kV [49]

Fig. 6b, the Paschen curves exist in a wide pressure range up to AP for all d values. These results prove that, as d increases, the breakdown curves are shifted not only to higher U_B but also simultaneously to higher pressures (see Fig. 7).

3.3 Stable Discharge Glow and Optimal Operation Regions

Figure 7 shows the pressure-driven stable discharge glow (SDG) regions of GDED with ZC for D = 22 mm, d = 50 and 250 μm, and pressures p = 4–760 Torr. To maintain stable homogeneous discharge in air media, the amplitude of the applied voltage, d and cathode materials are important factors. I-V characteristics determine the following cell parameters: (1) breakdown voltage U_B; (2) the disturbed voltage

(U_d) at which the SDG is disturbed; and (3) the corresponding limiting current I value which we recorded. It should be noted that the U_B demonstrates the Paschen curve. U_d is a voltage value of the density distribution of electronic and ionic currents, when the field in the gap is not disturbed by the influence of space charge. Considering these figures for the ZC at different discharge gaps, the following conclusions can be reached: When d increases, all pressure ranges of SDG become narrower. For instance, ΔU_{SDG} is 666 V for $d = 50$ μm while this value is 155 V for $d = 250$ μm at AP. The main reason for this voltage difference is the enhanced effect of electric field. The largest SDG region was obtained for $d = 50$ μm. The stable discharges up to atmospheric pressure operate with a ZC in air for all d and D. At the same time, when the ZC diameter increased, the SDG expanded. To put it more clearly, the range of SDG disturbed voltage and the breakdown voltage obtained are 646 V for $D = 9$ mm $d = 50$ μm, and 666 V for $D = 22$ mm $d = 50$ μm at atmospheric pressure. Because the volume of zeolite consisting of channels and pores is increased, the number of pores and channels participating in the gas discharge increased. Therefore, the number of ions increased in these regions, and as a result, the conductivity increased. The range of SDG can be expanded by increasing the conductivity of the ZC, and the conductivity can be adjusted by the applied voltage and pressure. Water plays an important role for the stability of CL framework. When the pressure increases, the interaction between the water and charge-balancing cations increases. Hence, cation mobility increases due to the decrease in the cation-lattice attraction. In this way, the cations contribute more to transport.

As it is known, a high gas p means many collisions in the plasma (i.e., a short collision mean free path, compared to the discharge length), leading to an efficient energy exchange between the plasma species. However, a low gas p results in only a few collisions in the plasma (i.e., a long collision mean free path compared to the discharge length). Atmospheric pressure glow discharges are exceptions to this rule. The reason is that not only does the p play a role, but also the discharge length or the d [79]. Generating a discharge at AP is based on the Paschen law ($p \times d$). The difference between the minimum of the Paschen curve and breakdown voltage in atmospheric pressure increases as d increases. When d increased, the minimum dots shifted to lower pressure while the nanoporous zeolite D was kept constant. Paschen minimum was 100 Torr for $d = 50$ μm while this value was decreased to 18 Torr for $d = 250$ μm at $D = 22$ mm. According to $U_B = f(pd)$, d is increased and p is decreased.

The reduction of the d enhanced the electric field, which reduced ignition voltage further. By optimization of the gas pressure, a minimum ignition voltage and operation voltage were obtained. Due to the small gap between the electrodes, higher pressure was required to exceed the Paschen minimum [80]. At the same time, the 'left side' of the Paschen curve exists for all electrode distances. It should be noted that when d increased, U_B is shifted to higher values (i.e., U_B values are 709 V for $d = 50$ μm and 1795 V for $d = 250$ μm at AP). This shift of the breakdown curves to higher values with the increase of the d is associated with the growth of the losses of charged particles on the lateral walls of the discharge tube due to the diffusion across the electric field [81]. In conclusion, by increasing the d, the U_B increases and the lowest value of U_B is obtained for the smallest d (see also Fig. 6b). Also, the widest and homogeneous SDG at AP is obtained for $d = 50$ μm. Therefore, $d = 50$ μm and $D = 22$ mm were the optimal operating conditions in our experiments.

Figure 8a shows the plot of conductivity multiplied by temperature with pressure for CL. The conductivity of CL was measured in the pressure range of 8–760 Torr. The obtained conductivity results can be divided into three different regions (R). In $R1$, the conductivity increased rapidly in the pressure range 8–66 Torr. Factors which may be responsible for the increased conductivity are an increased interaction between water and the mobile charge-balancing ions, an increase in the protonic conduction contribution, and/or an increase in ion mobility with increasing pressure. At atmospheric pressure, the conductivity value decreases rapidly at 600 V while this drop is smaller at 800 and 1000 V, because the U_B is 709 V for $D = 22$ mm and $d = 50$ μm. It has been shown that as the degree of zeolite hydration increases, the conductivity also increases [82]. An increase in the proportion of H_2O molecules available for hydrating the mobile ions could enhance ionic conduction. Polar H_2O molecules associate with the charge-balancing ions to facilitate the motion of the mobile ions in the external applied electric field. The pressure effect is to break up the H_2O clusters, thereby liberating more individual H_2O molecules to move throughout the framework. This allows the polar H_2O molecules to interact more strongly with the mobile charge-carrying ions, thereby reducing the strength of ion bonding with framework O atoms. This increased interaction of the H_2O molecules with the ions could result in an increase in conductivity in the low pressure region. Because of volume reduction on pressure increase and the concomitant changes in ion framework interaction, one could speculate that applied pressure promotes an increase in diffusion between the channels and cavities and/or creates new pathways for the ions to contribute to conduction. Either one of these effects, acting individually or in combination, would give rise to an increase in conduction with applied pressure. The pressure enhancement of the conductivity could prove useful in specific high-pressure solid-state applications [83].

In $R2$, the conductivity is approximately constant with pressure from 66 to 290 Torr. In this region, a plateau is formed. This plateau's width has expanded with the increase of applied voltage. As the applied voltage is increased further, this direction will show an increasing trend. If pressure increases, the conductivity decreases ($R3$). At atmospheric pressure, the conductivity value decreases rapidly at 600 V while this drop is smaller at 800 and 1000 V. Because the U_B is 709 V for $D = 22$ mm and $d = 50$ μm at AP. As shown in Fig. 6, ionic conduction at 600 V is a dominant transport mechanism in pre-breakdown mode. At the same time, when the applied voltage is increased, conductivity increases for all pressure ranges. The residual p and U_0 are important factors when working with porous materials such as zeolites. Because the zeolite structure highly composed of H_2O and density of H_2O in the nanopores is highly depending on these two parameters.

In conclusion, the zeolite conductivity decreases with pressure increase, and conduction takes place by ions. However, conductivity increased with increase of pressure up to 100 Torr which is not quite usual situation as shown in Fig. 8a. At the same time, a visible variation is formed in the slopes of pressure-driven conductivity graph with the U_0. In addition to the factors mentioned above as reasons for these situations, due to both the high voltage applied between electrodes in the gas discharge cell and the porous structure, electronic conduction also contributed to zeolite total conduction, i.e., we assume that the conduction was formed by the ions and electrons. The best approach to confirm this assumption is to examine the dynamics of the current behavior (see its time depended characteristic).

The similarity of the transport properties of both systems (i.e., with porous ZC and polished surface of the GaAs SC) gives us ground to suggest that the linear electrical characteristics of current distributions are similar as well. Therefore, we can judge the spatial structure of the current in both systems by studying the linear electrical characteristics observed in the GDED (Fig. 8b). Comparison of two different cells with porous ZC and GaAs SC indicated that the presence of porous ZC in the gas discharge region had significantly enhanced the character of the charge transport in our system. It is shown that under the experimental conditions, the discharge gap played only a passive role and was not responsible for the observed phenomena. It is found that active properties of porous ZC that manifest themselves at high electric fields play an important role in the observed conductivity mechanism [71].

Figure 8b is a plot of the current as a function of pressure for GDED with ZC and GaAs semiconducting cathode (SC) at $U_0 = 1$ kV and 0.8 kV. As the residual gas pressure changes, the current values change up to AP. We established that the current and GDLE increase in large pressure range when porous zeolite plates are used as a discharge electrode instead of a GaAs SC because ZC is not uniform, so the electric field becomes higher. Hence, we consider that it is easy to produce the initial electrons and channel of discharge. Based on the above results, it can be concluded that a steady gas discharge can be initiated in the nanopores of a ZC. Since nanopores in a ZC go through, they contain a gas with a small amount of electrons and ions in the absence of an electric field. These charge carriers provide a steady electric field through current. When electric fields cause the exponential run of the I-V characteristic, the amount of electrons and ions grows and a gas discharge arises. They are the nanopores of GDLE emerging on the zeolite surface, so stable, uniform, and homogeneous plasma is observed. The experimental results suggest that the amount of current carriers increases through the ionization of the gas contained in the nanopores rather than through the ionization of the pore wall material, because the current disappears at low pressures.

In AP and under electric field excitation, mechanisms of field emission, electron-impact ionization, and *SEE* are responsible for inducing gas micro/nano-discharge [43, 84, 85]. It is known that contributions of these mechanisms in total discharge depend on the Paschen law ($p \times d$). DC voltages can induce gas discharge[86–89], and the U_B is reached when the generation rate of species becomes higher than their loss rate. At this point, an avalanche increase in the number of charged particles results in the formation of a plasma region. Hence, two distinct operation regimes of pre-breakdown and post-breakdown can be considered in discharge problems.

I-V characteristics of GDED with ZC at AP are given in Fig. 9, and the inset in Fig. 9 shows the time dependence of pre-breakdown current at 600 V. Experimental results of the current dynamic behavior during the time range of 150 s across various

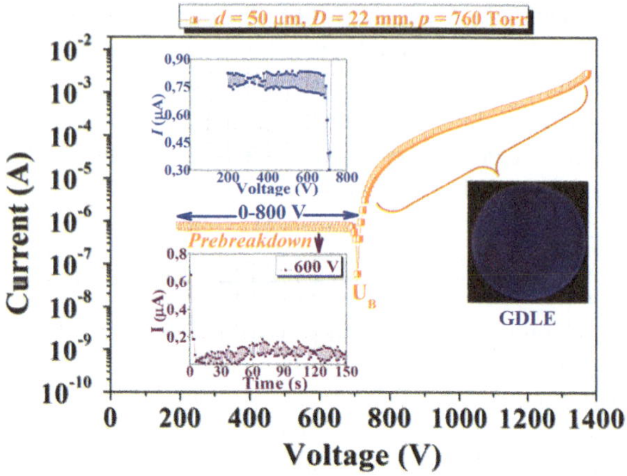

Fig. 9 I-V characteristics and the time dependence of pre-breakdown current at 600 V in the GDED with ZC and gas discharge light emission through the discharge cell [49]

regions of voltages indicated that the current value and stability varied significantly with U_0. Gas (i.e., atmospheric air) and zeolite were electrically insulated to a specific voltage value, and cations (especially Na^+, K^+, Ca^{2+}, Mg^{2+}) were important in the pre-breakdown mode and ionic conduction (cations motion) which were responsible for non-monotonic current oscillations with microampere amplitude (see inset of I-V characteristics in Fig. 9). Since cations are much heavier than the electrons, they move slower and respond later to an applied field. Cations can also have an important contribution specifically in the pre-breakdown regime, where the density of ions is orders of magnitude larger than that of the electrons due to their lower mobility.

In the case of high resistivity of the ZC, the ionizing effect of the active components of the discharge on the nanoporous dielectric, and consequently low density of carriers equilibrium, the generation of carriers in a ZC under the influence of a gas discharge plasma plays an important role [90]. This carrier generation (about 100 eV electrons and ions) occurs in a very thin skin layer. The carriers, aided by the field, then penetrate deep into the interior of the ZC, where they can modulate the conductance. The following process results in a local increase of current value. Modulation of the bulk of the ZC and reducing its resistance increase the current in the plasma, intensity of the GDLE, and the flux of the ionizing particles, which in turn reduce the resistance of the ZC to an even greater degree within this range.

I-V characteristics of GDED with ZC at AP are given in Fig. 10, and the inset in Fig. 10 shows the time dependence of the breakdown current at three different U_0. The current values were decreasing monotonously with time at 734 V. The current in the conventional cell with dielectric materials always decreased on a time scale, which is a peculiarity of ionic conductivity [27]. When the applied voltage was raised (at 834 V), the current values become monotonous. As an electron moves from one side of the ZC to the other side, ionic transport needs to occur to maintain charge neutrality. In Ref. [91] the authors proposed that the Na^+ ion from the sacrificial electron donor accompanies the electron and keeps the charge balance. As in the case of pressure-driven behavior, we suggest that when the electric field increases, H_2O

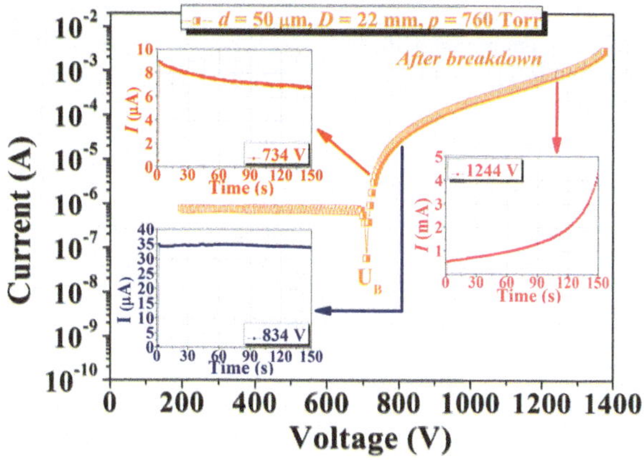

Fig. 10 I-V characteristics and the time dependence of post-breakdown current in the GDED with ZC at different feeding voltages[49]

dissociation could increase. The density and structure of water in the pore are highly field dependent [92]. When the separation of H_2O into OH^- and H^+ ions increases, Na^+, K^+, Ca^{2+}, Mg^{2+} cations transport decreases simultaneously and equilibrium behavior occurs at a specific voltage range. So, the electric current in zeolites is caused by movement of positive metal ions in the pore space, and it depends on the content of the H_2O molecules in the pores. At this stage, ionization in the gas started to increase and electrons appeared with collisions in porous zeolite bulk and discharge gap [93]. In other words, when the electric field increased, electrons became a part of a total transport mechanism. After the dominant conduction mechanism transition from ionic to electronic, when the applied voltage is further increased (at 1244 V), current increase reached considerably high magnitude. This situation was not peculiar to ionic transport. Ionization phenomena increased in this region because of both high-voltage application to gas discharge gaps and porous structure of the cathode. Therefore, collision ionization was greater. Consequently, electronic conduction also contributed to total zeolite conduction [1]. Therefore, we suggest that the new mixed transport mechanisms (i.e., electronic and ionic) of aluminosilicate materials occur when zeolite plates are used as a cathode in GDED at AP. It is of importance to have knowledge in peculiarities of these mixed conductivity mechanisms, in order to provide stable and controllable operation of GDED (Fig. 11).

In conclusion, ionic transport predominated in large discharge gaps and low feeding voltage values (i.e., low electric field) while electronic transport predominated in small discharge gaps and high feeding voltages. At the same time, it was shown that ions can have important role in pre-breakdown modes and ions' contribution to charge transport decreases with applied voltage. Also it should be noted that in the GDED with ZC, some challenges must be overcome such as sustaining wide SDG region (see Fig. 7), lower voltages are required for gas breakdown (see Fig. 6b), and easy to form uniform plasma throughout the reactor volume (see the inset in Fig. 9) when working at atmospheric pressure.

Thus, novel results concerning the transport properties of the GDED with zeolite cathode at atmospheric pressure were investigated. Ionic transport predominated in

Fig. 11 **a** XRD diffractogram of unmodified (N1) and **b** Ag°-modified (N4) porous zeolite plates [120]

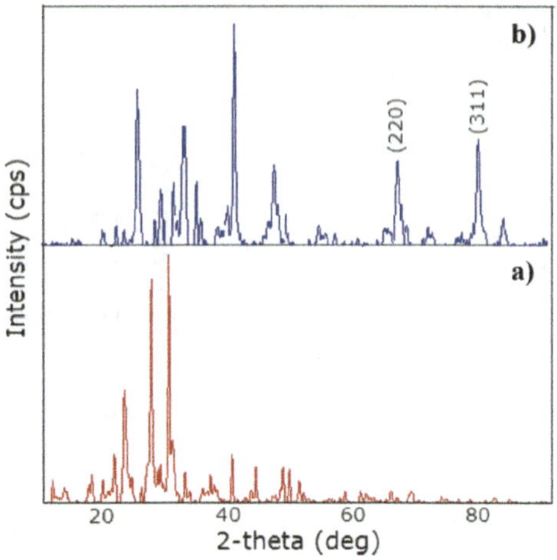

large discharge gaps and low feeding voltage values (i.e., low electric field), while electronic transport predominated in small discharge gaps and high feeding voltages. At the same time, it was shown that ions can have important role in pre-breakdown modes and ions' contribution to charge transport decreases with applied voltage. These results indicate that the main source of changes in the dominant transport mechanisms from ionic to electronic is the electric field. This is because the zeolite structure is highly composed of water, and the density and structure of water in the nanopores are highly electric field dependent. When the field increases, H_2O dissociation could increase, and therefore, ionic transport is decreased. Ionization in the discharge gap increased with increase of feeding voltage and decrease of gas discharge gap. Consequently, electrons appeared with collisions in zeolite bulk and discharge gap. Therefore, the electronic processes in this two-layer cell are controlled by the properties of the porous ZC and the gas discharge plasma layer. In this way, electronic conduction also contributed to zeolite total conduction. Thus, the new mixed transport mechanisms (i.e., electronic and ionic) are well interpreted in terms of enhanced electric field effect in GDED at AP. With this new approach, the ionic and electronic transport mechanisms and their interactions which are essential in enhancing applications in microdischarge devices with nanoporous zeolite cathodes were interpreted. It has been observed that as the electric field increases, charge carriers are transported, and therefore, conductivity of ZC and stable discharge glow regions are increased.

At the same time, pressure-driven charge transport in nanopores of a ZC was studied to evaluate the effect of water. For hydrated zeolite samples, effects of pressure-induced conductivity changes such as in cation-lattice and cation-water bonding may also be significant. When the gas pressure is relatively low (dehydrated zeolites), it has a pronounced effect on the ionic charge transport, which should be attributed to the diffusion mobility of cation charge carriers inside the channels. Thus, in the GDED with ZC to overcome some challenges such as sustaining wide

SDG region, lower voltages are required for gas breakdown and for easy formation of uniform plasma throughout the reactor volume, with working at atmospheric pressure.

The obtained results are shown in the nanoporous zeolites can be used in plasma devices, optoelectronics, field-emission displays, and energy storage devices, functional fillers in composites, catalysis, and biomedical applications. Understanding of the nanoporous materials transport properties is necessary for further expansion of their application areas. Therefore, due to the abundance and considerable low cost of natural CL, GDED with ZC, it may be a preferred and efficient device for application in generating and sustaining stable, uniform, and homogeneous non-thermal AP plasma.

4 Peculiarities of the Dielectric Response of the Silver-Modified Zeolite Porous Microstructure

Nanoporous zeolites have three-dimensional framework structures. The zeolite structure is built from TO_4 (T = Si, Al) tetrahedrals which are linked together to form a three-dimensional framework of interconnecting pore and channel structure. The conductivity of zeolites is usually associated with ionic conductivity [94, 95]. Charge is carried by means of the motion of mobile charge-balancing ions in the anion framework. Hydrated zeolites are also known as proton conductors and hydroxyl ion conductors [96]. It is usually presumed that zeolites do not act as electronic conductors [83]. Furthermore, dielectric materials such as zeolites have the ability to store energy upon the application of external electric field [97]. Specific electronic properties of nanopores in zeolite structure make zeolites good candidate materials in the electronic industry so that silicon-based technologies could be replaced or supplemented by nanoporous zeolite materials. The dielectric response of zeolites is one of the key parameters for applications. Such knowledge about zeolites has become important because of their potential application in electrochemical devices.

Conductivity and relaxation properties of various types of zeolites exchanged with monovalent or divalent cations have been extensively studied and results published in the literature [96, 98–101]. In particular, it has been shown that both ionic conductivity and dielectric relaxation are due to cation jumps. The analysis of these studies shows that they are often carried out in different experimental conditions, so the obtained results are difficult to compare. It is generally admitted that the dielectric relaxation and conduction phenomena observed at frequencies below 100 MHz (the so-called intermediate frequencies) are related to the movements of cations in the large cavities and channels in which the mentioned cations can be introduced and where they can move between preferred positions. The polarization effects are visible in the frequency dependence of imaginary part of complex permittivity $\varepsilon'' = f(f)$ or in frequency dependence of loss tangent tg $\delta = f(f)$ creating one or more peaks. These peaks can be characterized by critical frequency and the intensity (maximum). It is well known that the absorbed molecules or exchanged cations influence the conductivity and the relaxation properties of zeolites [96, 98–100]. In general, there are the following aspects which must be taken into account:

- chemical composition of the zeolite with respect to the relation $n(\mathrm{Si})/n(\mathrm{Al})$,
- properties and influence of the exchanged cations,
- the contents of the humidity,
- the temperature of the sample [101].

To optimize the conductivity and the relaxation properties of zeolites, the knowledge of dielectric properties is needed. Under various conditions, a number of works are devoted to study on dielectric spectra of zeolites. In Ref. [102], authors studied the dependence of the dielectric properties of the zeolite on the type of primary ion or alkaline cations incorporated in a zeolite, i.e., ion controlling and ion-migration polarization. At the same time, the temperature dependence of the dielectric permittivity at different frequencies is studied. It is established that the experimental results are satisfactorily described by a model of two-component system of the zeolite-air pores. Effect of water on the dielectric properties of the natural zeolite Ca-clinoptilolite has been studied in Ref. [103]. It is shown that relaxation of water and its various concentrations influenced the dielectric spectra in the clinoptilolite-water system. It has been established that the contribution of water on the dielectric properties of the zeolite, i.e., water bound in the pores and water in the free volume, is different. In Ref. [104], it was shown that modification of the natural zeolite clinoptilolite by some ions increases its dielectric permittivity, and at the same time, ion exchange by other ions decreases the dielectric permittivity. This may be observed when comparing it with permittivity to unmodified samples. In a previous study [105], the dielectric properties at room temperature of natural zeolite powders and plates have been studied. In our earlier work [106], dielectric spectra of natural zeolite clinoptilolite are studied at room temperature and 85% air humidity (at atmospheric pressure). The measurements were carried out on samples of high density (natural plate) and low density (unpressed powder). We have shown that the frequency dependence of both real and imaginary parts of the dielectric permittivity is characterized by approximately the same relaxation time of about 10^{-5} s. Moreover, the dielectric spectra are determined by fluctuations of alkali metal ions associated with the water molecules inside the zeolite pores. In Ref. [107], the dielectric permittivity of natural clinoptilolite has been measured for the first time in the wide frequency range up to microwave frequencies ($\omega = 4.5 \times 10^{10}$ rad/s). On the basis of these measurements, an approximation of the frequency dependence of the dielectric constant in the range 10^2–10^6 Hz is deduced. In Ref. [108], dielectric spectra of natural zeolite composites prepared by using zeolite and silicon powders are studied. The maximum dielectric permittivities were detected depending on the concentration of Si powder at different frequencies. It was found that there is a stable decrease of the dielectric response with increasing concentrations of Si if the concentration of Si powder is more than 9%.

In microelectronics, conductivity of the natural zeolites is usually associated with ionic conductivity [109, 110]. The aluminosilicate framework of zeolites has a negative charge, which is balanced by the cations of alkali-earth metals and water molecules connected weakly with the pores and cavities of the framework. Hydrated zeolites are also known as proton conductors [96]. Normally, as a rule, it is suggested that electronic conduction [83] is not realized in zeolites. In addition, materials such as zeolites have the ability to store energy upon application of an external electric field [97]. Special electrical properties of the nanopores of the zeolite structure make them good candidate materials in microelectronics [111], gas sensors [112], solar cells [113], functional fillers in composites [114], and plasma light sources with low-energy consumption [115]. Thus, it should be noted that the zeolite is actively used as an element of the electrical circuit. However, a systematic study of the zeolite as an electrical circuit element has not yet been carried out. Clinoptilolite is the most abundant natural zeolite and is characterized by large, intersecting, open channels of ten- and eight-membered tetrahedral rings [112]. The channels are occupied by ion-exchangeable cations and water molecules. Keeping this in mind, we have investigated the frequency dependence of the dielectric permittivity, conductivity of natural zeolite, and its modifications. In this study, the abovementioned measurements for unmodified and silver-modified zeolite plates with different amounts of silver ion (Ag^+) and silver nanoparticles (Ag^0) are performed. We chose the silver-modified nanozeolite because it is widely used in various fields of microelectronics, medicine, and biology [116–119].

In recent years, different studies are conducted for natural zeolite modified by ions and nanoparticles of various metals, not only for medical purposes, but also as an element of the electrical circuit. Silver is a very useful material in microelectronics due to its excellent electrical and thermal conductivities, photosensitivity, and antimicrobial properties. These processes can be profitably used when the metal concentrations in nanoporous zeolite are sufficiently high. Taking into account all the above, we have considered it of great interest to assess the ability of locally available natural zeolite, clinoptilolite, for increasing its dielectric permittivity, and optimization of conditions for its maximum ion-exchange modification. The introduction of Ag metals into the framework potentially allows the modification of the electrical characteristics of the framework itself, yielding mixed conductors. A promising use of zeolite-like mixed conductors is in energy applications, e.g., electrical energy storage, where an electrically conducting framework is required [106]. Yet, uses for electrically conducting zeolite-like frameworks can be envisioned in other important energy applications, in ion-exchange membrane technologies and catalysis. A study of the electronic and ionic conduction mechanisms helps demarcate such applications. Therefore, this work can be considered a preliminary study to conclude that silver-modified clinoptilolite is a suitable and efficient material for improving the electrical conductivity by the adsorption of Ag^+ from aqueous solution and its applications as an element of the electric circuit. Thus, we have prepared four types of samples:

1. N1—is unmodified natural ZC plate;
2. N2—is Ag^+-modified zeolite plate with silver ions and nanoparticles;
3. N3—is Ag^0-modified zeolite plate prepared in a 0.02 M $AgNO_3$ solution;

4. N4—is Ag^0-modified zeolite plate prepared in 1 M $AgNO_3$ solution.

The most informative characteristics of the element of an electrical circuit are to measure the frequency dependence of impedance and capacitance (and calculate from them the real and imaginary parts of the dielectric permittivity) of the sample. Therefore, we conducted our study to the significant influence of Ag modification on the electrical characteristics of nanoporous zeolite plates.

4.1 Characterization of Ag-Modified Zeolite

It is evident that the modification does not significantly affect the position of the diffraction peaks of the zeolite, but the relative intensity of the peaks is reduced and additional peaks observed in the range of 60°–90° ($2\theta = 65.680$ and 78.58°), which confirmed the presence of silver in the sample. The average crystallite size for the studied zeolite was calculated from the broadening of the reflections in the diffraction patterns by the Debye–Scherrer equation, and it was 8–10 nm. The morphology, distribution, and elemental composition of the samples were *analyzed* using scanning electronic microscope (SEM). Measurements were carried out in the secondary electron mode and the backscattered electron mode.

4.2 Theoretical Model

Denote the thickness and dielectric permittivity of the zeolite as d_1,ε_1, and the thickness and dielectric permittivity of the gap through d_2 and ε_2. Also denote s as the structure area, ω the frequency of the external harmonic field, $\varepsilon_0$the dielectric permittivity of the vacuum, and σ as the zeolite conductivity. The complex integrated resistance of the zeolite-gap structure has the following form [121].

$$R = \frac{d_1}{s}\frac{1}{\sigma + i\omega\varepsilon_{ij}\varepsilon_1} + \frac{d_2}{s}\frac{1}{i\omega\varepsilon_0\varepsilon_2} \tag{7}$$

The first term is the complex integrated resistance of the zeolite, and the second is that of the gap, since they are connected in series, therefore the total resistance is the sum of both of them. From the ratio between the complex integrated resistance and the complex permittivity is follows

$$i\omega\left(\varepsilon' - i\varepsilon''\right) = \frac{d_1 + d_2}{\mathrm{SR}} \tag{8}$$

Taking into account Eq. (7), we obtain

$$\varepsilon' = \varepsilon_2 \frac{d_1 + d_2}{d_2} \frac{1 + (\omega\tau)^2 x(1 + x)}{1 + (\omega\tau)^2 (1 + x)^2} \tag{9}$$

$$\varepsilon'' = \frac{d_1 + d_2}{d_1} \frac{\omega\tau}{1 + (\omega\tau)^2 (1 + x)^2} \tag{10}$$

Here, $x = \frac{\varepsilon_1 d_1}{\varepsilon_2 d_2}$ and $= \frac{d_1 \varepsilon_2}{d_2 \sigma}$. From Eq. (9), we determine the variation limits of the real part of the dielectric permittivity:

$$\varepsilon_2 \frac{d_1 + d_2}{d_2} \cdot \frac{1}{1 + \frac{\varepsilon_2 d_1}{\varepsilon_1 d_2}} \leq \varepsilon \leq \varepsilon_2 \cdot \frac{d_1 + d_2}{d_2} \tag{11}$$

From Eq. (11), we determine the ratio of the highest to the lowest value of the measured dielectric permittivity

$$\frac{\varepsilon_{\max}}{\varepsilon_{\min}} = 1 + \frac{\varepsilon_2 d_1}{\varepsilon_1 d_2} \tag{12}$$

If we ignore the difference in dielectric permittivities (ε_1 and ε_2), even when the thickness of the dielectric gap on the order of, or less than zeolite plate thickness, then ratio of the maximum value to the minimum value of the dielectric permittivity is 10, i.e., we observed the change of the dielectric permittivity in Fig. 12. Equation (9) also explains the increase in dielectric permittivity with increasing conductivity. From the fact that Eq. (9) is a decreasing function from $\tau = \frac{d_1 \varepsilon_2}{d_2 \sigma}$, which itself is inversely proportional to conductivity, it follows that the real part of the dielectric permittivity is increasing function of conductivity. Figure 5a, b shows that the capacity is the largest for those zeolite samples which have the lowest resistance. Thus, in the framework of the proposed model, we are able to explain the trend in the frequency dependence of the dielectric permittivity (or capacity) in Fig. 12a. Let us now consider the frequency

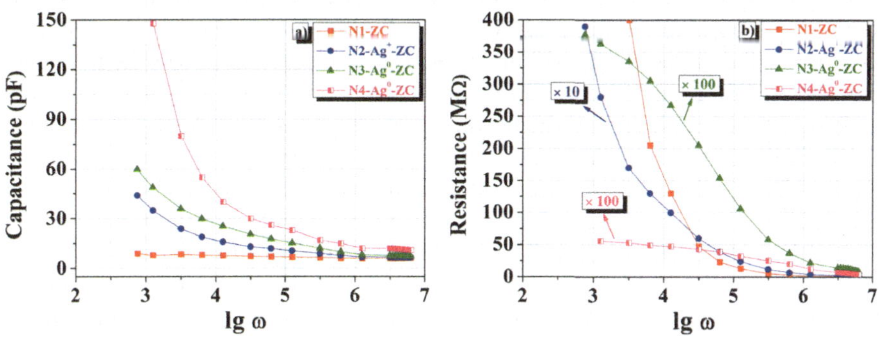

Fig. 12 Frequency dependence of the capacitance **a** and conductivity **b** of the zeolite plates with different amounts of silver ions and silver nanoparticles, respectively: 1—N1–ZC; 2—N2–Ag^{+}–ZC; 3—N3–Ag^{0}–ZC; 4—N4–Ag^{0}–ZC [120]

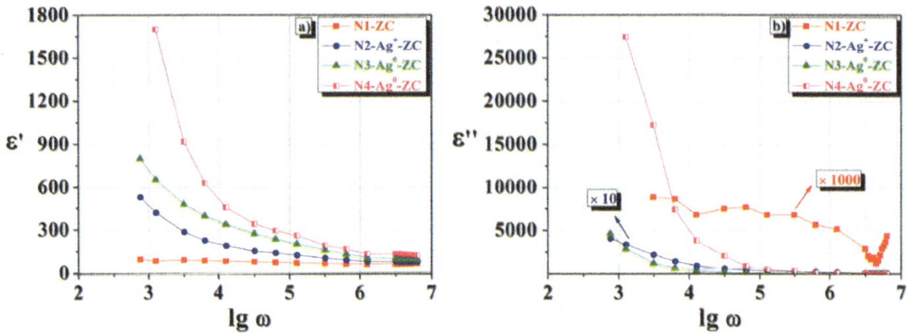

Fig. 13 Frequency dependence of the real **a** and imaginary **b** parts of the dielectric permittivity of the unmodified and silver-modified zeolite plates with different amounts of silver ion and silver nanoparticles, respectively: 1—N1-ZC; 2—N2–Ag^+–ZC; 3—N3–Ag^0–ZC; 4—N4–Ag^0–ZC [120]

dependence of the resistance $R(\omega)$ for the whole system, i.e., magnitude which is measured experimentally. This value is determined from Eq. (7):

$$R(\omega) = \frac{d_1}{s\sigma}\left[\frac{1}{\left(\omega\frac{d_1}{s\sigma}\frac{\varepsilon_2\varepsilon_0}{d_2}\right)^2} + \left(1 + \frac{\varepsilon_1 d_2}{\varepsilon_2 d_1}\right)^2\right] \tag{13}$$

Equation (13) shows that, firstly, the measured resistance decreases with increasing concentration of silver (as the silver concentration increases the conductivity σ of the sample increases), and secondly, $R(\omega)$ the measured resistance decreases as a function of frequency. Both of these phenomena are observed in our experiment (see Fig. 12b). At low frequencies, this dependence is related to the pumping of the electric field from the sample volume to the contact. In the case of high frequencies, the electric field remains in the volume of the sample.

Frequency dependence of the capacitance and conductivity of the ZC for all samples are presented in Figs. 12 and 13. According to our data, in the sequence content of silver in the zeolite volume is increased. Figure 12a shows the frequency dependence of the capacity for natural and silver-modified zeolite samples. Firstly, the capacitance decreases with increasing frequency, and the secondly, capacitance consistently increases by increasing the content of silver in the sample volume. Figure 12b presents the results of measured resistance for all abovementioned samples, respectively. It is seen that for all samples with increasing frequency, we observed decrease resistance, at the same time by increasing the content of silver in silver-modified zeolite plate resistance decreases. The observed phenomenon can be explained by considering the fact that with increasing content of silver, the conductivity increases and the sample itself is separated from one of the electrodes by a dielectric gap. Indeed, from Eqs. (9) and (11), it follows that at low frequencies, when the field is mostly concentrated near electrode in a narrow gap, the measured capacity is the large capacity of the gap. At high frequencies when the field during the period of an external electric field does not have time to move to the gap and remains in the sample volume, we measured small capacity of the sample. The ratio of these capacities (or ratio recalculated from these dielectric permittivities) is determined by Eq. (12). If we assume that the gap thickness is 10–100 times smaller than the sample thickness, we obtained also the same ratio for the largest and smallest values of capacitance (or dielectric permittivity) as is observed in Figs. 12 and 13.

It is noticed that if the conductivity is equal to zero then as it follows from Eq. (9), the measured value of the dielectric permittivity (or capacitance) does not depend on the frequency, because field redistribution between the sample volume and the gap is not observed. Consequently, the value of the dielectric permittivity (or capacitance) is different at different frequencies. Therefore, with increasing of the conductivity, the measured value of the dielectric permittivity (or capacitance) increases at low frequencies. Especially, this is observed in Figs. 12a and 13a, and the largest value of the capacity is obtained for samples with a high content of silver. As can be seen from Fig. 12b, measured resistance value decreases with increasing frequency. This is consistent with Eq. (13). From Eq. (13), it also follows that in the absence of gap (formally this is observed when the gap thickness d_1 approaches zero), the resistance is not dependent on the frequency (or it is pure resistance), and with increasing the bulk conductivity σ, the resistance at low frequencies is reduced. This can be seen in Fig. 12b that with increasing content of silver, the resistance decreases.

4.3 The Effect of the Ag Concentration on the Conduction

The metal uptake is attributed to different mechanisms of ion-exchange and adsorption processes [119]. During the ion-exchange process, metal ions move through the pores of the zeolite and channels of the lattice, and they replace exchangeable cations (mainly Na or Ca) and additionally exchange with protons of surface hydroxyl groups. Diffusion was faster through the pores and retarded when the ions moved through the smaller diameter channels. The ion-exchange processes in zeolites are affected by several factors such as concentration and nature of cations, pH, and crystal structure of the zeolite [122, 123]. Earlier experiments conducted with different initial Ag^+ concentrations show that the adsorption capacity of zeolite increases with the initial concentration of Ag^+ ions [124]. It can be concluded that percentage adsorption for Ag^+ decreases with increasing metal concentration in aqueous solutions. These results indicate that energetically less favorable sites become involved with increasing metal concentrations in the aqueous solution. The metal uptake can be attributed to different mechanisms of ion-exchange and adsorption processes as it is concerned in some previous work [125].

Table 2 shows the frequency dependence of the electric resistance of ZC and Ag^0–ZC which are loaded with different amounts of silver ions/nanoparticles at atmospheric pressure. Figure 12 shows that the dielectric response of Ag-modified clinoptilolite is monotonically dependent on the Ag concentration in the porous zeolite. With increasing Ag concentration, resistance of the porous zeolites decreases

Table 2 Resistance change of the unmodified and silver-modified zeolites for different frequencies

Resistance (M) at atmospheric pressure				
f (Hz)	N1–ZC	N2–Ag^+ ZC	N3–Ag^0 ZC	N4–Ag^0 ZC
1000	205	13	3.05	0.49
5000	47	6	2.05	0.43

and the capacity increases. Usually, zeolite resistance exhibits such behavior, because with increasing Ag concentration, the static conductivity of the sample increases. Another behavior is observed in dependence of capacity on the Ag concentration. Such behavior of this dependence is unusual, as well as its correlation with the conductivity (with increasing zeolite resistance the capacity decreases). However, the main contradiction is anomalously large values of the dielectric permittivity, calculated from the measured capacitance values (see Fig. 13). Turning to a discussion of these results, we note that in zeolites, free carriers are ions which are not able to penetrate the metal electrode. Therefore, some studies discuss field depolarization produced in materials with ionic conductivity upon application of an external electric field. For consistent accounting roles, depolarization fields should take into account the capacity of the metal electrode–dielectric interface gap. If the potential drop across this electrode–dielectric interface gap is small, compared with an external applied voltage, the depolarization fields are small and can be ignored. In the opposite case, the redistribution of the voltage between the gap and the sample volume can be the determining factor in the interpretation of the obtained results. Therefore, in this study for an explanation of the observed experimental results, we propose a simple model, consisting of an ion crystal which is separated from the electrode by thin dielectric gap.

4.4 Peculiarities of the Dielectric Properties

Dielectric behavior of the unmodified and silver-modified zeolite plates with different amounts of silver ion and silver nanoparticles at room temperature is shown in Fig. 13. Note that both parts of the permittivity change with frequency in a qualitatively similar way for both unmodified and Ag^+-modified zeolites, namely they decrease exponentially with increasing frequency. For sample N4, which was ion-exchanged for a 1 M $AgNO_3$ solution, the values of ε' and ε'' are considerably lower than for the unmodified one. This means that the introduction of Ag^+ ions decreases the concentration of dipoles, which are responsible for the dielectric response of the material. Note that the dielectric response in zeolite is formed largely by dipoles so that the values of dielectric permittivities characterize dipoles irrespective of their concentration. Such a change in the behavior of permittivities compared with the original sample may be related to Ag^+ induced structural modifications in the unit cell of the zeolite lattice. Ag^+ concentration creates electric dipoles with characteristic parameters differing from those of dipoles in the original material [126].

As mentioned earlier [108], the conductivity measured in natural clinoptilolite pellets is superposed of AC and DC conductivities. That means that they are both active. The measured value tends toward its static limit at a low frequency. Because of electrode–dielectric interface effects, it can be difficult to measure the DC conductivity directly. There is a simple method to find the static value of the conductivity σ using a complex impedance plot, but the electrode material plays a significant role [127]. In Ref. [126], the frequency dependences of the imaginary part of permittivity of silver-modified zeolitic material show the behavior typical for hopping conductivity mechanism of ionic conductor. No polarization characteristic of Cu-modified zeolite at used intermediate frequencies of the measuring AC signal (10^1–10^5 Hz) was observed. It was pointed out in Ref. [128] that the samples prepared in the form of pressed pellets are often not convenient for dielectric absorption measurement because the loss peak can be covered with high conduction losses. The relaxation process caused by the migration of the ions appeared at very low temperatures (lower than 273 K). When the temperature decreased, the conductivity dropped as well and the relaxation peak would appear at the $\varepsilon'' = f(f)$ plot. This relaxation is characterized by an absorption peak (temperature-dependent intensity and distribution of the relaxation times). No polarization peak was found at the room temperature, either.

Thus, the prepared by ion-exchange method, silver-modified and unmodified nanoporous zeolite plates were tested at room temperature in this study. Depending on the concentration of solutions (or Ag content), the samples with different amounts of silver and silver nanoparticles loading of the zeolite plates are obtained. Furthermore, the XRD pattern of silver-modified nanoporous zeolite showed the existence of Ag nanoparticles incorporated to clinoptilolite samples. It was sufficient for studying the effect of silver ions and silver nanoparticles on dielectric spectra of Ag-modified zeolite porous microstructure. It was found, for the first time to our knowledge, that the capacity value of the Ag-modified zeolite is abnormally high and the more, the greater the conductivity of the sample. Furthermore, it was found that for all Ag-modified zeolite samples, the resistance decreases with increasing frequency. All of the observed peculiarities of the dielectric response are due to the redistribution of the electric field between the zeolite sample volume and dielectric gap near surface of the electrode. It was found that the observed anomalously high values of dielectric permittivity are determined by the capacity of the near-electrode gap.

5 Conclusions

Non-thermal plasma characteristics in the GDED with a ZC were experimentally investigated up to atmospheric pressure (AP) in air medium. Porous zeolite cathodes are used for plasma light source applications with very low supply voltage and power consumption. The microstructures and performance of the ZCs are found to depend sensitively on the strong electric field in the nanopores, which supplies an efficient electron multiplication and related excitation of gas atoms. The use of gas discharge gap with nanoporous ZC gives rise to considerable increase in the plasma emission (PE) intensity. This device might find an application in generating and sustaining a stable, uniform, and homogeneous non-thermal AP plasma. The unique electronic properties of microporous zeolites make them also good candidate materials in the electronic industry so that silicon-based technologies could be replaced or supplemented by microporous zeolite-like materials. Furthermore, it was found that for all Ag-modified zeolite samples, the resistance decreases with increasing frequency. All of the observed peculiarities of the dielectric response are due to the redistribution of the electric field between the zeolite sample volume and dielectric gap near surface of the electrode. It was found that the observed anomalously high values of dielectric permittivity are determined by the capacity of the near-electrode gap. We also believe that in modification of the ZC as well as the geometry effect in a non-thermal plasma reactor, it is possible to build rather sensitive large emitting area plasma light source with internal emission amplification. Therefore, we confirm that the microporous zeolite is a suitable electrode material for GDED and can serve as a source of UV radiation if pressure and electric field are sufficiently high.

References

1. K. Koseoglu, M. Ozer, B.G. Salamov, Plasma Process. Polym. **11**, 1018 (2014)
2. T. Yokoyama, M. Kogoma, T. Moriwaki, S. Okazaki, J. Phys. D: Appl. Phys. **23**, 1125 (1990)
3. F. Massines, A. Rabehi, P. Decomps, R.B. Gadri, P. Segur, C. Mayoux, J. Appl. Phys. **83**, 2950 (1998)
4. B.G. Salamov, Y. Çiftci, K. Çolakoglu, IEEE Trans. Plasma Sci. **32**, 2093 (2004)
5. O. Sakai, Y. Kishimoto, K. Tachibana, J. Phys. D **38**, 431 (2005)
6. A.V. Phelps, Z. Petrovic, Plasma Sources Sci. Technol. **8**, R21 (1999)
7. A.A. Kudryavtsev, L.D. Tsendin, Tech. Phys. Lett. **28**, 621 (2002)
8. H. Willebrand, Y. Astrov, L. Portsel, S. Teperick, T. Gauselmann, Infrared Phys. Technol. **36**, 809 (1995)
9. B.G. Salamov, K. Colakoglu, S. Altindal, M. Ozer, J. Phys. III. **7**, 927 (1997)
10. B.G. Salamov, S. Altindal, M. Ozer, K. Colakoglu, E. Bulur, Eur. Phys. J. Appl. Phys. **2**, 26 (1998)
11. B.G. Salamov, S. Ellialtioglu, B.G. Akinoglu, N.N. Lebedeva, L.G. Paritskii, J. Phys. D: Appl. Phys. **29**, 628 (1996)
12. B.G. Salamov, M. Özer, M. Kasap, S. Altındal, J. Phys. D. **32**, 682 (1999)
13. V. Soghomonian, J.J. Heremans, Appl. Phys. Lett. **95**, 152112 (2009)
14. A.V. Eletskii, Phys. Usp. **40**, 899 (1997)
15. Y.V. Gulyaev, L.A. Chernozatonskii, Z.J. Kosakovskaja, N.I. Sinitsyn, G.V. Torgashov, Y.F. Zakharchenko, J. Vac. Sci. Technol. B **13**, 435 (1995)
16. N.N. Lebedeva, V.I. Orbukh, Ch.A. Sultanov, Tech. Phys. **55**, 565 (2010)
17. J.T.H. Tsai, H.C. Ko, Appl. Phys. Lett. **88**, 013104 (2006)
18. G. Gottardi, E. Galli, *Natural Zeolites* (Springer, Berlin, 1985).
19. C. Senaratne, J. Zhang, M.D. Baker, C.A. Bessel, D.R. Rolison, J. Phys. Chem. **100**, 5849 (1996)
20. A. Giaya, R.W. Thompson, R. Denkewicz, Microporous Mesoporous Mater. **40**, 205 (2000)
21. K. Koseoglu, M. Ozer, S. Ozturk, B.G. Salamov, Jpn. J. Appl. Phys. **53**, 086203 (2014)

22. H.Y. Kurt, B.G. Salamov, J. Electron. Mater. (2019) doi https://doi.org/10.1007/s11837-019-03956-0
23. B.G. Salamov, K. Colakoglu, S. Altındal, Infrared Phys. Technol. **36**, 661 (1995)
24. A. Mizuno, IEEE Trans. Dielec. Electr. Insul. **7**, 615 (2000)
25. A. Jaworek, A. Krupa, T. Czech, J. Phys. D **29**, 2439 (1996)
26. K. Hensel, V. Martisovits, Z. Machala, M. Janda, M. Lestinsky, P. Tardiveau, A. Mizuno, Plasma Process. Polym. **4**, 682 (2007)
27. J. Pawlat, P. Terebun, M. Kwiatkowski, B. Tarabova, K. Kucerova, Z. Machala, M. Janda, K. Hensel, Plasma Chem. Plasma Process. **39**, 627 (2019)
28. J.F. Roman-Zamorano, M. Flores-Acosta, H. Arizpe-Chavez, F.F. Castillon- Barraza, M.H. Farias, R. Ramirez-Bon, J. Mater. Sci. **44**, 4781 (2009)
29. A. Fred Mumpton, *Mineralogy and Geology of Natural Zeolites* (Fairford, GLOS, United Kingdom 2018)
30. A. Khataee, S. Bozorg, S. Khorram, M. Fathinia, Y. Hanifehpour, S.W. Joo, Ind. Eng. Chem. Res. **52**, 18225 (2013)
31. E. Chmielewska, E. Samajova, J. Kozac, Turk. J. Chem. **26**, 281 (2002)
32. T. Armbruster, Am. Mineral. **78**, 260 (1993)
33. S.V. Lankin, V.V. Yurkov, Перспективные Материалы. **5**, 59 (2006)
34. U. Simon, M.E. Franke, Micropor. Mesopor. Mater. **41**, 1 (2000)
35. H. Kabashima, H. Einaga, S. Futamura, IEEE Trans. Indust. Appl. **39**, 340 (2003)
36. R. Burlica, K.Y. Shih, B.R. Locke, Indust. Eng. Chem. Res. **49**, 6342 (2010)
37. M.A. Malik, K.H. Schoenbach, J. Phys. D: Appl. Phys. **45**, 132001 (2012)
38. A. Kumada, S. Okabe, K. Hidaka, J. Phys. D: Appl. Phys. **42**, 095209 (2009)
39. V. Bloshchitsyn, Review of surface discharge experiments (2010). https://arxiv.org/abs/1005.5044v1
40. M.A. Malik, S. Xiao, K.H. Schoenbach, J. Hazard. Mater. **209**, 293 (2012)
41. G.N. Tsikrikas, A.A. Serafetinides, J. Phys. D: Appl. Phys. **29**, 2806 (1996)
42. S.S. Yang, S.M. Lee, F. Iza, J.K. Lee, J. Phys. D: Appl. Phys. **39**, 2775 (2006)
43. Y.P. Raizer, *Gas Discharge Physics* (Springer, Berlin, 1991).
44. A. Bulletti, L. Capineri, M. Materassi, B.D. Dunn, IEEE Trans. Electron. Packag. Manuf. **30**, 115 (2007)
45. N. De Geyter, R. Morent, C. Leys, L. Gengembre, E. Payen, Surf. Coat. Technol. **201**, 7066 (2007)
46. Y. Sadiq, M. Ozer, B.G. Salamov, J. Phys. D: Appl. Phys. **41**, 045204 (2008)
47. G.J.M. Hagelaar, M.H. Klein, R.J.M. Snijkers, G.M.W. Kroesen, J. Appl. Phys. **89**, 2033 (2001)
48. V.I. Orbukh, N.N. Lebedeva, S. Ozturk, Ş. Uğur, B.G. Salamov, Optoelectron. Adv. Mater. Rapid Commun. **6**, 947 (2012)
49. K. Koseoglu, B.G. Salamov, Plasma Process. Polym. (2015). doi https://doi.org/10.1002/ppap.201500066
50. V.I. Orbukh, N.N. Lebedeva, S. Ozturk, B.G. Salamov, Superlattice Microstruct. **54**, 16 (2013)
51. E. Krogh Anderson, J.G. Krogh Anderson, E. Skou, in *Chemistry of Solid State Materials*, vol. 2 edited by P. Colomban (Cambridge University Press, Cambridge 1992)
52. R.A. Secco, M. Rütter, Y. Huang, Tech. Phys. **45**, 1447 (2000)
53. D.A. Faux, W. Smith, T.R. Forester, J. Phys. Chem. **101**, 1762 (1997)
54. S.V. Goryainov, R.A. Secco, Y. Huang, H. Liu, Phys. B **390**, 356 (2007)
55. P. Bruggeman, F. Iza, P. Guns, D. Lauwers, M.G. Kong, Y.A. Gonzalvo, C. Leys, D.C. Schram, Plasma Sourc. Sci. Technol. **19**, 015016 (2010)
56. M. Laroussi, F. Leipold, Int. J. Mass Spectrom. **233**, 81 (2004)
57. I. Fridovitch, Annu. Rev. Biochem. **64**, 97 (1995)
58. J. Shen, Q. Sun, Z. Zhang, C. Cheng, Y. Lan, H. Zhang, Z. Xu, Y. Zhao, W. Xia, P.K. Chu, Plasma Process. Polym. **12**, 252 (2015)
59. I. Amber, R.O. Odekhe, Y.S. Sanusi, J. Mech. Eng. Res. **4**, 142 (2012)

60. M. Ramos, R.L. Espinoza, M.J. Horn, A. P. F. Leite, *ISES Solar World Congress*. Göteborg Sweden, 14–19 June 2003
61. A. Arbuznikov, V. Vasilyev, A. Goursot, Surf. Sci. **397**, 395 (1998)
62. D.W. Breck, *Zeolites Molecular Sieves* (Malabar, R.E. Krieger Pub., 1984), pp. 392–410
63. D.X. Liu, P. Bruggeman, F. Iza, M.Z. Rong, M.G. Kong, Plasma Sourc. Sci. Technol. **19**, 025018 (2010)
64. Y.S. Akishev, A.A. Deryugin, V.B. Karalnik, I.V. Kochetov, A.P. Napartovich, N.I. Trushkin, Plasma Phys. Rep. **20**, 511 (1994)
65. A. Van Deynse, N. De Geyter, C. Leys, R. Morent, Plasma Process. Polym. **11**, 117 (2014)
66. R.H. Perry, D. Green, *Perry's Chemical Engineers' Handbook* (McGraw-Hill, New York, 1997).
67. J.J. Thomson, *Conduction of Electricity Through Gases* (Cambridge University Press, Cambridge, 1933).
68. M.J. Druyvesteyn, F.M. Penning, Rev. Mod. Phys. **12**, 267 (1940)
69. J. Meek, J. Craggs, *Electrical Breakdown of Gases* (Wiley, New York, 1978).
70. D. Carvalho, S. Ghosh, R. Banerjee, P. Ayyub, Nanotechnology **19**, 445713 (2008)
71. K. Koseoglu, M. Özer, S. Ozturk, B.G. Salamov, Jpn. J. Appl. Phys. **53**, 086203 (2014)
72. H. Jacobs, J. Freely, F.A. Brandt, Phys. Rev. **88**, 492 (1952)
73. J.M. Millet, J.J. Lafon, Phys. Rev. A. **52**, 433 (1995)
74. S. Watanabe, T. Watanabe, K. Ito, N. Miyakawa, S. Ito, H. Hosono, S. Mikoshiba, Sci. Technol. Adv. Mater. **12**, 034410 (2011)
75. N.L. Yasnopol'skii, A.E. Shabel'nikova, Fiz. Tverd. Tela. (Leningrad) **10**, 103 (1968) (Sov. Phys. Solid State **10**, 75) (1968)
76. N.L. Yasnopol'skii, A.E. Shabel'nikova, A.P. Balashov, N.S. Lozhkina, in *Research in Radio Engineering and Electronics* (in Russian), Part 2 (Moscow, 1975), p. 67
77. P.M. Shikhaliev, Tech. Phys. Lett. **24**, 752 (1998)
78. I.M. Bronshtein, B.S. Fraiman, *Secondary Electron Emission* (in Russian) (Nauka, Moscow, 1969), p. 408
79. A. Bogaerts, E. Neyts, R. Gijbels, J. van der Mullen, Spectrochim. Acta B. **57**, 609 (2002)
80. V.I. Orbukh, N.N. Lebedeva, S. Ozturk, Ş Uğur, B.G. Salamov, Optoelectron. Adv. Mater. Rapid Commun. **6**, 947 (2012)
81. V.A. Lisovskiy, S.D. Yakovin, V.D. Yegorenkov, J. Phys. D: Appl. Phys. **33**, 2722 (2000)
82. B. Soltabayev, H. Kurt, S. Acar, B.G. Salamov, J. Electronic Materials, **47** (2019). doi https://doi.org/10.1007/s11664-019-07210-w
83. M.D. Rutter, R.A. Secco, Y. Huang, Chem. Phys. Lett. **331**, 189 (2000)
84. M.A. Lieberman, A.J. Lichtenberg, *Principles of Plasma Discharges and Materials Processing* (Interscience Publication, Wiley, 2005).
85. A. Semnani, A. Venkattraman, A. Alexeenko, D. Peroulis, Appl. Phys. Lett. **102**, 174102 (2013)
86. M. Klas, S. Matejcik, B. Radjenovic, M. Radmilovic-Radjenovic, Phys. Scr. **83**, 045503 (2011)
87. F. Iza, J.A. Hopwood, IEEE Trans. Plasma Sci. **31**, 782 (2003)
88. A. Semnani, A. Venkattraman, A. Alexeenko, D. Peroulis, Appl. Phys. Lett. **103**, 063102 (2013)
89. A. Semnani, D. Peroulis, Appl. Phys. Lett. **105**, 253105 (2014)
90. B.G. Salamov, B.G. Akinoglu, N.N. Lebedeva, J. Phys. D: Appl. Phys. **32**, 2068 (1999)
91. H. Lee, P.K. Dutta, J. Phys. Chem. B. **106**, 11898 (2002)
92. J. Dzubiella, R.J. Allen, J.-P. Hansen, J. Chem. Phys. **120**, 5001 (2004)
93. K. Koseoglu, I. Karaduman, M. Demir, M. Ozer, S. Acar, B.G. Salamov, Superlattice Microstruct. **81**, 97 (2015)
94. I.M. Kalogeras, A. Vassilikou-Dova, Defect Diffus. Forum **164**, 1–36 (1998)
95. H.Y. Kurt, B.G. Salamov, JOM The Minerals, Metals & Materials Society (2019). doi https://doi.org/10.1007/s11837-019-03955-1
96. E. Krogh Anderson, I.G. Krogh Anderson, E. Skou, in: P. Colomban (Ed.), *Chemistry of Solid State Materials Series* (Cambridge University Press, Great Britain, 1992), p. 210

97. K. Praveena, S.R. Murthy, Int. J. Emerg. Technol. Adv. Eng. **3**, 363 (2013)
98. S. Ozturk Koç, S. Galioglu, B. Akata Kurç, E. Koç and B.G. Salamov, J. Electron. Mater. **47**(5), 2791 (2018)
99. P. Tabourier, J.C. Carru, J.M. Wacrenier, Zeolites **3**, 50 (1983)
100. P. Tabourier, J.C. Carru, J.M. Wacrenier, J. Chim. Phys. **87**, 43 (1990)
101. J.C. Carru, P. Tabourier, J.M. Wacrenier, J. Chim. Phys. **88**, 307 (1991)
102. S.V. Baryshnikov, S.V. Lankin, E.V. Stukova, V.V. Yurkov, Sovremenniye Naukoemkiye Tehnologii (2004), pp. 26–27
103. L.I. Gafarova, D.V. Sheds, I.V. Lunev, Y. Gusev, Struktura I Dinamika Molekularnih Syst. **10**(1), 354 (2003)
104. S.V. Lankin, Physica **N2**, 90854 (2011)
105. S. Ozturk, K. Koseoglu, V.I. Orbukh, G.M. Eyvazova, A.H. Muradov, N.N. Lebedeva, B.G. Salamov, Optoelectron. Adv. Mater. Rapid Commun. **8**, 733 (2014)
106. V.I. Orbukh, G.M. Eyvazova, A.H. Muradov, N.N. Lebedeva, B.G. Salamov, Transaction of ANAS of sci. Phys. and Astron. **33**(5) 49 (2013)
107. A.H. Muradov, V.I. Orbukh, G.M. Eyvazova, N.N. Lebedeva, Ch.G. Akhundov, A.M. Mamedov, Int. J. Scient. Techn. Res. **1** (9) 55 (2015). ISSN 2422-8702
108. S. Ozturk Koc, V.I. Orbukh, G.M. Eyvazova, N.N. Lebedeva, B.G. Salamov, Superlattices Microstruct. **9**, 269 (2016)
109. M. Kalogeras, A. Vassilikou-Dova, Defect Diffus. Forum **164**, 1–36 (1998)
110. F.J. Jansen, R.A. Schoonheydt, in *Advances in Chemistry Series*. ed. by W.M. Meier, J.B. Uytterhoeven (American Chemical Society, Washington, 1973), p. 96
111. M. Alvaro, J.F. Cabeza, D. Fabuel, H. Garcia, E. Guijarro, J.L.M. de Juan, Chem. Mater. **18**, 26 (2006)
112. K. Sahner, G. Hagen, D. Schonauer, S. Reib, R. Moos, Solid State Ionics **179**, 2416 (2008)
113. P. Alienzar, S. Valeneia, A. Kozma, H. Garcia, Chem. Phys. Chem. **8**, 1115 (2007)
114. K. Suresh, B.V. Siddaramaiah. M.B. Shagan, K.S. Manjula, C. Banganatbaiah, G.V. Narasimha Rao, B. Basavalingu, K. Byrappa, J. Polym. Res. **17**, 135 (2010)
115. S. Ozturk, M. Ozer, E. Bulur, B.G. Salamov, IEEE Trans. Plazma Sci. **43**, 1330 (2015)
116. S. Galioglu, M. Isler, Z. Demircioglu, M. Koc, E. Vocanson, N. Dectouches, R. Turan, B. Akata, Micropor. Mezopor. Mater. **159**, 1–8 (2012)
117. I. Tiscornia, S. Irusta, P. Pradanos, C. Teller, I. Coronos, J. Santamaria, J. Phys. Chem. C **111**, 4702 (2007)
118. Z. Cai, T.L. Coordrich, K.S. Ziemer, J. Warrywocla, A. Jr, Sacco. Appl. Catal. B: Environ. **102**, 323 (2011)
119. J.R. Macdonald, Ann. Biomed. Eng. **20**, 289 (1992)
120. U. Bunyatova, S. Ozturk Koc, V.I.Orbukh, G.M.Eyvazova, Z.A.Agamaliev, N.N. Lebedeva, I.C. Koçum, B.G. Salamov, M. Ozer, Superlattice. Microstruct. **98**, 295 (2017)
121. V.I. Orbukh, G.M.Eyvazova, N.N.Lebedeva, M.B.Muradov, B.G. Salamov, Trans. Azerb. Nat. Acad. Sci. Phys. Astron. **35**(5) 51 (2015)
122. M. Majdan, S. Pikus, M. Kowalska-Ternes, A. Gladysz-Plaska, H. Skrzypek, W. Kazimierczak, J. Molec. Struct. **657**, 47 (2003)
123. E. Torracca, P. Gali, M. Pansini, C. Colella, Micropor. Mesopor. Mater. **20**, 119 (1998)
124. S. Ozturk Koc, K. Koseoglu, S. Galioglu, B. Akata, B.G. Salamov, Micropor. Mesopor. Mater. **223**, 18 (2016)
125. E. Erdem, N. Karapınar, R. Donat, J. Colloid. Interface. Sci. **280**(2), 309 (2004)
126. V. Sály, S. Kocálka, Chem. Papers. **50**(6), 328 (1996)
127. J.E. Bauerle, J. Phys. Chem. Solids **30**, 2657 (1969)
128. B.G. Salamov, G.M. Eyvazova, V.I. Orbukh, N.N. Lebedeva, Z.A. Agamalıev, J. Low Dimensional Syst. **1**(1), 19 (2017)

Physical Properties and Ethanol Response of Sprayed In_2S_3:Sn Films

Mabrouk Kraini and Noureddine Bouguila

Abstract In this study, tin-doped indium sulfide thin films (In_2S_3:Sn) at different molar ratios of Sn:In (0–1% by mol in solution) were deposited on glass substrates by chemical spray pyrolysis method. The films were investigated by X-ray diffraction, optical absorption, Raman and photoluminescence spectroscopies, field emission scanning electron microscopy, energy dispersive X-ray spectroscopy and atomic force microscopy. The structural properties reveal that the In_2S_3:Sn thin films have a polycrystalline cubic structure and the average crystallite size increases from 16.3 to 25.5 nm. The surface morphology of the films is continuous and free of cracks. The average surface roughness and the root-mean square roughness increase from 13.12 to 31.65 nm and from 16.14 to 39.39 nm, respectively, with increasing of Sn:In molar ratio. Raman studies reveal the presence of modes of vibrations related to In_2S_3 phase, with no signature of secondary phases. The transmission coefficient is about 65–70% in the visible region and 70–90% in near-infrared region. The optical band gap values of In_2S_3:Sn for allowing direct transitions were found to be in the range 2.68–2.80 eV. The refractive index of In_2S_3:Sn thin films has decreased from 2.45 to 2.37 and the k values are in the range 0.02–0.25 for all wavelengths. Defects-related photoluminescence properties were also discussed. Effect of tin doping on ethanol sensing response was also investigated. A slight enhancement in response is shown for In_2S_3:1%Sn sample.

Keywords Sn-doped In_2S_3 · Spray pyrolysis · Structural properties · Raman and photoluminescence spectroscopy · Optical properties · Ethanol sensing

1 Introduction

Indium sulfide (In_2S_3) is one of the potential materials that can be used for many applications. This includes photovoltaic [1, 2], photoelectrochemical solar cells [3],

M. Kraini (✉) · N. Bouguila
Faculté des Sciences de Gabès, Laboratoire de Physique des Matériaux et des Nanomatériaux Appliquée à l'environnement, Université de Gabès, Cité Erriadh, 6072 Zrig, Gabès, Tunisia
e-mail: mabrouk.karini@gmail.com

S. J. Ikhmayies et al. (eds.), *Advances in Optoelectronic Materials*, Advances in Material Research and Technology,
https://doi.org/10.1007/978-3-030-57737-7_8

electronic [4], optical [1, 5], acoustic [6] ,and dry cells [7], which are displayed [8, 9] as photocatalysts for dye degradation [10], water splitting [11] and gas sensor [12]. This is mainly due to its chemical stability, wide energy band gap, and controllable electrical properties [13]. In photovoltaic solar cells, In_2S_3 has been proved to be a promising buffer layer due to its lower toxicity compared to that of CdS [14].

Due to the above important properties, Several techniques have been utilized to synthesize In_2S_3 such as chemical spray pyrolysis (CSP) [15–21], ultrasonic dispersion [22], chemical bath deposition (CBD) [23], physical vapor deposition [24], Hydrothermal [25], etc. Among these methods, spray pyrolysis is interesting because it allows preparing large area In_2S_3 thin films at low cost and it is easy to incorporate different dopants.

In_2S_3 belongs to the *III–VI* group of semiconductors, which is found to crystallize into three different structural forms, defective cubic structure α, β and γ [26, 27]. β-In_2S_3 being the stable phase at room temperature, crystallizes in a normal spinel structure with a high degree of tetrahedral and octahedral vacancy sites [2, 22, 23].

Few works have been reported to study the effect of tin doping on In_2S_3 thin films on their structural, morphological, optical, and electrical properties [28–33]. Furthermore, a precise study on the properties of doped films is essential for the doping under 1% of Sn into In_2S_3 films which are useful for photovoltaic and optoelectronic applications.

In this work, we report the influence of Sn doping (less than 1% by mol) into In_2S_3 on the structural, morphological, optical properties and ethanol sensing response of In_2S_3 films. We successfully synthesized Sn-doped In_2S_3 films using a simple and low-cost spray pyrolysis method to control the Sn content. This method may be used for the preparation of many new materials for electronics and photovoltaic devices.

2 Experimental Details

The In_2S_3 thin films were deposited by CSP using an aqueous solution of indium (III) chloride ($InCl_3$), thiourea ($CS(NH_2)_2$) in a ratio of S:In = 2.5, and Tin (IV) chloride ($SnCl_4$). The tin-to-indium molar ratio (Sn:In) is varied from 0 to 1% in the spray solution. Compressed air was used as a carrier gas at a flow rate of 6 l/min. The nozzle-to-substrate distance was approximately 25 cm, and the spraying time was around 15 min. The substrate temperature was maintained at 350 °C within an accuracy of ±5 °C. The solution was sprayed on glass substrate. The obtained In_2S_3 thin films thickness was approximately 300 nm.

3 Characterization of the Films

Structural properties of the films are determined by the Bruker D8 Advance diffractometer (*XRD*) using Cu Kα radiation (λ = 1.5406 Å) at diffraction angles varying between 10° and 70°. Raman spectroscopy measurements were carried out in a Renishaw inVia Reflex Spectrometer system, employing an argon laser (514 nm) as an excitement source. Room temperature photoluminescence (PL) measurements were carried with a Jobin–Yvon micro-Raman spectrometer (T 64,000) using an argon laser excitation wavelength of 488 nm with a power of 3 mW. PL spectra were collected through a confocal optical microscope coupled with a holographic grating (600 grooves mm^{-1}) with a resolution less than 0.5 cm^{-1} by a charge coupled device (CCD) camera detector.

A Shimadzu UV 3101 PC spectrophotometer was used to obtain the optical transmittance and absorbance spectra. Morphological properties were determined by field emission scanning electron microscope (*Zeiss FESEM Ultra PLUS*) equipped with an accessory of energy dispersive spectroscopy (EDS) and atomic force microscope using XE-100 instrument (*Park Systems Corporation*).

To investigate ethanol sensing response, In_2S_3 films were tested in a cylindrical chamber ($R = 10$ cm, $h = 10$ cm) using the experimental setup shown in Fig. 1.

The dilution of ethanol vapor in dry air was achieved using two-arm gas-flow devices. Two mass flow controllers allowed the flow rate of the dry air to be controlled from 0 to 1 l min^{-1} in both arms (d_1, d_2). The carrier gas sent in the first arm passes through a bottomed flask containing the ethanol vapor equilibrated with 200 cm^3 of the liquid, while air dry without ethanol vapor is sent along the second arm. Then both flows mix at the extremity of the two arms. The bottomed flask was placed in a water bath at a fixed temperature T_{vap}. By varying d_1 and d_2 ($d_1 + d_2$ *was kept constant at* 1 l min^{-1}), diverse concentrations of ethanol can be obtained and computed using

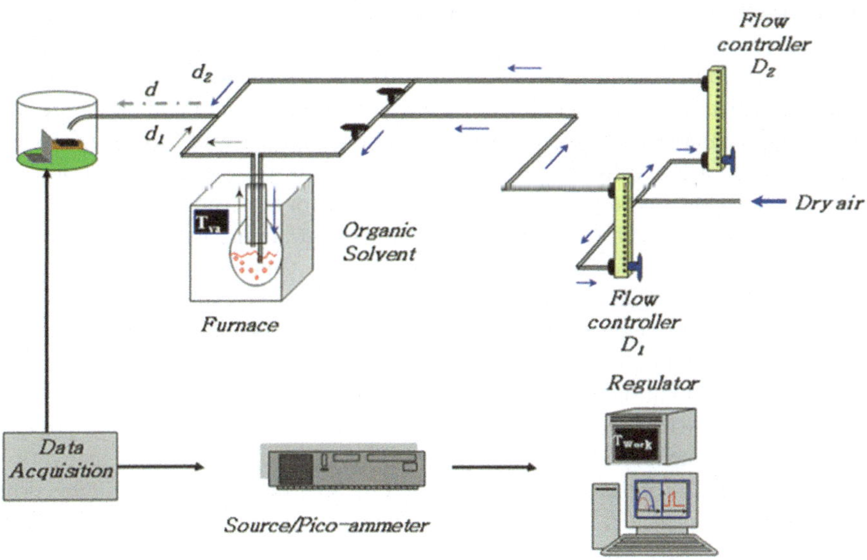

Fig. 1 Experimental setup for ethanol gas detection

the formula [12]:

$$[C](\mathrm{ppm}) = \left(\frac{xd_1}{(1+x)d_1 + d_2}\right) \times 10^6 \tag{1}$$

where x is the molar fraction of the vapor in the bottomed flask at T_{vap}, given by:

$$x = \frac{P_{\mathrm{vap}}}{P_{\mathrm{atm}}} \tag{2}$$

with P_{vap} the partial pressure of the vapor at a specified temperature T_{vap} and P_{atm} the atmospheric pressure.

The total flow charged with ethanol vapor was blown on the sensor for 5 min. Then, the test chamber was purged for 15 min. Samples were polarized at 0.5 V with an (HP4140B) source/pico-ammeter. Alcohol sensing test of the films was carried out using DC electrical characterization.

4 Results and Discussion

4.1 Structural Characterization

Figure 2 shows the XRD patterns of In_2S_3:Sn films doped with various values of molar ratio Sn:In from 0 to 1%. Seven pronounced XRD peaks are provided, corresponding to the planes (111), (220), (311), (222), (400), (511), and (440) [34, 35] according to joint committee on powder diffraction standards (JCPDS) card no. 32-0456, indicating that In_2S_3:Sn films has a polycrystalline cubic structure. No signature of secondary phases corresponding to Sn impurity is observed. This result is similar to that found by Yan et al. [29], Kilani et al. [30], and Maha et al. [32]. However, Sall et al. [31] reported tetragonal structure for their films and Mathew et al. [33] found that In_2S_3:Sn films obtained by in-situ Sn doping in the spray solution are amorphous. The spectra show evident amorphous "humps", at low angles in all diffraction patterns. These data are characteristic of the glass substrate [34, 35].

Tin atoms introduced into the crystalline structure can occupy different positions. They can be substituting the indium atoms in the In_2S_3 matrix or inserted into the vacant tetrahedral cationic sites. However, the ionic radius Sn^{4+} (0.69 Å) is lower than that of In^{3+} (0.80 Å) for octahedral coordination [36]; therefore, it is possible that Sn^{4+} ions systematically occupy the positions of the In^{3+} ions in the network.

From the results presented in Table 1, we note that the value of the angle corresponding to the peak (311) increased first with increasing the doping level up to 0.5%, then decreased. This decrease can be attributed to a lattice contraction (Table 1). The insertion of tin (IV) ions in the In_2S_3 material may be the responsible of this lattice deformation.

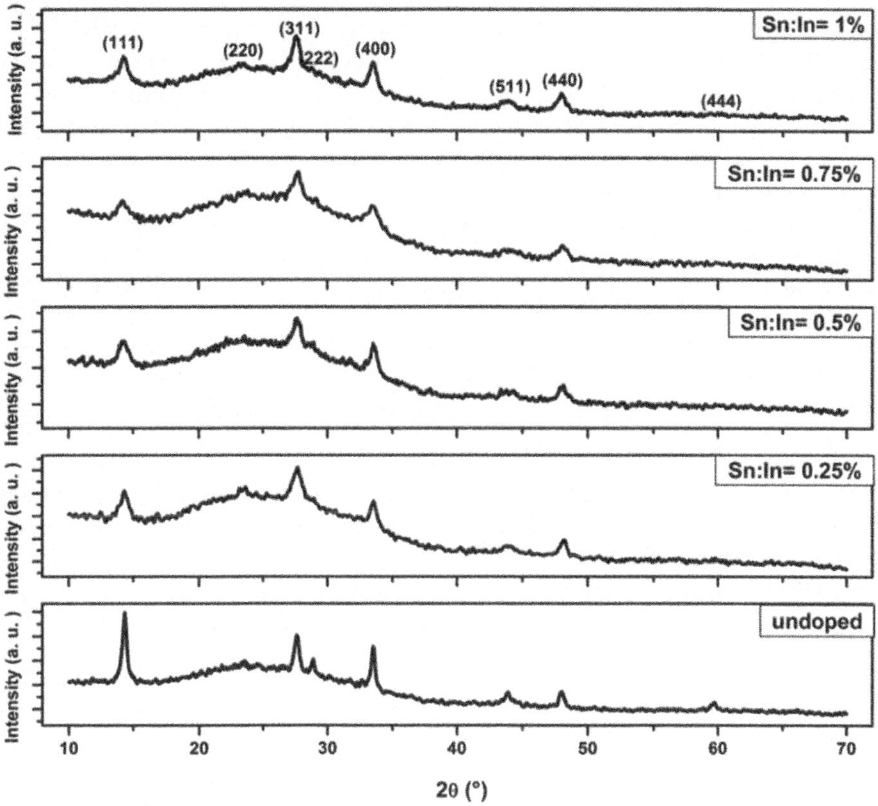

Fig. 2 XRD patterns of In_2S_3:Sn thin films

Table 1 Position of peak (311), full width at half maximum $\Delta(2\theta)$, crystallite size (D), dislocation density (δ) and microstrain (ε) as a function of molar ratio Sn: In

Sn:In	2θ (°)	B (°)	D (nm)	a (Å)	V (Å^3)	δ (10^{11} lines cm^{-2})	ε (10^{-3})
Undoped	27.614	0.35	23.3	10.710	1228.6	1.8	6.2
0.25%	27.622	0.50	16.3	10.707	1227.6	3.7	8.8
0.5%	27.660	0.45	18.1	10.692	1222.6	3.0	7.9
0.75%	27.651	0.40	20.4	10.696	1223.7	2.4	7
1%	27.626	0.32	25.5	10.705	1227	1.5	5.6

The average crystallite size was estimated from the (311) main peak by employing the Scherrer's formula [37]

$$D = \frac{0.9\lambda}{\beta \cos\theta} \quad (3)$$

where β is the size broadening in radians, obtained as $\beta = \sqrt{\beta_e^2 - \beta_0^2}$, where β_e is the full width at half maximum measured from the film and β_0 is the instrumental broadening (1.5×10^{-3} rd), λ is the wavelength of X-ray, θ is the Bragg angle.

The values of average crystallite size are listed in Table 1. It can be seen that the crystallite size increases with Sn concentration. Therefore, increasing the concentration of tin favors the crystal growth and improves the texture.

From the XRD patterns, we can observe a shift in position of the peak (311) that can be due to the presence of uniform strains in the samples.

The microstrain of the samples can be calculated from Stokes–Wilson equation [38]

$$\varepsilon = \frac{\beta}{4 \tan \theta} \tag{4}$$

Note that the microstrain varies inversely with the crystallite size, more the crystallite size increases more the grain boundary density decreases, resulting in improving the crystalline quality and consequently the reduction of microstrain. The microstrain decreases from 6.2×10^{-3} to 5.6×10^{-3} (Table 1). This evolution is probably due to the occupation of substitution sites by Sn ions in the lattice.

Dislocation density (δ) that is a measure of the amount of defects in the crystal structure is defined as the length of the dislocation lines per unit volume and it can be found [39, 40] by

$$\delta = \frac{1}{D^2} \tag{5}$$

The calculated values of dislocation density for In_2S_3:Sn thin films as a function of Sn:In molar ratio are tabulated in Table 1. The values of δ decreases from 3.7×10^{11} to 1.5×10^{11} lines cm^{-2} with increasing of Sn:In..This behavior can be explained by the change of the crystallites size (D) with doping concentration. [34, 35, 41].

4.2 Morphological Characterization

The surface morphological of In_2S_3:Sn films was performed out using FESEM measurements. FESEM images of undoped and Sn-doped In_2S_3 films are shown in Fig. 3 The surface morphology of the films depends on the Sn:In molar ratio. The surface of the samples revealed continuous films with no cracks and voids and it is clearly seen also from the FESEM micrographs that the films are dense without pinholes and perfectly covering the entire substrates for all Sn:In molar ratios.

The composition of In_2S_3:Sn thin films which are detected by EDS are summarized in Table 2. The EDS analysis confirmed the presence of In, S, and Sn elements in the films and the atomic percentage of these elements was given in Table 2. It has been observed that upon increasing the Sn concentration, the percentage of In in the films

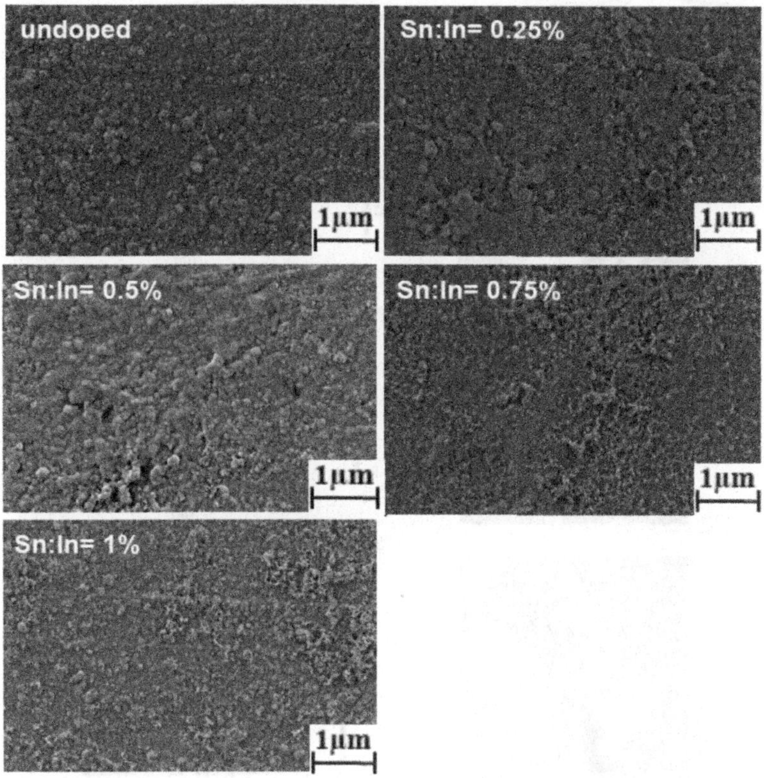

Fig. 3 FESEM images of In_2S_3 thin films for different Sn doping

Table 2 Chemical composition of In_2S_3:Sn thin films for different molar ratios Sn:In determined by EDS

Sn:In	S (at.%)	In (at.%)	Sn (at.%)	S:In	Sn:In
Undoped	60.76	39.24	0	1.54	0
0.25%	60.72	38.94	0.34	1.56	0.008
0.5%	61.34	38.05	0.61	1.61	0.016
0.75%	61.43	37.74	0.83	1.62	0.022
1%	61.97	37.10	0.93	1.67	0.025

was decreased slightly, whereas the percentage of S was almost constant. It would be presumed that Sn doping in the films was substitutional.

The surface morphology of the layers was studied by non-contact atomic force microscopy (NC-AFM) technique. Figure 4 shows two dimensions (2D) and three dimensions (3D) NC-AFM topographic images (2 µm × 2 µm) of In_2S_3:Sn films for different Sn:In molar ratios.

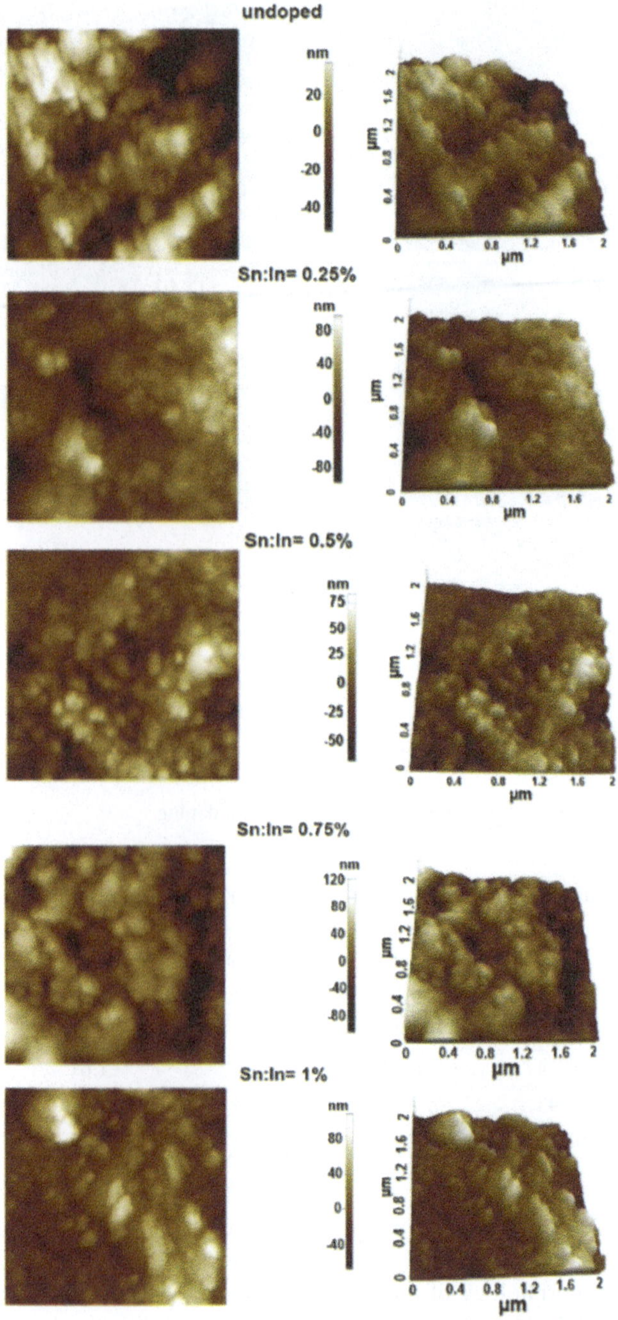

Fig. 4 NC-AFM images of In_2S_3 thin films for different Sn doping

Table 3 Surface roughness parameter values for In_2S_3:Sn films for different molar ratios

Sn:In	R_a (nm)	R_q (nm)	R_{ku}	R_{sk}
Undoped	13.12	16.14	2.58	0.230
0.25%	22.19	27.85	3.07	0.028
0.5%	24.68	30.06	2.55	0.225
0.75%	28.18	34.16	2.58	0.181
1%	31.65	39.39	3.18	−0.660

From these images, we can see that the layers are rough showing a relief at least more perturbed by clusters that represent rounded crystallites and spread over the entire surface. The average surface roughness (R_a) and the root-mean square roughness (R_q), skewness (R_{sk}) , and kurtosis (R_{ku}) for samples are summarized in Table 3.

We note that R_a and R_q increase from 13.12 to 31.65 nm and 16.14 to 39.39 nm, respectively. This can be correlated with the variation of the average grain size as a function of tin concentration. R_{sk} and R_{ku} both parameters indicate that the surface of the layers is composed of a very inhomogeneous structure having a distribution in the form of scattered patterns. Thus, the increase of the roughness of the films favors the adsorption of oxygen on the surface of the grains to form bonds then acting as electron traps [42]. The increase in the films roughness with tin doping concentration was confirmed by an increase in the grain size as calculated from XRD data.

4.3 *Optical Characterization*

4.3.1 Transmittance and Absorbance Spectra

The spectra of absorbance and transmittance measured in the wavelength range 300–2400 nm of the layers of In_2S_3:Sn for various molar ratios Sn:In are shown in Fig. 5a, b, respectively.

We can notice the presence of very low contrast of interference fringes due to reflections on the surface of the layer, the layer/substrate interface and the diffusion of light favored by low grain sizes, indicative of inhomogeneous layers or very thin layers. The measured transmission shows three distinct zones; a low absorption for wavelengths where transmission is the highest and the other of high absorption for shorter wavelengths where the transmission decreases rapidly to near zero values. The region of strong absorption is the fundamental absorption ($\lambda < 450$ nm) in the film due to the interband electronic transitions. The change in absorbance in this region is exploited to determine the gap. Absorption front of the layers moves slightly with the rate value of tin doping to low wavelength. The layers are transparent in the visible range (65–70%) and in the near-infrared (70–90%) with a linear absorption edge in the neighborhood of the ultraviolet (UV) region.

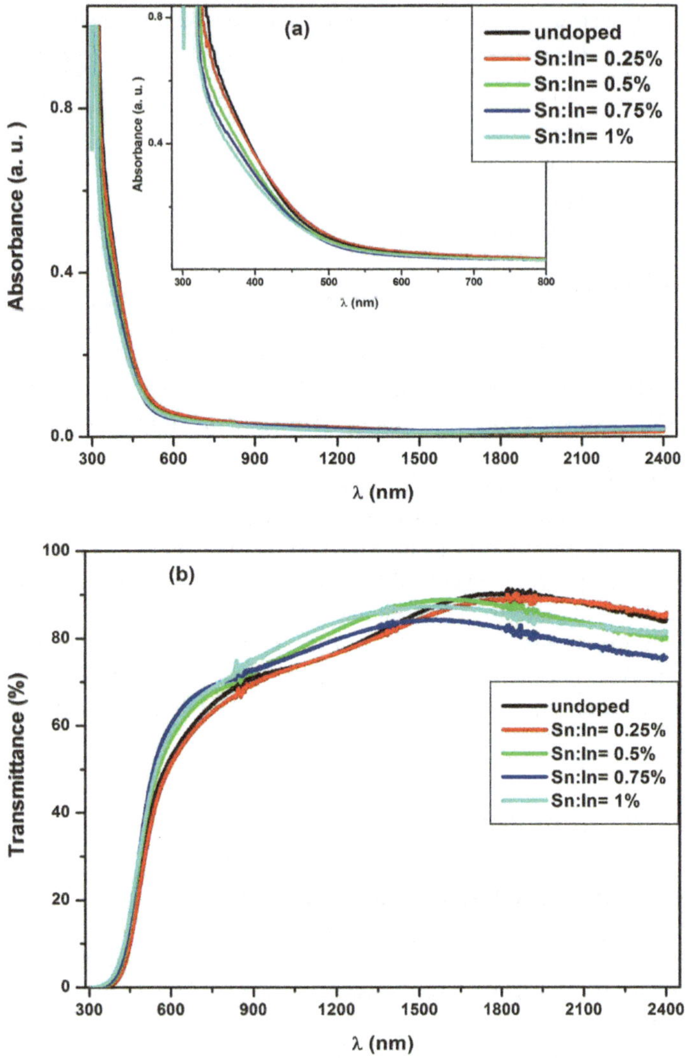

Fig. 5 Absorbance (**a**) and transmittance (**b**) spectra of In_2S_3:Snthin films

4.3.2 Absorption Coefficient and Extinction Coefficient

The absorption coefficient (α) of the films can be estimated by the relation [43]:

$$\alpha = \frac{2.303}{d} A \quad (6)$$

where $d = 300$ nm is the thickness of the prepared thin film, A is the absorbance.

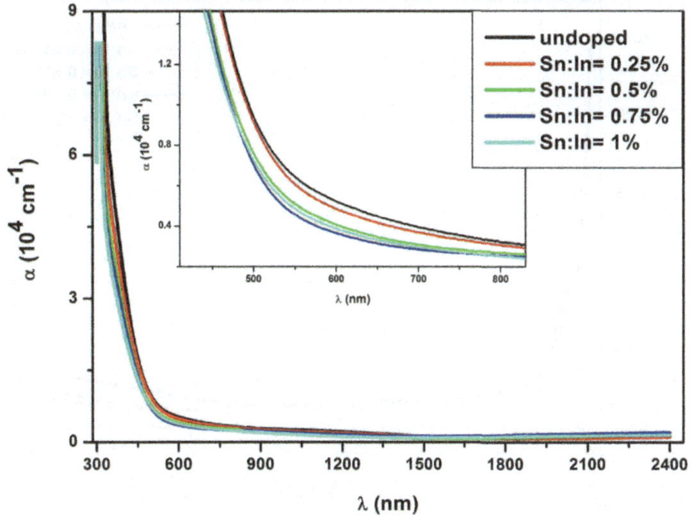

Fig. 6 Absorption coefficient of In_2S_3:Sn thin films

The variation of the absorption coefficient of In_2S_3:Sn thin films with the wavelength for various molar ratios Sn:In is shown in Fig. 6.

We can notice that α in general decreases with increasing of wavelength. It is observed a strong absorption in the UV-visible for the different molar ratios. In this zone, the absorption coefficient has a maximum value ranging from 6×10^4 to 9×10^4 cm^{-1} for the wavelength of 330 nm. The absorption becomes negligible in the visible and infrared (IR) regions. It was also noted that the absorption edge was not sharp for the doped samples indicating absorption in longer wavelength region also. This might be due to the introduction of shallow donor levels due to the doping. These results are consistent with literature [21, 44, 45]. The higher values of the absorption are attributed to the fact that the incoming photons have the sufficient energy to excite the electrons from the valence band to the conduction band. The absorption decreases in the higher wavelength region and this decrease corresponds to the reduction in the photon's energy.

The extinction coefficient k is given by the following expression [46]:

$$k = \frac{\lambda \alpha}{4\pi} \tag{7}$$

On Fig. 7, we present the evolution of the extinction coefficient k as a function of the wavelength of the Sn-doped In_2S_3 films.

The values of k are high in the zone of strong absorption or the extrinsic zone of absorption; then, decrease in the zone of weak absorption. The k values are in the range 0.02–0.25 for all wavelengths. Indeed, the k values vary slightly depending on

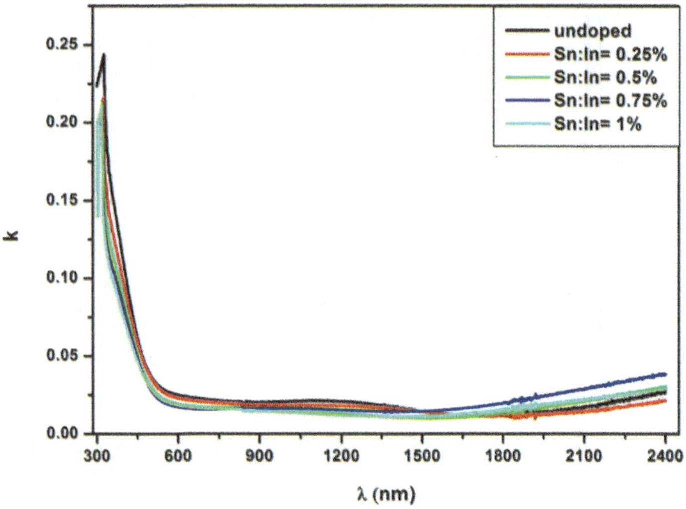

Fig. 7 Extinction coefficient of In_2S_3:Sn thin films

the amount of tin in the layers. In the visible, the low value of k means that these layers are transparent as shown by the transmittance spectra of.

4.3.3 Optical Energy Gap, Urbach Energy, and Refractive Index

The optical energy band gap "E_g" is determined using the following relation, which is valid for direct band gap semiconductors [21]:

$$(\alpha h\nu)^2 = A(h\nu - E_g) \tag{8}$$

where A is a constant, h is the Planck constant, υ is the frequency. Figure 8 shows the variation of $(\alpha h\nu)^2$ against $h\nu$. The point of the extrapolation of the linear part that meets the abscissa will give the value of the band gap energy of the material.

The absorption coefficient near the band edge shows an exponential dependence on photon energy and this dependence is given as follows [47–49].

$$\alpha = \alpha_0 \exp\left(\frac{h\nu}{E_U}\right) \tag{9}$$

$$E_U = \left[\frac{d\ln\alpha}{dh\nu}\right]^{-1} \tag{10}$$

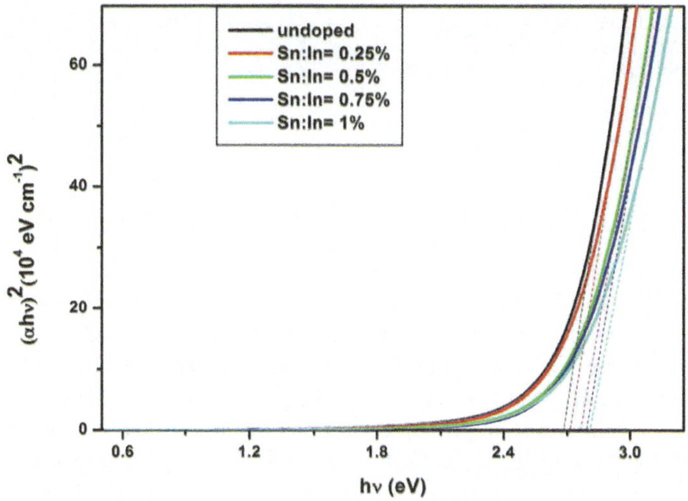

Fig. 8 Variation of $(\alpha h\nu)^2$ versus photon energy for In_2S_3:Sn films

where α_0 is a constant, E_U is the Urbach energy refers to the width of the exponential absorption edge deduced from inverse of the slope of the curve ln(α) versus $h\nu$, as shown in Fig. 9.

The values of the energy gap and Urbach are summarized in Table 4.

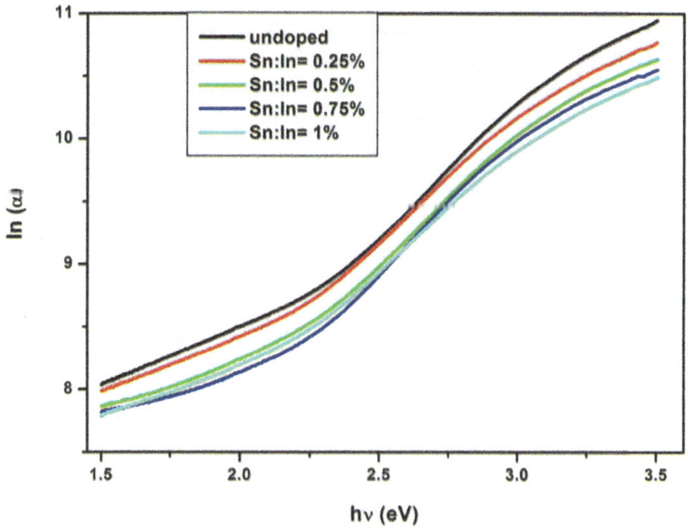

Fig. 9 Variation of ln(α) versus photon energy for In_2S_3:Sn films

Table 4 Optical constants of In_2S_3:Sn thin films as function of the molar ratio Sn:In

Sn:In	E_g (eV)	E_u (eV)	n
Undoped	2.68	0.480	2.45
0.25%	2.71	0.460	2.44
0.5%	2.75	0.448	2.43
0.75%	2.78	0.440	2.42
1%	2.8	0.476	2.41

We note that the optical gap increases from 2.68 to 2.80 eV. These results are similar to that found in the literature [30–32] who reported that doping tin has the effect of increasing the band gap of the In_2S_3 material. The determined value of the optical gap is smaller than that reported by Mathew et al. [33]. This difference can be explained by the different growth conditions of these films. As can be seen, the addition of dopant tends to increase the optical gap. This phenomenon is related to the increased number of carriers resulting from the occupation of interstitial sites by dopant atoms because they represent the main native donors in movies and is explained by the effect Burstein-Moss [50, 51] which is observed in naturally n-type semiconductors. On the other hand, when the concentration of dopant increases, the Fermi level moves toward the conduction band. Therefore, due to filling of the conduction band, transitions are made between the valence band and the Fermi level located in the conduction band between the valence band and the bottom of the conduction band. Such an increase of optical energy gap is useful for photovoltaic applications since it increases the collection of photons in the ultraviolet range [52].

The E_u value varies in the range of 0.440–0.476 eV. The tin atoms occupy the vacant site of indium and consequently the crystallinity improves, which reduces disorder. These results clearly confirm the analysis with XRD, where a correlation between strain and disorder in the different layers was observed. As it can be seen, the increase in strain is accompanied by a high disorder in the films.

The refractive index of a semiconductor increases with decreasing energy band gap or vice-versa. Various empirical and semi-empirical rules are available to correlate n and E_g. Among them, the Herve–Vandamme relationship for n and E_g is [53]:

$$n^2 = 1 + \left(\frac{A}{E_g + B} \right)^2 \tag{11}$$

where A and B are constants ($A = 13.6$ eV and $B = 3.4$ eV).

The calculated n and E_g values of In_2S_3 thin films in the present work are obeying the above said relations. The refractive index is related to the electronic polarizability of ions and the local field inside the optical materials. It is visible in Table 4 that the refractive index of In_2S_3:Sn thin films is decreased from 2.45 to 2.37 with increasing Sn:In molar ratio which may be attributed to the variation in packing density of the films. Indeed, this reduction is accompanied by a small increase in transmission. These results are comparable to that found in the literature [44, 54, 55].

4.4 Raman and Photoluminescence Spectroscopy Analysis

Figure 10 shows the Raman spectra of In_2S_3:Sn films for different Sn:In molar ratios. Multi-peak fitting of the Raman spectra is used to obtain the knowledge of the positions and the full width at half maximum (FWHM) of the Raman peaks. Red lines in Fig. 10 show the fitting results.

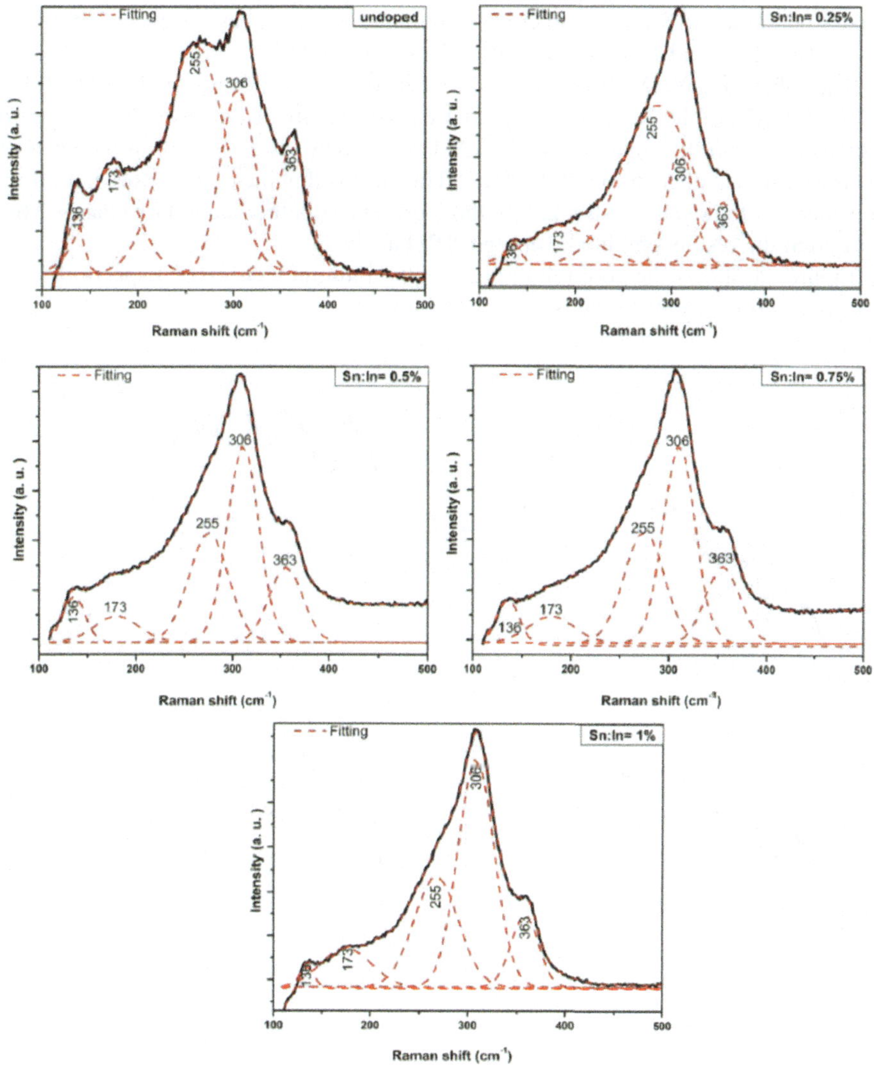

Fig. 10 Raman spectra of In_2S_3:Sn thin films for different molar ratios Sn:In

Table 5 FWHM of peak 306 cm^{-1} of the samples

Sn:In	Undoped	0.25%	0.5%	0.75%	1%
FWHM (cm^{-1})	57	41	35	28	24

According to Fig. 10, modes obtained of β-In_2S_3 by Raman analysis confirms well know the β-In_2S_3 phase of our samples with a main mode at 306 cm^{-1} is ascribed to the symmetric stretching vibrating mode of tetrahedral InS_4 [34, 56–58]. Others modes belonging to β-phase and corresponding to A_{1g} at 363 cm^{-1}, E_g at 255 cm^{-1}, F_{2g} at 163 cm^{-1} and 173 cm^{-1} are also observed [59, 60].

The values of FWHM of the tetrahedral peak obtained by fitting are listed in Table 5. The FWHM of the Raman peak decreases significantly, from 57 to 24 cm^{-1} with increasing of Sn:In molar ratio. The FWHM decrease is a sign of an increase of the crystalline quality of the In_2S_3 films. This means that the doping does not affect considerably the sample's structure without presence of other secondary phases. This supported the results obtained from the XRD analysis.

Figure 11 shows the room temperature photoluminescence (PL) spectra of the β-In_2S_3:Sn films for different Sn:In molar ratios.

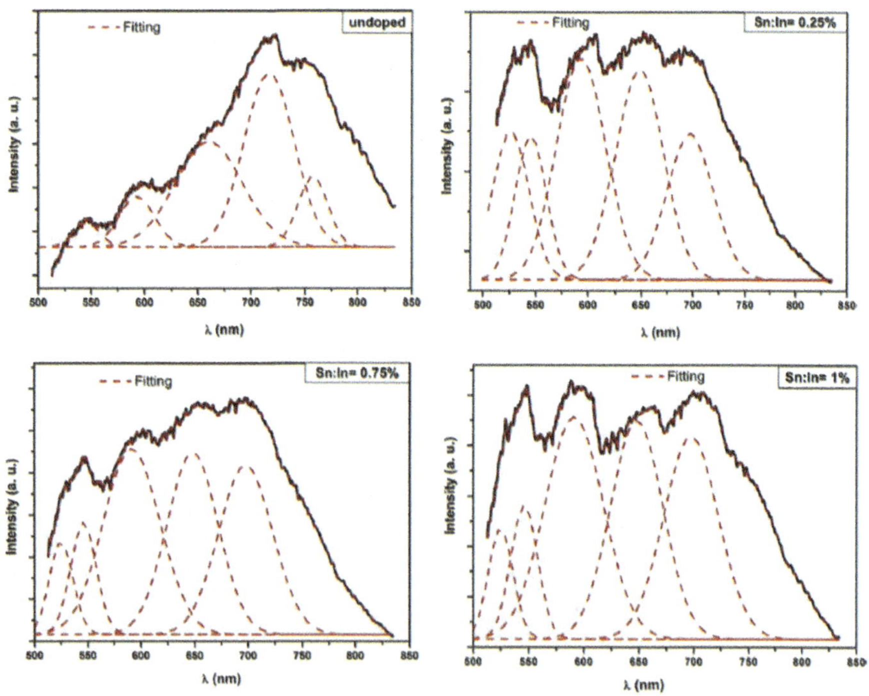

Fig. 11 Room temperature photoluminescence spectra of In_2S_3:Sn thin films for different molar ratios Sn:In

PL spectra reveal the visible peaks centered at 529, 537, 597, 650, 703, and 758 nm. Red lines in Fig. 10 show the fitting results. The peak at 529 and 537 nm (green emission) may be associated with the transition from a sulfur vacancy (V_S) level to the In vacancy (V_{In}) level [34, 61, 62]. Mathew et al. [63] and Chen et al. [9] reported that the green emission from In_2S_3 was attributed to the donor level formed by indium interstitials. The peak at 597 nm (orange emission) is assigned to the luminescence ions In^{3+} [64]. A shoulder at 758 nm accompanies two peaks centered at 650 and 703 nm (red emissions). These emissions arise as a transition from indium interstitial (In_i) donors to oxygen in vacancy of sulfur (O_{Vs}) acceptors [34, 62].

4.5 *Ethanol Detection*

It is known that morphology and microstructure affect the performance of gas sensors [65]. Thus, the increase of the roughness favors gas adsorption on the surface of the film. The increase in roughness with the tin doping concentration is remarkable for an Sn1% doping. In this section, we will focus on this sample. In_2S_3: Sn1% film was studied in needs of its response magnitude as a function of ethanol concentration and working temperature using DC electrical characterization. Results are compared with those published in our previous work for undoped In_2S_3 sensor.

4.5.1 Ethanol Concentration Effect

In_2S_3:Sn1% film is exposed to different concentrations of ethanol with dry air as a carrier gas. Figure 12 illustrates the transient current versus time for the sensors at a fixed working temperature of 350 °C.

Five responses and recovery curves are shown corresponding to ethanol gas concentration ranging from 250 to 1250 ppm. We note that during the exposure to ethanol vapor, the current through sensor increased quickly and got to equilibrium value with time. When target gas was removed, current regained its initial value. This behavior shows an increase of electrical current after the introduction of reducing gases is consistent with the sensing mechanism of *n*-type semiconductors [66].

Besides, the response of the sensor to a given concentration of the gas is proportional to the change in the conductance of the material on exposure to gas. Here, the sensing response "*S*" of the sensor is defined by [12, 67]:

$$S = \frac{I_{\mathrm{EtOH}}}{I_{\mathrm{Dry\ air}}} \tag{12}$$

where I_{air} and I_{EtOH} are, respectively, the current intensities before and after the introduction of ethanol vapor.

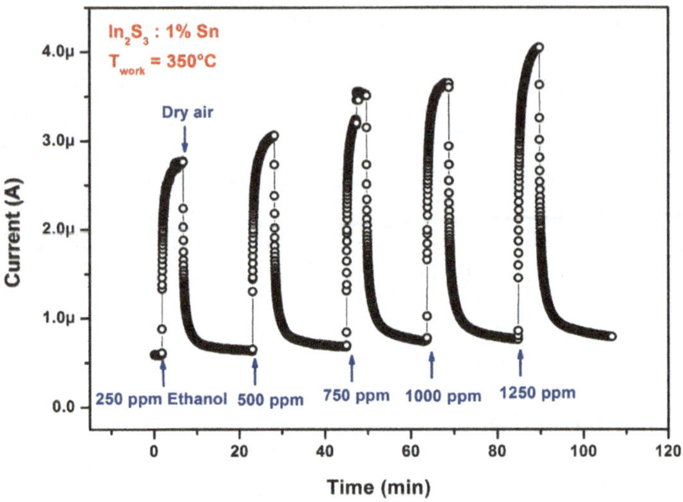

Fig. 12 Sensing transients of doped In_2S_3:Sn1% sensor for 250–1250 ppm ethanol at 350 °C

Figure 13 shows response curves versus ethanol concentration at an operating temperature of 350 °C of In_2S_3:Sn1% and undoped In_2S_3 response [12].

We note that response increases by increasing ethanol concentration and tends to saturation. The responses obtained for In_2S_3: Sn1% and undoped In_2S_3 films augment from 4.1 to 5.6 and from 1.4 to 3.6, respectively, with rising concentration

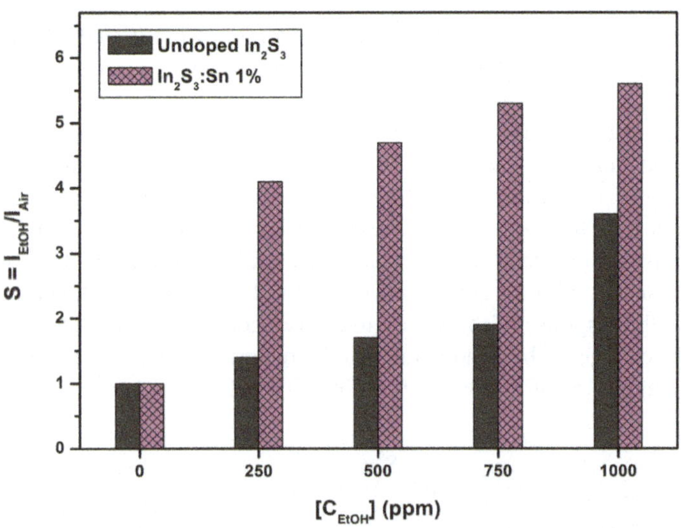

Fig. 13 Response of sensors based on undoped In_2S_3 and doped In_2S_3: Sn1% sensors at 350 °C versus ethanol concentrations

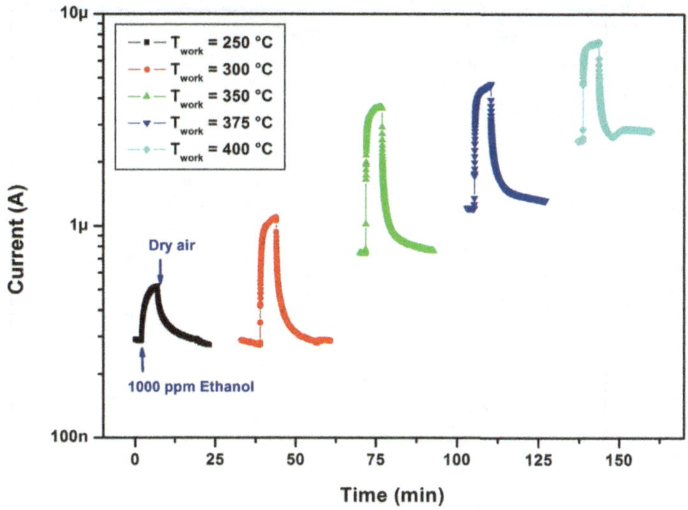

Fig. 14 Sensing transients of doped In_2S_3:Sn1% sensor to 1000 ppm ethanol at operating temperature ranging from 250 to 400 °C

from 250 to 1000 ppm. This comparison shows a slight improvement of the response to ethanol In_2S_3 sensor for a tin doping of 1%. In fact, the enhancement in sensing response of the latest sensor may be attributed to the highest surface roughness of the material which increases active surface area and abundant adsorption sites.

4.5.2 Working Temperature Effect

In order to study the working temperature effect on the sensor response, measurements were performed under 1000 ppm ethanol concentration at different operating temperatures varying from 250 to 400 °C. Figure 14 shows transient current intensity during alternating exposure to ethanol and dry air for In_2S_3:Sn1% sample.

Figure 15 illustrates response curves versus operating temperature of In_2S_3:Sn1% and undoped In_2S_3 response [12]. Thin films reveal a maximum of response at optimal temperature surroundings 350 °C. It is clear that the best sensing response is obtained for the doped In_2S_3:Sn1% film. Such a consequence corroborates result mentioned in the previous paragraph.

5 Conclusions

In summary, In_2S_3 thin films doped by tin were carried out by CSP technique with different Sn:In molar ratios. The structural, morphological, and optical properties

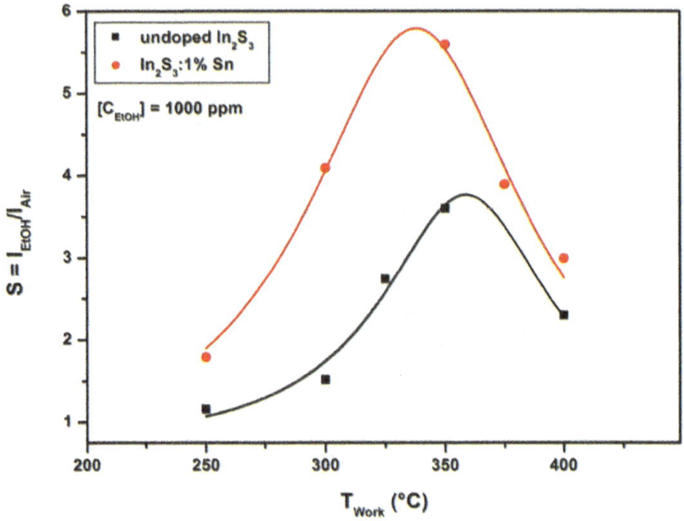

Fig. 15 Response of sensors based on undoped In_2S_3 and doped In_2S_3: Sn1% sensors to 1000 ppm ethanol as a function of the operating temperature

of In_2S_3:Sn thin films have been studied in this work. The prepared films are polycrystalline and exhibit a cubic β-In_2S_3 phase. The grain size increased from 16.3 to 25.5 nm with Sn doping. The phonon vibration modes determined by Raman spectroscopy confirm also the presence of the In_2S_3 phase in our samples. According to FESEM, the surface morphology of the films is continuous and free of cracks. R_a and R_q increased from 13.12 to 31.65 nm and 16.14 to 39.39 nm, respectively. The room temperature PL spectra of In_2S_3:Sn films for different Sn:In molar ratios reveal the presence of green, orange, and red emissions centered at 529 nm, 537, 597, 650, 703, and 758 nm. In_2S_3:Sn films exhibit transparency over 65–70% in the visible region and 70-90 % in near-infrared region. The optical band gap is found to vary in the range 2.68–2.80 eV for direct transitions. The refractive index of In_2S_3:Sn thin films is decreased from 2.45 to 2.37 and the k values are in the range 0.02–0.25 for all wavelengths and Sn:In molar ratio.

The In_2S_3:1%Sn sample with maximum roughness shows a slight improvement in ethanol response compared to undoped In_2S_3 mentioned in the previous publication. Indeed, the response increases from 1.4 to 4.1 for 250 ppm ethanol at an optimum temperature in the vicinity of 350 °C. The developed sensor shows promising applications in food control systems and can also be used to control drunk driving.

Acknowledgements This work was supported by Tunisian Ministry of Higher Education and Scientific Research. The authors are grateful to Hadi Sarra for her revision of the manuscript.

References

1. A.A. El Sbazly, D.A. Elhady, H.S. Metwally, M.A.M. Seyam, J. Phys.: Condens. Matter **10**, 5943 (1998)
2. M. Kilani, B. Yahmadi, N.K. Turki, M. Castagné, J. Mater. Sci. **46**, 6293 (2011)
3. K. Hara, K. Sayama, H. Arakawa, Sol. Energy Mater. Sol. Cells **62**, 441 (2000)
4. R. Nomura, S. Inazawa, K. Kanaya, H. Matsuda, Appl. Organomet. Chem. **3**, 195 (1989)
5. N. Kamoun, S. Belgacem, M. Amlouk, R. Bennaceur, J. Bonnet, F. Touhari, M. Nouaoura, L. Lassabatere, J. Appl. Phys. **89**, 2766 (2001)
6. S.H. Choe, T.H. Bang, N.O. Kim, H.G. Kim, C.I. Lee, M.S. Jin, S.K. Oh, W.T. Kim, Semicond. Sci. Technol. **16**, 98 (2001)
7. E. Dalas, L. Kobotiatis, J. Mater. Sci. **28**, 6595 (1993)
8. S. Yu, L. Shu, Y. Qian, Y. Xie, J. Yang, L. Yang, Mater. Res. Bull. **33**, 717 (1998)
9. W. Chen, J.O. Bovin, A.G. Joly, S. Wang, F. Su, G. Li, J. Phys. Chem. B **108**, 11927 (2004)
10. Y. He, D. Li, G. Xiao, W. Chen, Y. Chen, M. Sun, H. Huang, X. Fu, J. Phys. Chem. C **113**, 5254 (2009)
11. X. Fu, X. Wang, Z. Chen, Z. Zhang, Z. Li, D.Y.C. Leung, L. Wu, X. Fu, Appl. Catal. B **95**, 393 (2010)
12. R. Souissi, N. Bouguila, A. Labidi, Sens. Actuators B **261**, 522 (2018)
13. N. Barreau, S. Marsillac, J.C. Bernede, L. Assmann, J. Appl. Phys. **93**, 5456 (2003)
14. D. Braunger, D. Hariskos, T. Walter, H.W. Schock, Sol. Energy Mater. Sol. Cells **40**, 97 (1996)
15. N. Bouguila, M. Kraini, I. Najeh, I. Halidou, E. Lacaze, H. Bouchriha, H. Bouzouita, S. Alaya, J. Electron. Mater. **44**, 4213 (2015)
16. N. Bouguila, M. Kraini, A. Timoumi, I. Halidou, C. Vázquez-Vázquez, M.A. López-Quintela, S. Alaya, J. Mater. Sci.: Mater. Electron. **26**, 7639 (2015)
17. N. Bouguila, I. Najeh, N. Ben Mansour, H. Bouzouita, S. Alaya, J. Mater. Sci.: Mater. Electron. **26**, 6471 (2015)
18. N. Bouguila, M. Kraini, I. Halidou, E. Lacaze, H. Bouchriha, H. Bouzouita, J. Electron. Mater. **45**, 829 (2016)
19. N. Bouguila, A. Timoumi, H. Bouzouita, Eur. Phys. J. Appl. Phys. **65**, 20304 (2014)
20. N. Bouguila, A. Timoumi, H. Bouzouita, E. Lacaze, H. Bouchriha, B. Rezig, Eur. Phys. J. Appl. Phys. **63**, 2030 (2013)
21. M. Kraini, N. Bouguila, I. Halidou, A. Timoumi, S. Alaya, Mater. Sci. Semicond. Process. **16**, 1388 (2013)
22. N. Barreau, Sol. Energy **83**, 363 (2009)
23. H. Spasevska, C.C. Kitts, C. Ancora, G. Ruani, Int. J. Photoenergy **2012**, 1 (2011)
24. A. Akkari, C. Guasch, M. Castagne, N.K. Turki, J. Mater. Sci. **46**, 6285 (2011)
25. A. Timoumi, H. Bouzouita, B. Rezig, Thin Solid Films **519**, 7615 (2011)
26. S. Cingarapu, M.A. Ikenberry, D.B. Hamal, C.M. Sorensen, K. Hohn, K.J. Klabunde, Langmuir **28**, 3569 (2012)
27. L.J. Liu, W.D. Xiang, J.S. Zhong, X.Y. Yang, X.J. Liang, H.T. Liu, W. Cai, J. Alloys Compd. **493**, 309 (2010)
28. R.S. Becker, T. Zheng, J. Elton, M. Saeki, Sol. Energ. Mater. **13**, 97 (1986)
29. L.L. Yan, Y.J. Ling, C.S. Ying, L.P. Min, Chin. Phys. B. **24**, 078103 (2015)
30. M. Kilani, C. Guasch, M. Castagne, N.K. Turki, J. Mater. Sci. **47**, 3198 (2012)
31. T. Sall, M. Fahoume, B. Mari, M. Mollar, in *2014 International Renewable and Sustainable Energy Conference (IRSEC)* (IEEE, 2014), pp. 667–671
32. M.H.Z. Maha, M.M.B. Mohagheghi, H.A. Juybari, Thin Solid Films **536**, 57 (2013)
33. M. Mathew, M. Gopinath, C.S. Kartha, K.P. Vijayakumar, Y. Kashiwaba, T. Abe, Sol. Energy **84**, 888 (2010)
34. M. Kraini, N. Bouguila, I. Halidou, A. Moadhen, C. Vázquez-Vázquez, M.A. López-Quintela, S. Alaya, J. Electron. Mater. **44**, 2536 (2015)
35. M. Kraini, N. Bouguila, J. El Ghoul, I. Halidou, S.A. Gomez-Lopera, C. Vázquez-Vázquez, M.A. López-Quintela, S. Alaya, J. Mater. Sci.: Mater. Electron. **26**, 5774 (2015)

36. D.R. Lide, *Handbook of Chemistry and Physics, CRC Handbook of Chemistry and Physics*, 89th edn. (Taylor & Francis Group, New York, 2008), pp. 1923–1924
37. B.D. Cullity, *Elements of X-Ray Diffraction* (Addison-Wesley, Reading, MA, 1978).
38. A. Khorsand. Zak, W. H. Abd. Majid, M.E. Abrishami, R. Yousefi, Solid State Sci. **13**, 251 (2011)
39. K. Ravichandran, P. Philominathan, Sol. Energ. **82**, 1062 (2008)
40. V. Bilgin, S. Kose, F. Atay, I. Akyuz, Mat. Chem. Phys. **94**, 103 (2005)
41. P. Roy, S.K. Srivastava, Thin Solid Films **496**, 293 (2006)
42. K. Ben Bacha, A. Timoumi, N. Bitri, H. Bouzouita, Optik **126**, 3020 (2015)
43. G.B. Kamath, C.M. Joseph, C.S. Menon, Mater. Lett. **57**, 730 (2002)
44. S.P. Nehra, S. Chander, A. Sharma, M.S. Dhaka, Mater. Sci. Semicond. Process. **40**, 26 (2015)
45. M.H. Suhail, S.G. Kaleel, F.M. Yasser, APJR **1**, 80 (2014)
46. S. Rajeh, A. Mhamdi, K. Khirouni, M. Amlouk, S. Guermazi, Opt. Laser Technol. **69**, 113 (2015)
47. F. Urbach, Phys. Rev. **92**, 1324 (1953)
48. S. Ilican, Y. Caglar, M. Caglar, J. Optoelectron. Adv. Mater. **10**, 2578 (2008)
49. S. Ilican, Y. Caglar, M. Caglar, M. Kundakci, A. Ates, Int. J. Hydrogen Energ. **12**, 5201 (2009)
50. T.S. Moss, Proc. Phys. Soc. B **67**, 775 (1954)
51. E. Burstein, Phys. Rev. **93**, 632 (1954)
52. C. Guillén, T. Garcia, J. Herrero, M.T. Gutiérrez, F. Briones, Thin Solid Films **451**, 112 (2004)
53. P.J.L. Herve, L.K.J. Vandamme, J. Appl. Phys. **77**, 5476 (1995)
54. M.M. El-Nahass, B.A. Khalifa, H.S. Soliman, M.A.M. Seyam, Thin Solid Films **515**, 1796 (2006)
55. A. Timoumi, H. Bouzouita, B. Rezig, Aust. J. Basic Appl. Sci. **7**, 448 (2013)
56. L.L. Yan, Y.J. Ling, C.S. Ying, L.P. Min, Chin. Phys. B **24**, 078103 (2015)
57. Y. Xiong, Y. Xie, G. Du, X. Tian, Y. Qian, J. Solid State Chem. **166**, 336 (2002)
58. C. Guillen, J. Herrero, Thin Solid Films **510**, 260 (2006)
59. H. Tao, H. Zang, G. Dong, J. Zeng, X. Zhao, Optoelectron. Adv. Mater. Rapid Commun. **2**, 356 (2008)
60. H. Tao, S. Mao, G. Dong, H. Xiao, X. Zhao, Solid State Commun. **137**, 408 (2006)
61. R. Jayakrishnan, T.T. John, C.S. Kartha, K.P. Vijayakumar, T. Abe, Y. Kashiwaba, Semicond. Sci. Technol. **20**, 1162 (2005)
62. T.T. John, S. Bini, Y. Kashiwaba, T. Abe, Y. Yasuhiro, C.S. Kartha, K.P. Vijayakumar, Semicond. Sci. Technol. **18**, 491 (2003)
63. M. Mathew, R. Jayakrishnan, P.M. Ratheesh Kumar, C. Sudha Kartha, K.P. Vijayakumar, J. Appl. Phys. **100**, 033504 (2006)
64. Z.P. Ai, Opt. Mater. **24**, 589 (2003)
65. A.M. Azad, S.A. Akbar, S.G. Mhaisalkar, L.D. Birkefeld, K.S. Goto, J. Electrochem. Soc. **139**, 3690 (1992)
66. N. Barsan, U. Weimar, J. Electroceramic **7**, 143 (2001)
67. O. Lupan, V. Postica, J. Gröttrup, A.K. Mishra, N.H. de Leeuw, J.F.C. Carreira, J. Rodrigues, N. Ben Sedrine, M.R. Correia, T. Monteiro, V. Cretu, I. Tiginyanu, D. Smazna, Y.K. Mishra, R. Adelung, ACS. Appl. Mater. Interfaces **9**, 4084 (2017)

Magnetoelectricity in Ion-Implanted Ferroelectric Crystals

F. A. Mikailzade

Abstract In this chapter, the recent works on the synthesis of nanocomposite materials obtained by nanosized inclusions of magnetic 3D metals in the surface layer of various ferroelectric crystals by using ion-beam implantation technique, and the results of investigations of magnetoelectric effects in these structures have been reviewed. Remarkable shifts of FMR lines have been observed on applying DC electrical field on the samples. These results revealed a strong magnetoelectric coupling between the ion-synthesized magnetic nanoparticles and the ferroelectric matrix. Another evidences of the magnetoelectric coupling are magnetocapacitance effects observed clearly in 3D ion-implanted plates of ferroelectric crystals. These studies showed that ion-beam synthesized nanocomposite layers of ferroelectrics reveal ferromagnetic properties and strong magnetoelectric effects that makes them useful for magnetoelectronics applications.

Keywords Magnetoelectricity · Multiferroics · Ferromagnetism · Ferroelectrics · Nanomaterials

1 Introduction

In recent decades, there has been a great interest in multiferroic materials due to their attractive physical properties, multifunctionality and wide applications in various fields of information technology, radioelectronics, optoelectronics and microwave electronics. In these materials, the coupling interaction between ferroelectric and ferromagnetic substances could produce a magnetoelectric effect. This phenomenon provides extra functionality for their applications. However, the magnetoelectric coupling demonstrated in most known single-phase multiferroic materials is either weak or happens below room temperature, which limits their practical exploitations. It has been proposed that strong magnetoelectric coupling can be obtained by designing composites containing piezoelectric and magnetostrictive constituents,

F. A. Mikailzade (✉)
Department of Physics, Gebze Technical University, Gebze 41400, Kocaeli, Turkey
e-mail: faik@gtu.edu.tr

S. J. Ikhmayies et al. (eds.), *Advances in Optoelectronic Materials*,
Advances in Material Research and Technology,
https://doi.org/10.1007/978-3-030-57737-7_9

which can be electromagnetically coupled through stress mediation. In particular, multiferroic composites based on dispersion of ferromagnetic nanoparticles in ferroelectric and/or piezoelectric matrices typically reveal giant magnetoelectric effect.

The magnetoelectric effect, in its most general definition, describes the coupling between electric and magnetic fields in matter (i.e. induction of magnetization (***M***) by an electric field (***E***) or polarization (***P***) generated by a magnetic field (***H***)). Two independent events mark the birth of the ME effect. In 1888, Rontgen discovered that a moving dielectric became magnetized when placed in an electric field [1] , which was followed by observation of the reverse effect—polarization of a moving dielectric in a magnetic field 17 years later [2]. In 1894 when considering crystal symmetry, Curie predicted the possibility of an intrinsic magnetoelectric ME behaviour in some (non-moving) crystals [3].

The term 'magnetoelectric' was offered by Debye [4] a few years after the first (unsuccessful) attempts to demonstrate the static ME effect experimentally [5, 6]. In spite of Curie's early recognition of symmetry being a key issue in the search for ME behaviour, many decades passed until it was realized that the ME response is only allowed in time-asymmetric media [7]. Such violation of time-reversal symmetry can extrinsically occur through application of an external magnetic field [8] or movement as in the historic experiment conducted by Rontgen [1], or intrinsically in the form of long-range magnetic ordering. Dzyaloshinskii [9] was the first to show violation of time-reversal symmetry explicitly for a particular system (antiferromagnetic Cr_2O_3) . His work was soon followed by experimental confirmation of an electric field-induced magnetization [10, 11] and a magnetic field-induced polarization [12, 13] in Cr_2O_3 (see Fig. 1), both linear in the applied field.

Thus, Landau and Lifshits have pointed out the possible existence in magnetically ordered crystals of an equilibrium electric polarization proportional to the magnetic field intensity and of an equilibrium magnetization proportional to the

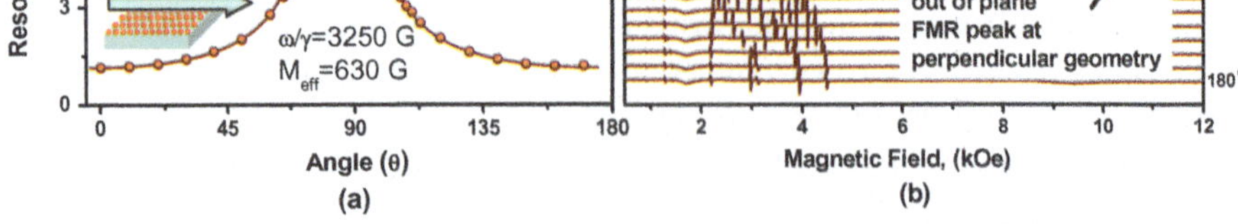

Fig. 1 **a** Angular dependence of the FMR resonance field in Co-implanted surface of $BaTiO_3$ in the out-of-plane geometry. The inset figure shows the coordinate system for FMR measurements. **b** The FMR spectra of the Co-implanted surface of $BaTiO_3$ at out-of-plane geometry [96]

electric field intensity (linear magnetoelectric effect) [7]. The linear magnetoelectric (ME) effect has been treated by Dzyaloshinskii [9] from the standpoint of magnetic symmetry. Magnetically ordered crystals showing a linear ME effect are commonly called magnetoelectrics. The experiments on Cr_2O_3 constituted a breakthrough in research on the ME effect. Nevertheless, it became immediately clear that technical applications, e.g. in the form of ME switches, were hampered by the small magnitude of the induced polarization or magnetization.

Thus, the magnetoelectric (ME) effect in its most general definition denominates the coupling between electric and magnetic fields in matter. A systematic progression of contributions to the ME effect is obtained from the expansion of the free energy (F) of a material. The magnetoelectric effect in a single-phase crystal is traditionally described in Landau theory by writing the free energy F of the system in terms of an applied magnetic field $\boldsymbol{H}$, whose ith component is denoted H_i, and an applied electric field $\boldsymbol{E}$ whose ith component is denoted E_i.

Note that this convention is unambiguous in free space, but that E_i within a material encodes the resultant field that a test particle would experience. Let us consider a non-ferroic material, where both the temperature-dependent electrical polarization $P_i(T)$ and the magnetization $P_i(T)$ are zero in the absence of applied fields, and there is no hysteresis. It may be represented as an infinite, homogeneous and stress-free medium by writing F under the summation convention in S.I. units as follows:

$$F(\boldsymbol{E},\boldsymbol{H}) = F_0 - \frac{1}{2}\varepsilon_0\varepsilon_{ij}E_iE_j - \frac{1}{2}\mu_0\mu_{ij}H_iH_j - \alpha_{ij}E_iH_j - \frac{1}{2}\beta_{ijk}E_iH_jH_k - \frac{1}{2}\gamma_{ijk}H_iE_jE_k - \cdots \quad (1)$$

Here F_0 is a ground state free energy. The subscripts refer to the three components of the vectors in spatial coordinates. E_i and H_i are the components of the electric field $\boldsymbol{E}$ and magnetic field $\boldsymbol{H}$, respectively. Here ε_0 and μ_0 are the dielectric and magnetic constants, ε_{ij} and μ_{ij} are the second-order tensors of dielectric and magnetic susceptibilities, β_{ijk} and γ_{ijk} are the third-order tensor coefficients, and most importantly, α_{ij} are the components of the tensor, which are designated as the linear magnetoelectric effect and correspond to the induction of polarization by a magnetic field or a magnetization by an electric field. The rest of the terms in (1) correspond to the high-order magnetoelectric effects parameterized by tensors β and γ.

In the present scheme, all magnetoelectric coefficients incorporate the field-independent material response functions $\varepsilon_{ij}(T)$ and $\mu_{ij}(T)$. The magnetoelectric effects can then easily be established in the form $P_i(H_j)$ or $M_i(E_j)$. The former is obtained by differentiating F with respect to E_i. A complementary operation involving Hi establishes the latter. One obtains:

$$P_i(\boldsymbol{E},\boldsymbol{H}) = -\frac{\partial F}{\partial E_i} = \varepsilon_0\varepsilon_{ij}E_j + \alpha_{ij}H_j + \frac{1}{2}\beta_{ijk}H_jH_k + \gamma_{ijk}H_jE_k + \ldots$$

and then setting $E_i = 0$

$$P_i = \alpha_{ij} H_j + \frac{\beta_{ijk}}{2} H_j H_k \tag{2}$$

and also;

$$M_i(\boldsymbol{E}, \boldsymbol{H}) = -\frac{\partial F}{\partial H_i} = \frac{1}{2}\mu_0\mu_{ij} H_j + \alpha_{ij} E_j + \beta_{ijk} E_i H_j + \frac{1}{2}\gamma_{ijk} E_j E_k + \cdots$$

setting $H = 0$ we get;

$$M_i = \alpha_{ij} E_j + \frac{1}{2}\gamma_{ijk} E_j E_k \tag{3}$$

The tensor α corresponds to induction of polarization by a magnetic field or of magnetization by an electric field which is designated as the linear ME effect. It is supplemented by higher-order ME effects like those parametrized by the tensors β and γ. The vast majority of research on the ME effect is devoted to the linear ME effect, and it is generally acceptable to omit the prefix 'linear' and simply to refer to the linear manifestation as the 'ME effect'.

Generally, the responses of the system to the action of an external field E and magnetic field H are described by the generalized susceptibility tensor

$$\chi = \begin{pmatrix} \chi^E & \chi^{EM} \\ \chi^{ME} & \chi^M \end{pmatrix}$$

Here we have

$$\chi_{ik}^E = \frac{\partial P_i}{\partial E_k}; \quad \chi_{ik}^{EM} = \frac{\partial P_i}{\partial H_k}; \quad \chi_{ik}^{ME} = \frac{\partial M_i}{\partial E_k}; \quad \chi_{ik}^M = \frac{\partial M_i}{\partial H_k};$$

are, respectively, the second-order tensors of the dielectric, magnetic and magneto-electric susceptibilities.

The components of the generalized susceptibility tensor can be expressed in terms of the derivatives of the thermodynamic potential with respect to the components of the electrical and magnetic fields.

$$\begin{vmatrix} \frac{\partial^2 F}{\partial E^2} & \frac{\partial^2 F}{\partial H \partial E} \\ \frac{\partial^2 F}{\partial E \partial H} & \frac{\partial^2 F}{\partial H^2} \end{vmatrix} > 0$$

$$\frac{\partial^2 F}{\partial E^2}\frac{\partial^2 F}{\partial H^2} - \frac{\partial^2 F}{\partial H \partial E}\frac{\partial^2 F}{\partial E \partial H} > 0$$

$$\left(\chi_{ik}^{ME}\right)^2 < \chi_{ik}^{E}\chi_{ik}^{M}$$

This condition defines the upper bound of the absolute magnitude of the ME susceptibility (χ_{ik}^{ME} can have either sign). This implies that we should expect the largest value of χ_{ik}^{ME} in crystals having large values of χ_{ik}^{E} and χ_{ik}^{M}, i.e. in ferroelectric–ferromagnets.

So, a multiferroic that is ferromagnetic and ferroelectric is liable to display large linear magnetoelectric effects. This follows because ferroelectric and ferromagnetic materials often (but not always) possess a large permittivity and permeability respectively, and α_{ji} is bounded by the geometric mean of the diagonalized tensors ε_{ii} and μ_{jj} such that:

$$\alpha_{ij}^2 \le \varepsilon_0\mu_0\varepsilon_{ii}\mu_{jj} \tag{4}$$

It represents a stability condition on ε_{ij} andμ_{ij}, but if the coupling becomes so strong that it drives a phase transition to a more stable state, then α_{ij}, ε_{ii}, μ_{jj} take on new values in the new phase. Note that a large ε_{ii} is not a prerequisite for a material to be ferroelectric (or vice versa); and similarly ferromagnets do not necessarily possess large μ_{jj}. For example, the ferroelectric KNO_3 possesses a small $\varepsilon = 25$ near its Curie temperature of 120 °C [14], whereas paraelectric $SrTiO_3$ exhibits $\varepsilon > 50{,}000$ at low temperatures [15]. Therefore, large magnetoelectric couplings need not arise in, or be restricted to, multiferroic materials.

2 Physical Properties of Ferroics and Multiferroics

2.1 *Multiferroic Materials*

As for the trends towards device miniaturization and high-density data storage, an integration of several functions into one material system has become highly desirable. Stemming from the extensive applications of magnetic and ferroelectric materials, it is natural to pursue a new generation of memories and sensing/actuating devices powered by materials that combine magnetism and ferroelectricity in an effective and intrinsic manner. The coexistence of several order parameters will bring out novel physical phenomena and offer possibilities for new device functions. The multiferroics addressed in this review represent one such type of materials, which do allow opportunities for humans to develop efficient control of magnetization or/and polarization by an electric field or/and magnetic field and to push their multi-implications. The novel prototype devices based on multiferroic functions may offer particularly high performance for spintronics, for example, reading the spin states, and writing the polarization states to reverse the spin states by electric field, to overcome the high writing energy in MRAMs.

Considering that little attention has been paid to multiferroicity until recently, it now offers us the opportunity to explore some important issues which have rarely been reachable. Although ferroelectricity and magnetism have been the focus of condensed matter physics and materials science since their discovery, quite a number of challenges have emerged in dealing with multiferroicity within the framework of fundamental physics and technological applications. There are, in principle, two basic issues to address in order to make multiferroicity physically understandable. The first is the coexistence of ferroelectricity (electric dipole order) and magnetism (spin order) in one system, since it was once proven extremely difficult for the two orders to coexist in a single material. Even so, the exploration of the microscopic conditions by which the two orders can coexist intrinsically in one system as a non-trivial problem has continued. Secondly, an efficient coupling between two orderings in the same multiferroic system (magnetoelectric coupling) seems to be even more important than their coexistence, because such a magnetoelectric coupling represents the basis for multi-control of the two orderings by either an electric field or magnetic field. Investigations have demonstrated that a realization of such strong coupling is even more challenging and, thus, the core of recent multiferroic researches. It should be mentioned here that most multiferroics synthesized so far are transitional metal oxides with perovskite structures. They are typically strongly correlated electronic systems in which the correlations among spins, charges/dipoles, orbitals and lattice/phonons are significant. Therefore, intrinsic integration and strong coupling between ferroelectricity and magnetism are essentially related to the multi-latitude landscape of interactions between these orders, thus making the physics of multiferroicity extremely complicated. Nevertheless, it is also clear that multiferroicity provides a more extensive platform to explore the novel physics of strongly correlated electronic systems, in addition to high TC superconductor and colossal magnetoresistance (CMR) manganites, etc. Since its discovery a century ago, ferroelectricity, like superconductivity, has been linked to the ancient phenomena of magnetism. Attempts to combine the dipole and spin orders into one system started in the 1960s, and some multiferroics, including boracites ($Ni_3BO_{13}I$, $Cr_3B_7O_{13}Cl$) fluorides ($BaMF_4$, M¼Mn, Fe, Co, Ni), magnetite Fe_3O_4, (Y/Yb)MnO_3 and $BiFeO_3$, were identified in the following decades. However, such a combination in these multiferroics has been proven to be unexpectedly tough. Moreover, a successful combination of the two orders does not necessarily guarantee a strong magnetoelectric coupling and convenient mutual control between them. Fortunately, recent work along this line has made substantial progress by discovering/inventing some multiferroics, mainly in the category of frustrated magnets, which demonstrate the very strong and intrinsic magnetoelectric coupling. Our theoretical understanding of this breakthrough is attributed to the physical approaches from various length scales/levels. Technologically, growth and synthesis techniques for high-quality single crystals and thin films become available.

2.2 *Single-Phase Multiferroics*

The term multiferroic was first used by H. Schmid in 1994 [16]. His definition referred to multiferroics as single-phase materials which simultaneously possess two or more primary ferroic properties. Today, the term multiferroic has been expanded to include materials, which exhibit any type of long-range magnetic ordering, spontaneous electric polarization and/or ferroelasticity. Working under this expanded definition, the history of magnetoelectric multiferroics can be traced back to the 1960s [17]. In the most general sense, the field of multiferroics was born from studies of magnetoelectric systems.

Multiferroics were grown for the first time in 1958, when magnetically active 3D ions were used to substitute ions with a noble gas shell in ferroelectrically distorted perovskite lattices [16, 17]. This leads us to ferroelectric antiferromagnetic compositions like $PbFe_{0.5}Nb_{0.5}O_3$ and $PbFe_{0.5}Ta_{0.5}O_3$ [18]. Up to now more than 80 single-phase multiferroics were grown either as a discrete composition or as a solid solution. Multiferroics have been the topic of various review papers with a focus on structure and materials science, phase diagrams, symmetries and theory [19–25].

Currently, four major crystallographic types of multiferroics can be distinguished as follows: (1) compounds with perovskite structure—$BiFeO_3$, $TbMnO_3$ and others [26–29]; (2) compounds with hexagonal structure, which include ferroelectric antiferromagnetic rare-earth manganites $RMnO_3$ with $R =$ Sc, Y, In, Ho, Er, Tm, Yb, Lu, Sr [30–36]; (3) Boracite compounds with the general formula $M_3B_7O_{13}X$ (M = Cr, Mn, Fe, Co, Cu, Ni) which are ferroelectric ferroelastic antiferromagnets [37–39]; 4) Orthorhombic $BaMF_4$ compounds (M = Mg, Mn, Fe, Co, Ni, Zn) [40–42].

Aside from these major types, a large number of multiferroics with different structures are known. Specific examples are discussed in the before mentioned review articles [16, 19, 21]. A systematic classification of symmetries-related types of ferroic ordering and compounds can be found in [23, 24]. Generally, magnetically driven multiferroics are insulating materials, mostly oxides, in which macroscopic electric polarization is induced by magnetic long-range order. A necessary but not sufficient condition for the appearance of spontaneous electric polarization is the absence of inversion symmetry. In these materials, inversion symmetry is broken by magnetic ordering. Such a symmetry breaking often occurs in so-called frustrated magnets, where competing interactions between spins favour unconventional magnetic orders. The microscopic mechanisms of magnetically induced ferroelectricity involve the polarization of electronic orbitals and relative displacement of ions in response to magnetic ordering.

However, finding a choice for multiferroic material is very limited because a few materials exhibit coexistence of ferroelectric and ferromagnetic properties at room temperature. There are very few natural multiferroic magnetoelectrics with both magnetic and electric polarizations existing or present in nature or synthesized in the laboratory, on the other hand, the observed ME effect for most such materials is too weak to be applicable [22]. The mechanisms and reasons for the existence of a very few magnetoelectric systems have been discussed in a number of works [23,

43–45]. According to these works, in perovskite compounds the transition metal ion can trigger two types of Jahn–Teller distortion, which is responsible for appearing ferroelectric polarization. A first-order Jan–Teller distortion occurs in the case of partially filled 3D orbitals, which retains the centre of symmetry, e.g. by elongation of the octahedron of ligands (as in the case of $LaMnO_3$ and $YTiO_3$). A second-order Jan–Teller distortion, which is weaker than the first-order one, requires an empty 3D shell for which a first-order Jahn–Teller distortion cannot occur and breaks the central symmetry by off-centre movement of the transition metal ion (examples are $BaTiO_3$ and PZT). Since only a partially filled 3D orbital can lead to magnetic ordering, whereas the breaking of central symmetry is a necessary condition for the formation of a spontaneous polarization, the conditions for ferroelectricity and (anti-) ferromagnetism are mutually exclusive. In an alternative approach [44], the incompatibility was explained by Hund's rule coupling that tends to keep the 3D spins parallel to one another. This mechanism breaks the strong covalent bonds that are necessary for ferroelectricity. In turn, magnetic ferroelectrics must be materials in which the double well potential provoking the ferroelectric distortion is not caused by the hybridization of transition metal ions in a noble gas configuration.

So, all these reasons and mechanisms led to presence of little number of single-phase multiferroic materials and a low value of magnetoelectric coefficient in these materials, which made them unsuitable for possible applications.

2.3 *Composite Multiferroics*

The first observation of the ME effect triggered a lot of excitement because of the obvious potential of the correlation between the magnetic and electric properties of matter for technical applications. However, the general weakness of the ME effect, the restricted number of compounds displaying it, the difficulties in developing useful applications and a limited understanding of the microscopic sources of ME behaviour led to a decline of research activities for about two decades.

Unfortunately, the magnetoelectric effect in single-phase compounds is usually too small to be practically applicable. The breakthrough in terms of the giant magnetoelectric effect was achieved in composite materials; for example, in the simplest case the multilayer structures composed of a ferromagnetic piezomagnetic layer and a ferroelectric piezoelectric layer [46–49]. Other kinds of magnetoelectric composites including co-sintered granular composites and column-structure composites were also developed [50–52]. In the composites, the magnetoelectric effect is generated as a product property of the magnetostrictive and piezoelectric effects, which is a macroscopic mechanical transfer process. A linear magnetoelectric polarization is induced by a weak AC magnetic field imposed onto a DC bias magnetic field. Meanwhile, a magnetoelectric voltage coefficient up to 100 Vcm^{-1} Oe^{-1} in the vicinity of electromechanical resonance was reported [46]. These composites are acceptable for practical applications in a number of devices such as microwave components, magnetic field sensors and magnetic memories. For example, it was reported recently that the magnetoelectric composites can be used as probes in scanning probe microscopy to develop a near-field room temperature scanning magnetic probe microscope.

The physical properties of a material that is formed from two or more single-phase compounds are determined by the properties of the constituents as well as by the interaction between them. Composites can be used to generate ME behaviour from materials which in themselves do not allow the ME effect. This is conveniently achieved by combining magnetostrictive and piezoelectric compounds. A magnetic field applied to the composite will induce strain in the magnetostrictive constituent which is passed along with the piezoelectric constituent, where it induces an electric polarization. The idea of synthesizing a composite medium displaying a ME effect was first formulated by Tellegen in 1948 [53], that is, more than a decade *before* the first observation of intrinsic ME behaviour in a single-phase compound. Tellegen suggested a suspension of macroscopic particles that carry both electric and magnetic dipole moments, an idea that, however, proved impossible to realize because of a matter of principle. Actually, van Suchtelen and co-workers [54–57] grew the first artificial ME material by combining ferroelectric piezoelectric $BaTiO_3$ and ferromagnetic piezomagnetic $CoFe_2O_4$ in eutectic composite by unidirectional solidification.

Because of the complex nature of the ME interaction between the constituents, the relation between the applied magnetic field and the voltage induced in the detection circuit is not simply linear as in the case of single-phase compounds. In composite materials, the ME effect is defined for a weak (up to 10 Oe) AC magnetic field applied in the presence of a large (up to 10 kOe) DC bias field with frequencies of the AC field between 100 Hz and 1 MHz. The voltage induced by the AC field is proportional to the AC field amplitude. Depending on the growth conditions, the composition used by van Suchtelen, van den Boomgard and others revealed ME voltage coefficients of up to 130 $mVcm^{-1}\ Oe^{-1}$.

Thus, even in the earliest experiments on composite magnetoelectrics, the ME response exceeds the largest values observed on single-phase compounds by more than an order of magnitude.

Hence, in normal practice, desired magnetoelectric effect is achieved by growing heterostructures of ferroelectric and magnetic materials. Realization of heterostructures with desired properties is not only difficult but also involves complicated lengthy procedures. Recently, attention to ME materials has been gradually drawn towards composite thin films. Compared to bulk composites, ME composite thin films exhibit unique advantages. Their composition and connectivity could be modulated at the microscopic scale, and the artificial thin-film heterostructures can thus be achieved, which have potential applications in all kinds of microdevices and integrated units such as microsensors, MEMS devices and high-density information storage devices. It has been identified that a number of magnetostrictive–piezoelectric composite structures show significant ME output qualified for potential applications.

Recently, efforts have been made to fabricate the artificial layers and tailor their structures for the suitability of multiferroics [52, 58–69]. These materials are in the form of either composites, superlattices or multilayers. Interestingly, the material made in the form of superlattices, whose structure consists of alternating ferroelectric and ferromagnetic layers, yielded unusual electrical and magnetic transport properties that cannot be obtained in either of their constituents. As a result of these studies of sandwich devices, practical applications of magnetoelectric effects now appear more feasible. Nevertheless, sandwich devices are intrinsically limited in the feature size and difficult to miniaturize. Therefore, further advances in new magnetoelectric composites are still desirable.

Recent research activities on magnetoelectric composites were discussed in [70]. It was mentioned that the novel applications of these structures include magnetic sensors, energy harvesters and power converters. The development of ME sensors based on the selection of material properties and fabrication methods has been discussed, which resulted in a new type of magnetic sensor with minute power consumption and extremely high detection performance, which is competitive to existing magnetic sensors. Different ME-based energy harvesters have been compared according to their directionality, bandwidth and sources of energy.

It has been noticed in [71] that in recent years significant progress was reported in the development of sensors of DC and AC magnetic fields with sensitivities on the order of a few pico-Tesla, in particular for possible use in medical imaging technologies. One could also utilize the composites for sensors of current, acceleration and speed of rotation. Multiferroic composites for high-frequency signal processing is another area of significant interest. Ferrites are materials of choice for devices such as resonators, filters and phase shifters for use in the frequency range from 1 to 110 GHz due to low losses. Such devices, however, require a source of variable magnetic field for tuning the device operating frequency. One could replace the ferrite with a ferrite–ferroelectric composite in order to realize the strain-mediated electric fieldtuning of the ferrite which would reduce the power requirements and facilitate miniaturization and integration with semiconductor devices. Voltage tuning of ferriteferroelectric resonators, filters and phase shifters has been reported in recent years. Magnetoelectric random access memories and multi-state memories were also explored. Ferrite–ferroelectric composites were used in devices called gyrators that are capable of direct conversion of current to voltage and may find applications in power electronics.

3 Materials and Experimental Methods for Investigations of the Novel Multiferroic Composites

3.1 Ferroelectric Crystals for Studies

There has been considerable research focused on the growth and characterization of perovskite ABO_3 oxide ferroelectric materials. The most typical one is barium titanate ($BaTiO_3$) , which is currently one of the most technologically interesting materials due to its remarkable properties, such as high dielectric constant, high optical transparency, ferroelectric, electro-optic and nonlinear optic properties [72–76]. This compound undergoes successive phase transitions to three phases on lowering the temperature. Above the Curie temperature of 403 K, the structure of barium titanate is cubic and paraelectric. Below the Curie point, the structure is slightly distorted to tetragonal structure. The tetragonal phase is stable between 278 and 403 K. The structure becomes orthorhombic below 278 K, and a further transition to rhombohedral structure occurs at about 183 K. The ferroelectricity has been observed in the three low-temperature phases [77, 78].

The ferroelectric transition in $BaTiO_3$ is generally considered to be the first-order displacive type. Respectively, the phase diagram of $BaTiO_3$ can be described in the frame of the phenomenological Landau–Ginzburg–Devonshire theory [78]. $BaTiO_3$ is a prototypical ferroelectric, an insulating solid whose macroscopic polarization can be reoriented by the application of an electric field. In the perovskite ferroelectrics, it is well known both experimentally and theoretically that the polarization is also strongly coupled to strain, and thus that properties such as the ferroelectric transition temperature and polarization magnitude are quite sensitive to external stress.

Besides, $BaTiO_3$ crystals, ceramics and thin films with perovskite structure have been studied extensively during the last few decades because of their proven superior electrical and optical properties. For barium titanate crystal, interest was mainly due to its photorefractive properties and other associated effects and applications, such as phase conjugation, volume holography for optical storage, two-beam coupling and optical computing [79].

The ternary compounds $TlInS_2$ and $TlGaSe_2$ belong to a group of ternary layered semiconductors with space group symmetry C_6^{2h} at room temperature [80, 81]. During the past two decades, there is a growing interest in these crystals due to the coexistence of ferroelectric and semiconductor properties. Structural anisotropy [81, 82], successive incommensurate and commensurate phase transitions [82–84], good optical and photoconductivity properties [85] of these compounds made them attractive for researchers and stimulated more detailed investigations of its physical properties using various experimental methods.

In their structural investigations, $TlInS_2$ and $TlGaSe_2$ were considered as the isostructural compounds [86–91]. According to X-ray diffraction measurements [92], their crystal structures consist of alternating two-dimensional metal–chalcogen layers. The layers are composed of A_4B_{10} polyhedron complexes, which represent a combination of four elementary AB_4 (A = In, Ga and B = S, Se) tetrahedra and are linked together by common chalcogen atoms at the corners. The elementary unit cell contains two partially disordered layers, containing successive rows of the tetrahedral complexes, which are turned away from each other by 90°. Each successive layer is shifted along the [010] direction by the length of the edge of the small AB_4 tetrahedron with respect to the upper layer. Monovalent Tl atoms are in trigonal prismatic cavities resulting from the combination of the A_4B_{10} tetrahedra into a layer.

It is well known that on cooling $TlInS_2$ and $TlGaSe_2$, crystals undergo a sequence of the structural phase transitions (PT), including the PT in the incommensurate (IC) and commensurate (C) ferroelectric phases [92–94]. According to X-ray diffraction experiments at room temperature [80–82], the initial paraelectric phase of $TlInS_2$

and $TlGaSe_2$ is characterized by the space symmetric group of C^6_{2h}. PT to IC phase is accompanied by soft mode condensation (at $T_i = 216$ K and 120 K, respectively, which are identified as IC phase transition temperatures) with the wave vector $\overrightarrow{q_i} = (\delta, \delta, 0.25)$ in the Brillouin zone, where δ is the incommensurate parameter. At $T_c \sim$ 201 K and 107 K, respectively, the δ value jumps to zero and the crystals $TlInS_2$ and $TlGaSe_2$ exhibit a structural phase transition to commensurate ferroelectric phase with the wave vector of $\overrightarrow{q_c} = (0, 0, 0.25)$. In this case, the spontaneous polarization lies in the plane of layers.

3.2 *Ion Implantation Technique*

Among different techniques for fabrication of composite materials, ion implantation has a number of advantages: easy control of the metal distribution and concentration; the availability of almost arbitrary metal–dielectric compositions; the ability to surpass the solubility limits constrained by the chemical and thermodynamic equilibrium of the host matrix and metalimpurities [95]. Besides, the ion implantation technique is ideally suited for fabrication of thin-film magnetic media and planar devices for magneto-sensor electronics. In the given context, the authors suggested high-fluency Co and Fe ion implantation into ferroelectric $BaTiO_3$, $TlInS_2$ and $TlGaSe_2$ crystals with perovskite structure to form magnetic nanoparticles in the near-surface region of the irradiated substrate aiming fabrication of a new multiferroic composite material.

Single-charged Co^+ and Fe^+ ions with energy of 40 keV were implanted into $BaTiO_3$ (*CrysTec GmbH*, Germany), $TlInS_2$ and $TlGaSe_2$ (synthesized at the Institute of Physics of Azerbaijan Academy of Sciences in Baku) single crystals at the fluency of 0.5×10^{17} ion/cm^2 and with ion current density of 8 μA/cm^2. The implantation was carried out at room temperature by using ion-beam accelerator *ILU-3* (Kazan Physical-Technical Institute of RAS) kept at residual vacuum of 10^{-5} Torr. Element composition and surface morphology of all samples were investigated using commercial scanning electron microscope *«Zeiss» EVO-50XVP* with energy-dispersive X-ray (EDX) spectrometer *Oxford INCA Energy 330*.

3.3 *Magnetic Measurement Techniques*

Magnetic resonance measurements were carried out by using Bruker EMX model X-band (9.8 GHz) spectrometer. A closed-cycle helium cryostat system and Lakeshore 340 model temperature controller were used in the measurements, which allowed to scan the temperature with a rate of about 0.2 K/min and to stabilize the temperature with accuracy better than 0.05 K. The measurements were performed in the temperature range of 10–300 K. The static magnetic field was varied in the range of 0–1600 mT. A goniometer was used to rotate the sample holder which is parallel to the microwave magnetic field and perpendicular to the applied static magnetic field. The measurements were performed in two different, in-plane and out-of-plane geometries. At the in-plane geometry, the sample was attached horizontally at bottom edge of sample holder, and the static magnetic field was scanned in the plane of the implanted surface. At the out-of-plane geometry, the sample was attached to the flat platform of sample holder, whereas the magnetic field of microwave lies in the film plane during measurement and static magnetic field is rotated from the sample plane to the surface normal. The field derivative of microwave power absorption (dP/dH) was recorded as a function of the DC field. To obtain intensities of electron magnetic resonance (EMR) and ferromagnetic resonance (FMR) signals, the double digital integration of the resonance curves was performed using Bruker WINEPR software package (Bruker Bio Spin Corporation, Billerica, MA 01,821 USA).

The static magnetization measurements were performed using VSM (PPMS, Quantum Design Corp.) for parallel orientations of the applied magnetic field with respect to the implanted surface plane in a wide temperature interval. The temperature dependences of the magnetization were measured in zero field cooled (ZFC) and field cooled (FC) regimes. For the ZFC measurements, samples were cooled in zero field to 10 K and then, the magnetization was recorded during heating of the sample up to 400 K in the magnetic field of 50 Oe applied parallel to the sample surface. For the FC measurements, the applied field of 50 Oe was kept constant during cooling to 10 K, and the magnetization was recorded during heating at the same constant magnetic field.

3.4 *Dielectric Constant Measurements*

The samples for capacitance measurements were oriented along the polar axis which lies in the cleavage plane (the morphology of crystals permits cleavage to plane parallel plates with mirror-like surfaces) . The plates were gently polished, cleaned and covered with silver paste. The dimensions of the electrodes were 3 × 0.3 mm^2 with an inter-electrode distance of 1 mm. The capacitance of the samples was measured using an agilent 4294 impedance analyzer. The dielectric measurements were performed at various frequencies (100 kHz–1 MHz) in the temperature range between 50 and 300 K. A superconducting vector magnet system used in the measurements allowed to apply the static magnetic field up to 5 T and scan the temperature with a rate of about 0.1 K/min and to stabilize the temperature with the accuracy better than 0.005 K. The temperature was measured by a Cernox™ temperature sensor placed close to the sample.

4 Ferromagnetic Resonance and Magnetization Studies

4.1 Co-implanted $BaTiO_3$ Crystal

Magnetic resonance measurements of Co-implanted $BaTiO_3$ were performed in [96]. In addition to paramagnetic signals, which were attributed to Fe impurities of $BaTiO_3$ substrate, a broad resonance signal at high field region of spectrum was observed (Fig. 1). When the external magnetic field is parallel to the implanted surface of $BaTiO_3$ crystal (parallel geometry), the observed signal intensity is very low according to the paramagnetic signals. During the rotation of sample as the magnetic field turns from the implanted surface towards the normal of implanted surface (perpendicular geometry), the broad signal moved to the high field region of spectrum and the intensity of this signal increased. The line width of this signal is approximately 350 Oe at perpendicular geometry. This broad resonance line is attributed to the ferromagnetic resonance signal due to the granular film composed of Co nanoparticles on the surface. Due to the size and shape distribution of particles on the surface gives very broad and low intensity FMR signal at parallel geometry. We did not observe anisotropic behaviour of this FMR line at in-plane geometry. We note that the observed FMR line dependence on the sample orientation is similar to that observed in the FMR of granular magnetic films [97]. In latter case, the resonance signal is due to the collective motion of particles magnetic moments, i.e. may be described approximately by the macroscopic magnetization of the granular layer as the whole system.

The value of the g-factor and effective magnetization was calculated from the variation of resonance field with the rotation angle of sample in the out-of-plane geometry as 2.1 and 630 Oe, respectively. The observed peculiarities of the ferromagnetic behaviour of Co-implanted $BaTiO_3$ may be attributed to the dipole–dipole interaction of magnetic cobalt nanoparticles formed as a result of high-influence implantation. This phenomenon (called magnetic percolation) is discussed in details in [97, 98]. The fact is that when the distance between the magnetic particles becomes comparable with their sizes, the dipole–dipole interaction couples the particle magnetic moments. As a result, the granular phase behaves as a ferromagnetic continuum with respect to the dipolar forces even without direct contact between the particles. The mechanism of such dipole–dipole interaction is discussed also in [99]; the authors considered a semi-quantitative model for dipolar field for regular array of closely separated spherical particles in granular magnetic layer. This model predicts the dipolar field exhibiting an anisotropic behaviour. Hence, the angular dependence of FMR spectra in such systems is qualitatively similar to that of a magnetic thin film.

In order to further investigate the magnetic properties of the Co-implanted $BaTiO_3$ we have performed temperature-dependent magnetization $M(T)$ measurements using a VSM magnetometer in field cooled (FC) and zero field cooled (ZFC) regimes. For ZFC measurements, the samples are cooled in zero field to 10 K, and the magnetization is recorded during heating the sample up to 400 K under the magnetic field of 100 Oe applied parallel to the sample surface. For FC measurements, the applied field of

100 Oe is kept constant during cooling to 10 K, and the magnetization is recorded on heating regime on applying the magnetic field of the same intensity. Figure 2 shows that the FC and ZFC curves diverge substantially from each other below 400 K, above which the coincidence of FC and ZFC curves takes place. So, the peculiarities of FC and ZFC curves reveal the presence of superparamagnetic behaviour at the temperatures above $T_b \sim 400$ K, which can be considered as 'blocking temperature'. So, the FC and ZFC curves show ferromagnetic-like behaviour up to high temperatures, and the behaviour of the magnetization is typical for magnetic granular systems with wide particle size distribution and strong magnetic interaction between particles [100].

Additionally, small peaks in both ZFC and FC curves at low temperatures are attributed to the interfacial magnetic moments of the particles that are 'frozen' [101] in certain directions at the temperatures lower than 40 K. In all measurements, the magnetic field is applied in the direction parallel to the implanted surface of $BaTiO_3$. The recorded $M(H)$ loops confirm the ferromagnetic-like behaviour. The coercive field decreases significantly by increasing the temperature and reaches to zero on approaching $T_b \sim 400$ K (Fig. 3). Thus, the results of the magnetization measurements also support the origin of observed ferromagnetic behaviour in granular ferromagnetic layer formed in a result of Co implantation of $BaTiO_3$.

4.2 *Fe-Implanted $BaTiO_3$ Crystal*

The results of magnetic resonance investigations of Fe-implanted $BaTiO_3$ samples are presented in Fig. 4. The measurements have been performed at room temperature for in-plane and out-of-plane geometries [101]. As it can be seen from the figure, magnetic resonance spectra contain narrow EPR lines originating from isolated paramagnetic Fe^{3+} centres located in Ti^{4+} sites, and broad FMR signal from ferromagnetic Fe nanoparticles of ion-implanted surface layer.

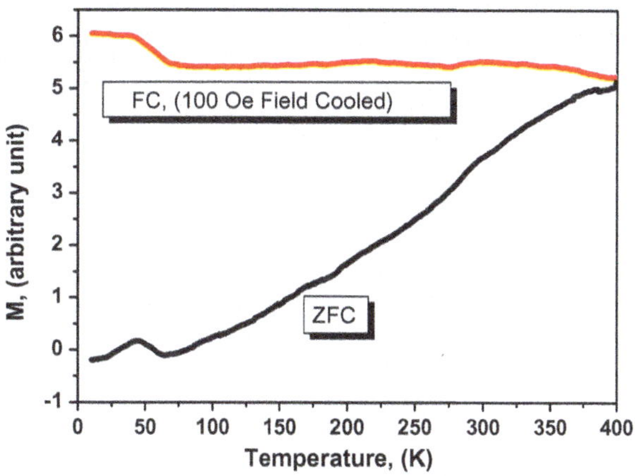

Fig. 2 Temperature dependence of the magnetization $M(T)$ of Co-implanted $BaTiO_3$ (fluency -1.0×10^{17} ion/cm^2) measured in FC and ZFC regimes [96]

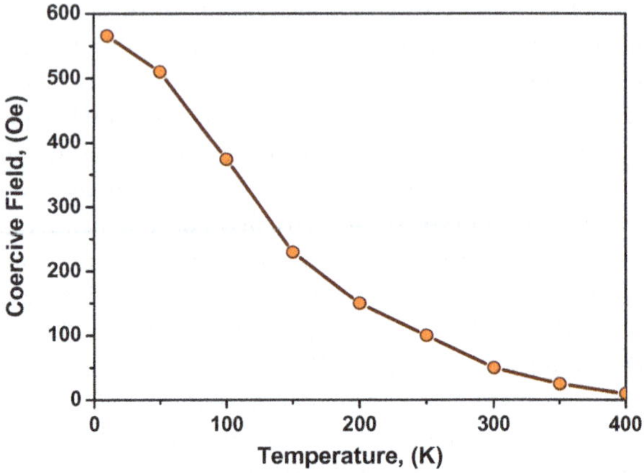

Fig. 3 Temperature dependence of the coercive field of Co-implanted $BaTiO_3$ (fluency - 1.0 × 10^{17} ion/cm^2) [96]

The angular evolutions of the EPR lines in the in-plane and out-of-plane geometries were modelled in the same way as in [96]. Based on this modelling the spectra were identified as coming from Fe^{3+} ion (S = 5/2, L=0) impurities substituting the host titanium (Ti^{4+}) sites in the body of the substrate plates and acquiring axially distorted ligand crystal field of cubic symmetry [98]. The magnetic resonance measurements of virgin substrates confirmed this conjecture.

In order to discuss the results of FMR measurements, it is worth mentioning that dependence of the FMR resonance field on the sample orientation is generally similar to that observed in the FMR of Co-implanted $BaTiO_3$ [96]. In this case, the resonance signal from a granular magnetic layer is considered as due to the collective motion

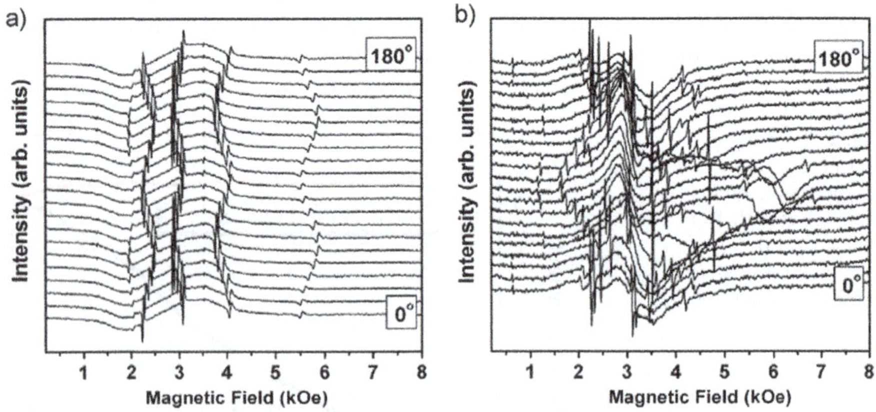

Fig. 4 Magnetic resonance spectra of the Fe-implanted $BaTiO_3$ plate measured at various orientations of the applied magnetic field in the in-plane (left) and out-of-plane (right) geometries [101]

of particles magnetic moments, i.e. may be described approximately by the macroscopic magnetization of the granular layer as the whole system. In this framework, it is possible to analyse the FMR signal as coming from a thin magnetic film with some effective values of the magnetization and the g-factor. Then, the resonance condition for the granular film at arbitrary orientation takes the same form as that for a continuous film. For the resonance condition, we used the classical resonance equation as we have made in [96]: The g-factor and effective magnetization have been calculated from variation of the resonance field with the rotation angle of the sample at the out-of-plane geometry (see Fig. 5) as 2.19 and 231 G, respectively.

We attribute the observed peculiarities of the FMR signal behaviour to dipole–dipole interaction of magnetic nanoparticles of iron formed as a result of the high-fluency implantation. The detailed discussion of the phenomenon is given in [97–99], it is called 'magnetic percolation', when the dipole–dipole interaction couples magnetic moments of ferromagnetic particles separated by distances comparable with their sizes. As a result, the granular phase behaves as a ferromagnetic continuum even without direct contact between the particles. The angular dependence of the FMR spectra in such systems becomes qualitatively similar to that of a magnetic thin film.

Figure 6 presents the temperature dependences of magnetization *M(T)* measured by increasing the temperature at low magnetic field of 50 Oe in beforehand zero field cooled (ZFC) and field cooled (FC) samples at 1 T. It is seen from the figure that the FC and ZFC curves diverge substantially from each other at temperature of about 150 K, above which the coincidence of FC and ZFC curves takes place. Wide peak at $T_B \sim 80$ K which can be considered as 'blocking temperature' for the ZFC curve indicates that the magnetic moments of the iron nanoparticles are 'frozen' in random directions at temperatures lower than T_B. In addition, the FC curve shows temperature-independent ferromagnetic-like behaviour at temperatures below 50 K. We used Fe bulk cubic anisotropy constant $K = 5.5 \times 10^4$ J m^{-3} at low temperature [102] to estimate an upper limit for iron particle sizes. Using well-known relation

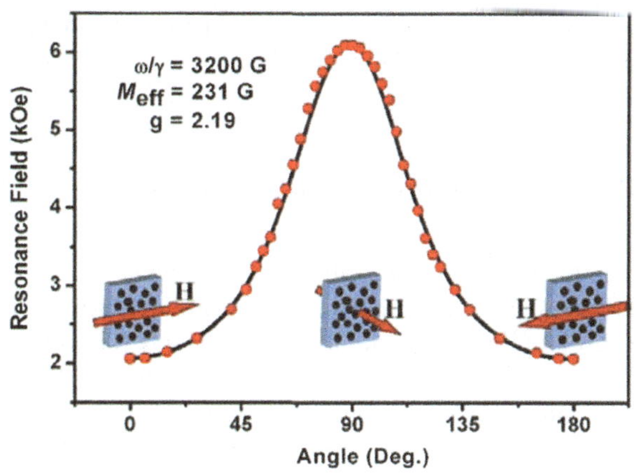

Fig. 5 Angular dependence of the ferromagnetic resonance field in the Fe-implanted $BaTiO_3$ plate measured at the rotation of magnetic field in the out-of-plane geometry [101]

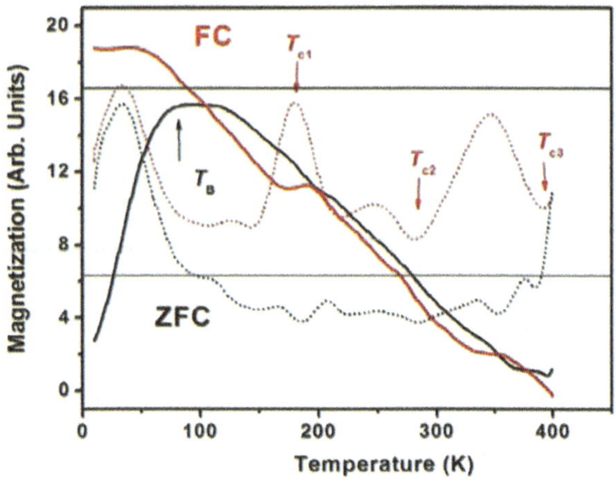

Fig. 6 Temperature dependences of the in-plane magnetization of the Fe-implanted $BaTiO_3$ sample measured in FC and ZFC regimes at magnetic field of 50 Oe. The dotted curves represent the temperature dependences of the first derivatives of $M(T)$ curves [101]

[100] $KV = 25\ k_B T_b$, where k_B is Boltzmann constant and V is iron nanoparticle's volume, the limiting value of a particle diameter was found to be about 10 nm that is in the agreement with TEM studies. Note that the observed FC and ZFC curves of magnetization are typical for magnetic granular systems with wide particle size distribution and strong magnetic interaction between the particles [100].

However, there are three peculiarities (the points of inflection) in the temperature dependences of magnetization as shown by three arrows in Fig. 6. The two clearly visible inflections of the ZFC and FC curves at $T_{c1} \sim 180$ K and $T_{c3} \sim 390$ K can be related to paraelectric cubic to tetragonal polar and from orthorhombic polar to rhombohedral polar phase transitions in $BaTiO_3$ crystal, respectively. The third, tetragonal polar to orthorhombic polar phase transition occurs at $T_{c2} \sim 280$ K and has much weaker impact on the magnetic moment, however, first derivatives of the $M(T)$ drawn by dotted curves of the corresponding colour clearly a feature induced by the structural phase transition. This observation indicates that there is a marked magnetoelectric coupling between the iron nanoparticles and the $BaTiO_3$ ferroelectric matrix upon para- to ferroelectric transition at about 390 K and upon low-temperature phase transition at about 180 K.

4.3 Co-implanted $TlInS_2$ and $TlGaSe_2$ Crystals

The results of magnetization measurements of Co-implanted $TlInS_2$ and $TlGaSe_2$ [103] are presented in Figs. 7 and 8. Figure 7 shows that FC and ZFC curves diverge substantially from each other below 400 K for both samples, above this temperature the coincidence of FC and ZFC curves takes place. So, the peculiarities of FC and ZFC curves reveal the presence of superparamagnetic behaviour at the temperatures above ~ 400 K, which can be considered as 'blocking temperature'. The results show ferromagnetic-like behaviour up to high temperatures, and the behaviour of

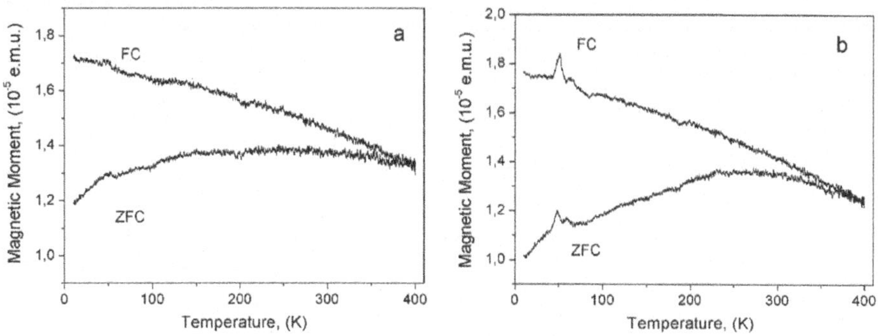

Fig. 7 **a** ZFC and FC temperature dependences of the magnetic moment of $TlInS_2$ **a** and $TlGaSe_2$ **b** crystal plates implanted with Co ions at the fluency of 1.0×10^{17} ion/cm^2 [103].

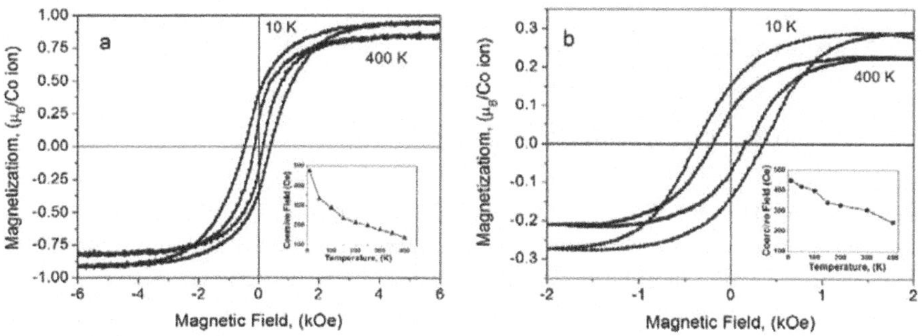

Fig. 8 Magnetic hysteresis loops of $TlInS_2$ **a** and $TlGaSe_2$ **b** crystal plates implanted with Co ions at the fluency of 1.0×10^{17} ion/cm^2 (the temperature dependences of the coercive field are presented in the insets) [103]

the magnetization is typical for magnetic granular systems with wide particle size distribution and strong magnetic interaction between particles [101, 102]. Additionally, small peaks in the both ZFC and FC curves at low temperatures are attributed to the interfacial magnetic moments of the particles that are 'frozen' [104] in certain directions at temperatures lower than 40 K.

Additionally, the studies of the magnetic field dependences of the magnetization of the samples also revealed ferromagnetic-like behaviour at low temperatures. In all measurements, the field is applied in the directions parallel to the implanted surface of the crystals. It has been obtained that the coercive field decreases significantly by increasing the temperature up to 400 K. In a result, the M(H) dependence exhibits a nonlinear behaviour with zero coercive field value at high temperatures that is peculiar to superparamagnetic systems.

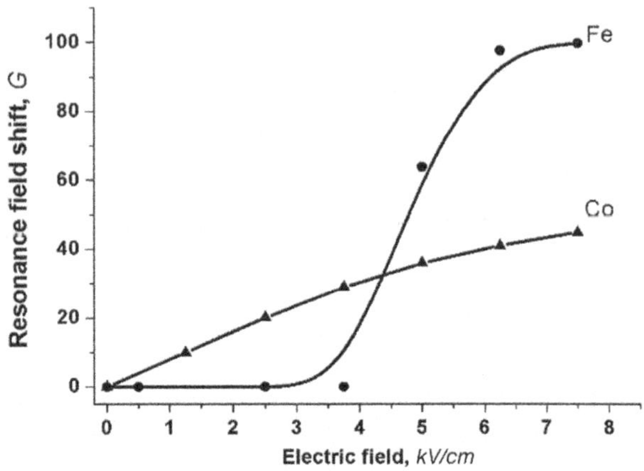

Fig. 9 Dependences of the shift in the resonance field of the FMR signal on the applied electric field for the $BaTiO_3$ samples modified by iron or cobalt ions [105]

5 Magnetoelectric Effects

5.1 Co- and Fe-Implanted $BaTiO_3$ Crystal

We investigated the behaviour of FMR signal under an action of external electric field to detect magnetoelectric coupling between magnetic nanoparticles and ferroelectric $BaTiO_3$ matrix in the samples under study [105]. The measurements were performed in out-of-plane geometry; both static electric and static magnetic fields were applied in the direction perpendicular to the implanted surface plane. It has been revealed that applying the electric field on the samples brings to remarkable decreases of ferromagnetic resonance field values. Figure 9 shows the dependences of amounts of the resonance field shifts on applied electric field. As it is seen from the figure, the dependences of the resonance fields on applied electric field exhibit nonlinear character. According to the classical Kittel's model [102] for the resonance field of FMR signal in a thin magnetic film, the observed shift can be caused only by a change in the magnetization of the sample. When the magnetic field is oriented perpendicular to the sample plane, the resonance field H_{res} is determined as follows:

$$H_{res} = \frac{\omega_{res}}{\gamma} + 4\pi M(E)$$

where ω_{res} is the resonance frequency (9.8 GHz), γ is the gyromagnetic ratio, M is the magnetization of the sample, and E is the external electric field. According to this expression, the observed shift of the resonance field of FMR signal indicates a decrease in the magnetization of the sample in an external electric field. So, taking into account the fact that the effective magnetization is observed to lie in the implanted surface plane, we observed transverse magnetoelectric effect, which includes the

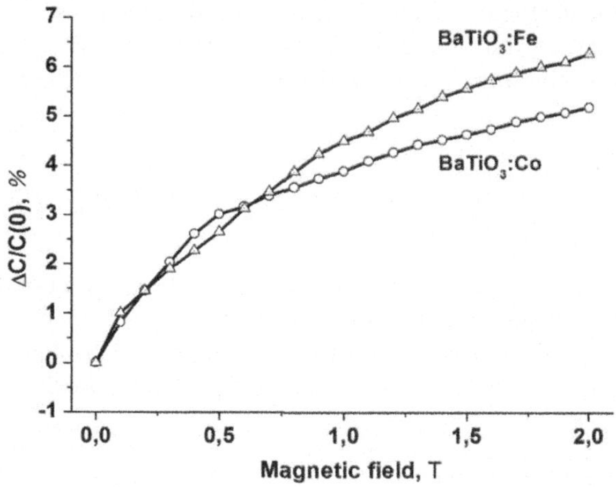

Fig. 10 Magnetic field dependence of the relative change in the capacitance Co- and Fe-implanted $BaTiO_3$ [105]

change of the value of effective magnetization in a result of increasing of electric field value applied in perpendicular direction.

Obviously, such shifts of FMR lines under electric field related to the changes of magnetizations indicate magnetoelectric coupling between the implanted cobalt (iron) particles and ferroelectric plate of $BaTiO_3$. We suppose that ion implantation leads to great mechanical strains in modified layer, and these strains provide good coupling between magnetic nanoparticles and $BaTiO_3$ matrix. Namely $BaTiO_3$ matrix changes its volume under an action of external electric filed due to piezoelectric effect. This leads to a change in the distance between cobalt nanoparticles. In the case of strong magnetic coupling between magnetic nanoparticles in Co (Fe) -implanted $BaTiO_3$ the displacement of cobalt (iron) nanoparticle results in changing of sample magnetization.

The study of the influence of applied magnetic field on dielectric properties of Co- and Fe-implanted $BaTiO_3$ crystal revealed a strong dependence of the dielectric properties of the samples on the external magnetic field. Figure 10 shows the dependences of the relative change in the capacitance of the parallel plate capacitors with a $BaTiO_3$ spacer on the applied magnetic field. The relative change in the capacitance, which is primarily determined by the dielectric constant of the spacer of the parallel plate capacitor, reaches 6% in an external magnetic field of 2 T for both samples. This is a very remarkable value, comparing with the data on various magnetoelectric composites given in the literature.

The observed phenomenon can be explained on the base of well-known interpretation of strong magnetoelectric coupling in ferroelectric–ferromagnetic composite structures. It is known from the literature that a strong ME effect could be realized in the composite consisting of magnetostrictive and piezoelectric constituents, so that an efficient magnetostrictive–piezoelectric coupling between the two phases is achieved. In this case, the magnetoelectric effect originates from the elastic interaction between the magnetostrictive and piezoelectric subsystems. In a magnetic field, the magnetostriction in the magnetostrictive phase gives rise to mechanical stresses that are transferred into the piezoelectric phase, owing to the piezoelectric effect, resulting in an electric polarization of the ferroelectric–piezoelectric phase.

5.2 Co-implanted $TlInS_2$ and $TlGaSe_2$ Crystals

The temperature dependences ε(T) of the real part of dielectric constant of pure and Coimplanted (at the ion concentration of 1.0×10^{17} ion/cm^2) $TlInS_2$ and $TlGaSe_2$ crystals measured on heating the samples are presented in Figs. 11 and 12 [106]. The measurements of capacitance were performed for the samples implanted with various

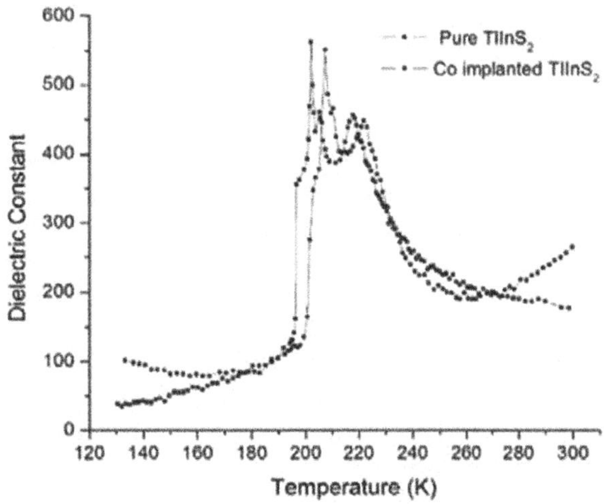

Fig. 11 Temperature dependences of dielectric constant of pure and Co-implanted (at the fluency of 1.0×10^{17} ion/cm^2) $TlInS_2$ crystals measured on heating the sample [106]

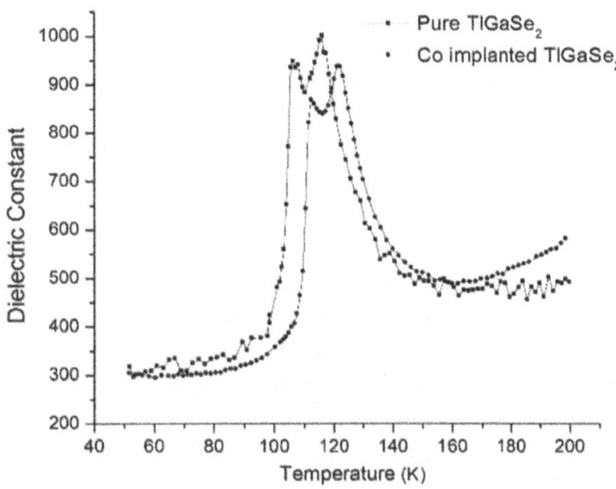

Fig. 12 Temperature dependences of dielectric constant of pure and Co-implanted (at the fluency of 1.0×10^{17} ion/cm^2) $TlGaSe_2$ crystals measured on heating the sample [106]

ion concentrations. As it is seen from the figures, the temperature dependences of the dielectric constant of pure $TlInS_2$ and $TlGaSe_2$ exhibit well-known anomalies at the well-known phase transition temperatures [92–94]. It is also seen that the implantation of the crystals brings to remarkable shifts of the phase transition points to high temperatures on heating the sample. It has been also revealed that these shifts are increased with increasing of ion concentrations. On the other side, the measurements of these dependences on cooling regime revealed the absence of any shifts of phase transition anomalies. Thus, the implantation of the samples brought to increase of the temperature hysteresis in $\varepsilon(T)$ dependences: the shifts of phase transition points in implanted samples are realized only on heating the samples.

It is known that the presence of incommensurately modulated structure in crystal leads to occurrence of long-lived metastable states in the temperature interval of the successive incommensurate and commensurate phase transitions. This causes the presence of the thermal hysteresis of the dielectric susceptibility, which has been observed for $TlInS_2$ and $TlGaSe_2$ in [107–109]. It was mentioned that many incommensurate systems, just above the incommensurate–commensurate phase transition point, consist of locally commensurate regions separated by narrow walls called discommensurations or solitons where the phase and amplitude of modulation change abruptly. Thermal hysteresis is usually attributed to defect-induced pinning of these discommensurations, as well as the domain-like structure of the ferroelectric phase, which prevents the crystal from reaching thermal equilibrium after the incommensurate–commensurate phase transition on decreasing the temperature.

The similar effect is observed on heating the ferroelectric crystals from low temperatures, which also may be qualitatively explained on the base of phenomenological theory of incommensurate systems. As it was shown in [110], the dielectric susceptibility of the crystal possessing incommensurate and commensurate phases consists of normal and 'anomalous' part. 'Anomalous' part of the dielectric constant arises due to movements of discommensurations in response to applied electric field and this part determines mainly a magnitude of the dielectric susceptibility in the temperature range of incommensurate–commensurate phase transitions. Then, an 'anomalous' part of the dielectric constant in crystals with incommensurate phases depends on the density of discommensurations. After existing the crystal in the ferroelectric state at low temperatures, it contains a minimum number of discommensurations. In this case, the domain walls are rigid, and it is difficult to reorient the domains under applied electric field. Low value of induced polarization will result to a lower value of the dielectric susceptibility. Heating the crystal will result to formation and evolution of discommensurations. But this process has high activation energy due to interaction with defects or other imperfections of the crystal structure and that is why it is comparatively embarrassed. Any increase of this interaction will lead to a delay of the transformation of the crystal into incommensurate phase, and we observe shifts of $\varepsilon(T)$ peaks to higher temperatures.

In our point of view, the same mechanism is responsible for appearing shifts of the phase transition points in Co-implanted $TlInS_2$ and $TlGaSe_2$ crystals on heating the samples and thus for increasing the temperature hysteresis of the dielectric susceptibility. The observed behaviour can be interpreted taking into account the interaction between the domain structure of the ferroelectric state of $TlInS_2$ and $TlGaSe_2$ substrates and ferromagnetic Co clusters of the composite surface layer, which can be attributed to some kind of magnetoelectric coupling. It is well known from the literature that a strong magnetoelectric coupling could be realized in ferroelectric–ferromagnetic composite structures, so that the magnetodielectric effect originates from the elastic interaction between the magnetostrictive and piezoelectric subsystems. According to this interpretation, the magnetostriction in the magnetostrictive phase gives rise to mechanical stresses that are transferred into the piezoelectric phase, owing to the piezoelectric effect, resulting in changes in the dielectric properties of the ferroelectric–piezoelectric phase [46]. Actually, this strong magnetoelectric coupling brings to lock-in of the domain walls and prevents their further transformation to incommensurate phase on increasing the temperature.

In order to continue detailed investigation of possible magnetodielectric effects, the temperature dependences of the real part of the dielectric susceptibility of Co-implanted $TlInS_2$ and $TlGaSe_2$ on applying the magnetic fields with various intensities were measured. The results are presented in Figs. 13 and 14. The magnetic field was applied in the direction perpendicular to the layers and implanted surfaces of the crystals, and the measurements were performed on heating the sample. As it is seen from the figures, the applying of the magnetic field brings to remarkable shifts of the phase transition points down to the values that were observed for pure samples. By the way, one must point out that no any shifts were observed again when the samples were measured in cooling regime.

Thus, the decreasing of the temperature hysteresis of dielectric constant of Co-implanted $TlInS_2$ and $TlGaSe_2$ crystals were observed in the temperature region of successive phase transitions on applying the magnetic field in the direction perpendicular to the implanted surface plane. This result also confirms our above-mentioned

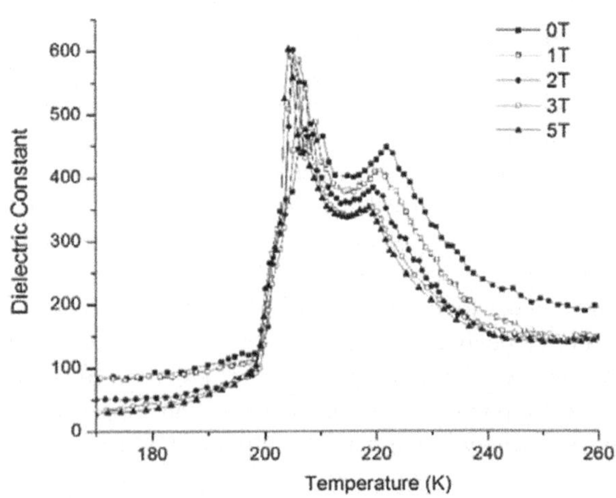

Fig. 13 Temperature dependences of dielectric constant of Co-implanted (at the fluency of 1.0×10^{17} ion/cm^2) $TlInS_2$ under various magnetic fields measured on heating the sample [106]

conclusion about magnetoelectric coupling between domain structures of ferroelectric and ferromagnetic constituents. The matter is that as mentioned in [104], Co implantation of $TlInS_2$ and $TlGaSe_2$ brings to formation of ferromagnetic surface layer in crystal structure, which behaves like to ferromagnetic thin granular film with characteristic shape anisotropy of magnetization, the direction of which lies in the implanted surface plane/layer plane. In this frame, above observed shifts of the phase transition temperatures can be interpreted as a result of longitudinal magnetoelectric coupling between ferroelectric and ferromagnetic domain walls, which are both aligned in the same direction. The application of the magnetic field perpendicular to this direction destroys the mentioned shape anisotropy, changes the magnetization direction and thus brings to weakening of the longitudinal magnetoelectric effect and shifting of the phase transition temperatures back to ones for pure samples.

6 Conclusions

Thus, high-fluency implantation of cobalt and iron ions into ferroelectric crystals results in the precipitation of implanted impurities in the form of magnetic metal nanoparticles. The irradiated ferroelectric samples exhibit properties of thin granular magnetic films. The observed shifts of the ferromagnetic resonance field values on applying the external electric field and the change in the dielectric constant of nanocomposite samples on applying the external magnetic field indicate that there is a magnetoelectric coupling between the Co and Fe nanoparticles and the ferroelectric matrix. This has opened up prospects for the use of the mentioned synthesized materials as new type of nanocomposite multiferroics.

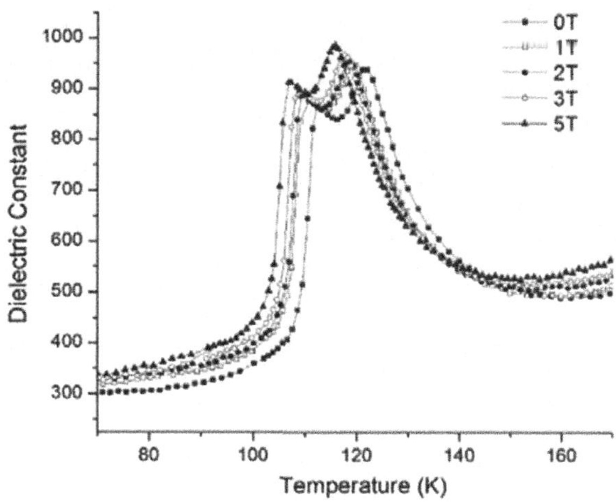

Fig. 14 Temperature dependences of dielectric constant of Co-implanted (at the fluency of 1.0×10^{17} ion/cm^2) $TlGaSe_2$ under various magnetic fields measured on heating the sample [106]

Acknowledgements The author is grateful to his collaborators—Rustam Khaibullin from Kazan Physical-Technical Institute (Russia), Sinan Kazan and Maksut Maksutoglu from Gebze Technical University (Turkey) for their great efforts and collaborations in joint research and investigations of novel nanocomposite multiferroic materials.

References

1. W.C. Rontgen, Ann. Phys. **35**, 264 (1888)
2. H.A. Wilson, Phil. Trans. R. Soc. Lond. A **204**, 121 (1905)
3. P. Curie, J. Physique **3**, 393 (1894)
4. P. Debye, Z. Phys. **36**, 300 (1926)
5. A. Perrier, A.J. Staring, Arch. Sci. Phys. Nat. **4**, 373 (1922)
6. A. Perrier, A.J. Staring, Arch. Sci. Phys. Nat. **5**, 333 (1923)
7. L.D. Landau, E.M. Lifshitz, *Electrodynamics of Continuous Media* (Pergamon Press, Oxford, 1960).
8. J.H. Van Vleck, *The Theory of Electric and Magnetic Susceptibilities* (Oxford University Press, London, 1932).
9. I.E. Dzyaloshinskii, Soviet Physics JETP **10**, 628–629 (1959)
10. D.N. Astrov, Soviet Physics JETP **11**, 708 (1960)
11. D.N. Astrov, Soviet Physics JETP **13**, 729 (1961)
12. G.T. Rado, V.J. Folen, Phys. Rev. Lett. **7**, 310 (1961)
13. V.J. Folen, G.T. Rado, E.W. Stalder, Phys. Rev. Lett. **6**, 607 (1961)
14. A. Chen, F. Chernow, Phys. Rev. **154**, 4 (1967)
15. M.A. Saifi, L.E. Cross, Phys. Rev. B **2**, 677 (1970)
16. H. Schmid, Ferroelectrics **162**, 317 (1994)
17. G.A. Smolenskii, I.E. Chupis, Sov. Phys. Usp. **25**, 475 (1982)
18. G.A. Smolenskii, A.I. Agranovskaya, V.A. Isupov, Sov. Phys. Solid State **1**, 149 (1959)
19. Y.N. Venevtsev, V.V. Gagulin, I.D. Zhitomirsky, Ferroelectrics **73**, 221 (1987)
20. G.A. Smolenskii, V.A. Bokov, V.A. Isupov, N.N. Krainik, G.H. Nedlin, Helv. Phys. Acta **41**, 1187 (1968)
21. Y.N. Venevtsev, V.V. Gagulin, Ferroelectrics **162**, 23 (1994)
22. S.M. Skinner, IEEE Transactions on Parts. Materials and Packaging **6**, 68 (1970)
23. H. Schmid, Int. J. Magn. **4**, 337 (1973)
24. K. Aizu, Phys. Rev. B **2**, 754 (1970)
25. M. Fiebig, T. Lottermoser, D. Meier et al., Nat Rev Mater **1**, 16046 (2016)
26. F. Kubel, H. Schmid, Acta Crystallogr. B **46**, 698 (1990)
27. M. Matsubara et al., Science **348**, 1112 (2015)
28. A.V. Kovalev, G.T. Andreeva, C.R. Acad, Sci. **256**, 1958 (1963)
29. K.H. Hellwege, A.M. Hellwege (eds.), *Numerical Data and Functional Relationships (Landolt-Börnstein, New Series, Group III*, vol. 16a (Springer, Berlin, 1981).
30. H. Sugie, N. Iwata, K. Kohn, J. Phys. Soc. Japan **71**, 1558 (2002)
31. C. Moure, M. Villegas, J.F. Fernandez, J. Tartaj, P. Duran, J. Mater. Sci. **34**, 2565 (1999)
32. B.B. Van Aken, J.W.G. Bos, R.A. de Groot, T.T.M. Palstra, Phys. Rev. B **63**, 125127 (2001)
33. M. Bieringer, J.E. Greedan, A.S. Wills, Appl. Phys. A **74**, S601 (2002)
34. N. Fujimura, H. Sakata, D. Ito, T. Yoshimura, T. Yokota, T. Ito, J. Appl. Phys. **93**, 6990 (2003)
35. M. Lilienblum et al., Nat. Phys. **11**, 1070 (2015)
36. C. Becher et al., Nat. Nanotechnol. **10**, 661 (2015)
37. Y.Y. Tomashpol'ski, Y.N. Venevtsev and V.N. Beznozdrev, Fiz. Tverd. Tela (in Russian) **7**, 2763 (1965)
38. E. Ascher, H. Schmid, D. Tar, Solid State Commun. **2**, 4 (1964)

39. H. Schmid, H. Rieder, E. Ascher, Solid State Commun. **3**, 327 (1965)
40. J.F. Scott, Ferroelectrics **24**, 127 (1980)
41. M. Eibschütz, H.J. Guggenheim, Solid State Commun. **6**, 737 (1968)
42. J. Wang, J.B. Neaton, H. Zheng, V. Nagarajan, S.B. Ogale, B. Liu, D. Viehland, V. Vaithyanathan, D.G. Schlom, U.V. Waghmare, N.A. Spaldin, K.M. Rade, M. Wuttig, R. Ramesh, Science **299**, 1719 (2003)
43. B.B. van Aken, T.T.M. Palstra, A. Filippetti, N.A. Spaldin, Nat. Mater. **3**, 164 (2004)
44. D.I. Khomskii, Bull. Am. Phys. Soc. C **21**, 2 (2001)
45. N.A. Hill, A. Filipetti, J. Magn. Magn. Mater. **242–245**, 976 (2002)
46. M. Fiebig, J. Phys. D: Appl. Phys. **38**, R123 (2005)
47. C.W. Nan, M.I. Bichurin, S.X. Dong, D. Viehland, J. Appl. Phys. **103**, 031101 (2008)
48. M. Zeng, J.G. Wan, Y. Wang, H. Yu, J.-M. Liu, X.P. Jiang, C.W. Nan, J. Appl. Phys. **95**, 8069 (2004)
49. S.X. Dong, J.R. Chen, J.F. Li, D. Viehland, Appl. Phys. Lett. **83**, 4812 (2003)
50. N. Cai, C.W. Nan, J.Y. Zhai, Y.H. Lin, Appl. Phys. Lett. **84**, 3516 (2004)
51. C.W. Nan, G. Li, Y. Lin, H. Chen, Phys. Rev. Lett. **94**, 197203 (2005)
52. H. Zheng, J. Wang, S.E. Lofland, Z. Ma, L. Mohaddes-Ardabili, T. Zhao, L. Salamanca-Riba, S.R. Shinde, S.B. Ogale, F. Bai, D. Viehland, Y. Jia, D.G. Schlom, M. Wuttig, A. Roytburd, R. Ramesh, Science **303**, 661 (2004)
53. B.D.H. Tellegen, Philips Res. Rep. **3**, 81 (1948)
54. J. van Suchtelen, Philips Res. Rep. **27**, 28 (1972)
55. J. Van Den Boomgaard, D.R. Terrell, R.A.J. Born, J. Mater. Sci. **9**, 1705 (1974)
56. J. Van Den Boomgaard, A.M.J.G. Van Run, J. Van Suchtelen, Ferroelectrics **14**, 727 (1976)
57. J. Van Den Boomgaard, A.M.J.G. Van Run, J. Van Suchtelen, Ferroelectrics **10**, 295 (1976)
58. K.-S. Chang, M.A. Aronova, C.-L. Lin, M. Murakami, M.-H. Yu, J. Hattrick-Simpers, O.O. Famodu, S.Y. Lee, R. Ramesh, M. Wuttig, I. Takeuchi, Appl. Phys. Lett. **84**, 3091 (2004)
59. C.-W. Nan, N. Cai, L. Liu, J. Zhai, Y. Ye, Y. Lin, J. Appl. Phys. **94**, 5930–5936 (2003)
60. G. Liu, C.-W. Nan, N. Cai, Y. Lin, J. Appl. Phys. **95**, 2660 (2004)
61. M. Trassin, J. Phys. Condens. Matter **28**, 033001 (2016)
62. J. Lian, F. Ponchel, N. Tiercelin, L. Han, Y. Chen, D. Remiens, T. Lasri, G. Wang, P. Pernod, W. Zhang, X. Dong, J. Appl. Phys. **124**, 064101 (2018)
63. R. Karthikeyan, M.K. Niranjan, J. Magn. Magn. Mater. **469**, 138 (2019)
64. X. Wang, Q. Yang, L. Wang, Z. Zhou, T. Min, M. Liu, N.X. Sun, Adv. Mater. **30**, 1803612 (2018)
65. P.F. Liu, J. Miao, Z.D. Xu, G. Jakob, Q. Liu, Z.Y. Ren, K.K. Meng, Y. Wu, J.K. Chen, X.G. Xu, Y. Jiang, Appl. Phys. Lett. **113**, 062401 (2018)
66. Z. Tang, B. Yang, J. Chen, Q. Lu, S. Zhao, J. Alloy. Compd. **772**, 298 (2019)
67. G. Singh, M.K. Singh, A. Kumar, S. Dussan, R.S. Katiyar, AIP Conf. Proc. **1731**, 080043 (2016)
68. X. Guo, D. Li, Li. Xi, Chin. Phys. B **27**, 9 (2018)
69. Kanchan Bala, Hakikat Sharma, and N. S. Negi, AIP Conference Proceedings **1731**, 080014 (2016)
70. Chung Ming Leung, Jiefang Li, D Viehland and X Zhuang. J. Phys. D: Appl. Phys. **51**, 263002 (2018)
71. D. Viehland, J.F. Li, Y. Yang, T. Costanzo, A. Yourdkhani, G. Caruntu, P. Zhou, T. Zhang, T. Li, A. Gupta, M. Popov, G. Srinivasan, J. Appl. Phys. **124**, 061101 (2018)
72. D. Roy, S.B. Krupanidhi, Appl. Phys. Lett. **61**, 2057 (1992)
73. K. Nashimoto, D.K. Fork, F.A. Ponce, J.C. Tramotana, Jpn. J. Appl. Phys. **32**, 4099 (1993)
74. K. Fujimoto, Y. Kobayashi, K. Yamamoto, K. Hirata, Y. Bando, Appl. Phys. Lett. **56**, 527 (1990)
75. H.A. Lu, L.A. Wills, B.W. Wessels, Appl. Phys. Lett. **64**, 2973 (1994)
76. R.A. McKee, F.J. Walker, E.D. Specht, G.E. Jellision Jr., L.A. Boatner, Phys. Rev. Lett. **72**, 2741 (1994)

77. M.E. Lines, A.M. Glass, *Principles and Applications of Ferroelectric and Related Materials* (Oxford University Press, London and New York, 1977).
78. F. Jona, G. Shirane, *Ferroelectric Crystals* (Pergamon, New York, 1962).
79. W. Ousi-Benomar, S.S. Xue, R.A. Lessard, A. Singh, Z.L. Wu, P.K. Kuo, J. Mater. Res. **9**, 970 (1994)
80. W. Henkel, H.D. Hochheimer, C. Carlone, A. Werner, S. Ves, H.G.V. Schnering, Phys. Rev. B **26**, 3211 (1982)
81. D. Müller, H. Hahn, Z. Anorg, Allg. Chem. **438**, 258 (1978)
82. D.F. McMorrow, R.A. Cowley, P.D. Hatton, J. Banys, J. Phys.: Condens. Matter **2**, 3699 (1990)
83. R.A. Aliev, K.R. Allakhverdiev, A.I. Baranov, N.R. Ivanov, R.M. Sardarly, Sov. Phys. Solid State **26**, 775 (1984)
84. A.A. Volkov, Y.G. Goncharov, G.V. Kozlov, S.P. Lebedev, A.M. Prokhorov, R.A. Aliev, K.R. Allakhverdiev, JETP Lett. **37**, 615 (1984)
85. J.A. Kalomiros, N. Kalkan, M. Hanias, A.N. Anagnostopoulos, K. Kambas, Solid State Commun. **96**, 601 (1995)
86. H. Hahn, B. Wellmann, Naturwissenschaften **54**, 42 (1967)
87. D. Müller, F.E. Poltmann, and H. Hahn, Z. Naturworsch. B **29**, 117 (1974)
88. T.J. Isaacs, J. Appl. Crystallog. **6**, 413 (1973)
89. T.J. Isaacs, R.H. Hopldns, J. Cryst. Growth **29**, 121 (1975)
90. T.J. Isaacs, J.D. Felchtner, J. Solid State Chem. **14**, 260 (1975)
91. N.M. Gasanly, A.F. Goncharov, N.M. Melnik, A.S. Ragimov, V.I. Tagirov, Phys. Status Solidi B **116**, 427 (1983)
92. F.M. Salaev, K.R. Allakhverdiev, F.A. Mikailov, Ferroelectrics **131**, 163 (1992)
93. R.A. Suleymanov, MYu. Seidov, F.M. Salaev, F.A. Mikailov, Sov. Phys. Solid State **35**, 177 (1993)
94. F.A. Mikailov, E. Başaran, T.G. Mammadov, MYu. Seidov, E. Şentürk, Phys. B **334**, 13 (2003)
95. G. Dearnaley, J.H. Freeman, R.S. Nelson, J. Stephen, *Ion Implantation* (North-Holland, Amsterdam, 1973).
96. S. Kazan, F. A. Mikailzade, A. G. Sale, M. Maksutoglu, M. Acikgoz, R. I. Khaibullin, N. I. Khalitov, Ju. I. Gatiiatova, and V. F. Valeev, Phys. Rev. B **82**, 054402 (2010)
97. G.N. Kakazei, A.F. Kravets, N.A. Lesnik, M.M. Pereira de Azevedo, Yu.G. Pogorelov, J.B. Sousa, J. Appl. Phys. **85**, 5654 (1999)
98. A.L. Stepanov, R.I. Khaibullin, B.Z. Rameev, A. Reinholdt, U. Kreibig, Tech. Phys. Lett. **30**, 151 (2004)
99. S. Tomita, K. Akamatsu, H. Shinkai, S. Ikeda, H. Nawafune, C. Mitsumata, T. Kashiwagi, M. Nagiwara, Phys. Rev. B. **71**, 180414(R) (2005)
100. J.L. Dormann, D. Fiorani, E. Tronc, Adv. Chem. Phys. **98**, 283 (1997)
101. M. Maksutoğlu, S. Kazan, N.I. Khalitov, V.I. Nuzhdin, R.I. Khaibullin, L.R. Tagirov, V.V. Roddatis, K.E. Prikhodko, F.A. Mikailzade, J. Magn. Magn. Mater. **373**, 103 (2015)
102. C. Kittel, *Introduction to Solid State Physics* (Wiley, New York, 1976).
103. F. Mikailzade, A.G. Şale, S. Kazan, R.I. Khaibullin, N.I. Khalitov, V.I. Nuzhdin, T.G. Mammadov, Solid State Commun. **152**, 407 (2012)
104. J. Martan, Mater. Sci. Eng., B **22**, L1 (1994)
105. N.I. Khalitov, H.M. Lyadov, V.F. Valeev, R.I. Khaibullin, I.A. Faizrakhmanov, E.N. Dulov, L.R. Tagirov, S.Z. Ibragimov, K.E. Prikhodko, V.V. Roddatis, M. Maksutoglu, S. Kazan, F.A. Mikailzade, Physics of Solid State **55**, 1187 (2013)
106. F. Mikailzade, M. Maksutoglu, R.I. Khaibullin, V.F. Valeev, V.I. Nuzhdin, V.B. Aliyeva, T.G. Mammadov, Phase Transitions **89**, 568 (2016)
107. F.A. Mikailov, E. Başaran, E. Şentürk, Solid State Commun. **122**, 161 (2002)
108. F.A. Mikailov, E. Başaran, E. Şentürk, L. Tümbek, T.G. Mammadov, V.P. Aliev, Phase Transitions **76**, 1057 (2003)
109. F.A. Mikailov, E. Başaran, E. Şentürk, L. Tümbek, T.G. Mammadov, V.P. Aliev, Solid State Commun. **129**, 761 (2004)
110. H. Mashiyama, H.-G. Unruh, J Phys C: Solid State Phys. **16**, 5009 (1983)

Index

S. J. Ikhmayies et al. (eds.), *Advances in Optoelectronic Materials*,
Advances in Material Research and Technology,
https://doi.org/10.1007/978-3-030-57737-7

D

F

N

Structural properties, 11, 26, 128, 211, 212